MATERIALS FOR ARCHITECTS AND BUILDERS

MATERIALS FOR ARCHITECTS AND BUILDERS

Third edition

ARTHUR LYONS

MA(Cantab) MSc(Warwick) PhD(Leicester) DipArchCons(Leics) Hon LRSA
Head of Quality, Principal Lecturer and Teacher Fellow, School of Architecture,
Faculty of Art and Design, De Montfort University, Leicester, UK

ELSEVIER

AMSTERDAM • BOSTON • HEIDELBERG • LONDON • NEW YORK • OXFORD
• PARIS • SAN DIEGO • SAN FRANCISCO • SINGAPORE • SYDNEY • TOKYO

Butterworth-Heinemann is an imprint of Elsevier

Butterworth Heinemann is an imprint of Elsevier
Linacre House, Jordan Hill, Oxford OX2 8DP, UK
The Boulevard, Langford Lane, Kidlington, Oxford OX5 1GB, UK
84 Theobald's Road, London WC1X 8RR, UK
Radarweg 29, PO Box 211, 1000 AE Amsterdam, The Netherlands
30 Corporate Drive, Suite 400, Burlington, MA 01803, USA
525 B Street, Suite 1900, San Diego, CA 92101-4495, USA

First Published 1997
Third edition 2007

British Library Cataloguing in Publication Data
Lyons, Arthur (Arthur R.)
    Materials for architects and builders. - 3rd ed.
    1. Building materials
    I. Title
    691

Library of Congress Cataloging-in-Publication Data
A catalog record for this book is available from the Library of Congress

    ISBN-13: 978-0-7506-6940-5
    ISBN-10: 0-7506-6940-3

For information on all Butterworth Heinemann publications
visit our web site at books.elsevier.com

Printed and bound in Italy
07 08 09 10 11 10 9 8 7 6 5 4 3 2 1

Working together to grow
libraries in developing countries

www.elsevier.com | www.bookaid.org | www.sabre.org

ELSEVIER   BOOK AID International   Sabre Foundation

CONTENTS

—

ABOUT THE AUTHOR

—

Dr Arthur Lyons is Head of Quality in the Faculty of Art and Design at De Montfort University, Leicester, UK. He also holds the position of principal lecturer and teacher fellow in building materials in the Leicester School of Architecture, Department of Product and Spatial Design, within the Faculty. He was educated at Trinity Hall, Cambridge, and Warwick and Leicester Universities in the fields of natural sciences and polymer science and has a postgraduate diploma in architectural building conservation. He has been a lecturer in building materials within schools of architecture and surveying for over thirty years. Arthur Lyons was honoured with life membership of the Leicestershire and Rutland Society of Architects in recognition of his services to architects and architecture. He continues his teaching and research role in building materials with students of architecture, architectural technology and interior design, in parallel with his senior faculty position.

PREFACE

—

Materials for Architects and Builders is written as an introductory text to inform students at undergraduate degree and national diploma level of the relevant visual and physical properties of a wide range of building materials. The third edition has been significantly enhanced by the incorporation of full colour images throughout, illustrating the materials and in many cases their use in buildings of architectural merit. The text includes the broad environmental debate with sections on energy saving and recycled materials. There are seventeen chapters covering the wide range of materials under standard headings. Each chapter describes the manufacture, salient properties and typical uses of the various materials, with the aim of ensuring their appropriate application within an awareness of their ecological impact.

European Standards are taking over from the previous British Standards, and for most key materials the European Norms have now been published. Generally, this has led to an increase in the number of relevant standards for building materials. However, in many cases, both the British and European Standards are current and are therefore included in the text and references.

New and rediscovered old materials, where they are becoming well integrated into standard building processes are described; other materials no longer in use are generally disregarded, except where increased concern for environmental issues has created renewed interest. The use of chemical terminology is kept to the minimum required to understand each subject area, and is only significantly used within the context of the structure of plastics. Tabulated data is restricted to an informative level appropriate to student use. An extensive bibliography and listed sources of technical information are provided at the end of each chapter to facilitate direct reference where necessary.

The text is well illustrated with over 250 line drawings and colour photographs, showing the production, appearance and appropriate use of materials, but it is not intended to describe construction details as these are well illustrated in the standard texts on building construction. Environmental concerns including energy-conscious design, and the effects of fire, are automatically considered as part of the broader understanding of the various key materials.

The text is essential reading for honours and foundation degree, BTEC and advanced GNVQ students of architecture, building, surveying and construction, and those studying within the broad range of built environment subjects, who wish to understand the principles relating to the appropriate use of construction materials.

Arthur Lyons
March 2006

ACKNOWLEDGEMENTS

—

I wish to acknowledge the assistance of my colleagues in the Leicester School of Architecture, Department of Product and Spatial Design, Faculty of Art and Design, De Montfort University, Leicester, for suggesting amendments and additions to the various chapters of this book for its third edition, especially Robert Sheen in the Resources Centre for sourcing considerable material. I wish to thank my wife, Susan, for her participation and support during the production of this work, also my daughters Claire and Elizabeth for their constant encouragement. I am indebted to the numerous manufacturers of building materials for their trade literature and for permissions to reproduce their published data and diagrams. I am grateful to building owners, architectural practices and their photographers for the inclusion of the photographs; to Her Majesty's Stationery Office, the Building Research Establishment, the British Standards Institute and trade associations for the inclusion of their material.

I should like to thank the following organisations for giving permission to use illustrations:

Aircrete Products Association (Fig. 2.3); Angle Ring Company Ltd (Fig. 5.10); Architectural Ceramics (Figs. 8.6, 8.9 and 8.10); Building Research Establishment (Figs. 2.3, 4.14 and 9.13) – Photographs from GBG 58, Digest 476 and IP 10/01; Baggeridge Brick plc (Figs. 1.16, 1.18 and 1.19); British Cement Association (Figs. 3.4, 3.8, 3.19 and 3.23); British Standards Institute (Figs. 2.8 and 5.26) – Permission to reproduce extracts from BS EN 771 Part 1: 2003 and BS 6915: 2001 is granted by BSI. British Standards can be obtained from BSI Customer Services, 389 Chiswick High Road, London W4 4AL. Tel: +44 (0) 20 8996 9001, email: cservices@bsi-global.com; CGL Comtec (Fig. 8.8); Construction Resources (Fig. 4.34); Copper Development Association (Figs. 5.21 – 5.23); Corus (Figs. 5.2, 5.4 – 5.7, 5.11 and 5.13); Glass Block Technology www.glassblocks.co.uk (Fig. 7.5); Hanson Brick Ltd. (Fig. 1.3); Ibstock Brick Ltd (Figs. 1.7 – 1.9, 2.9); Imperial Chemical Industries plc (Fig. 17.3); James & Son Ltd (Fig. 11.8); KME UK Ltd (Fig. 5.23); Lead Contractors Association (Figs. 5.25 and 5.27); Lead Sheet Association (Fig. 5.24); Lignacite Ltd (Fig. 2.7); Make Architects (Fig. 4.1); Marshalls plc (Fig. 2.14); The Metal Cladding and Roofing Manufacturers Association (Fig. 5.15); Metra Nonferrous Metals Ltd and Rheinzinc (Fig. 5.29); Monodraught (Figs. 14.5 and 14.6); Natural Stone Products Ltd (Fig. 9.9); Pilkington Glass Ltd (Figs. 7.6, 7.8, 7.9, 7.19 and 7.20); Pyrobel (Fig. 7.13); Ruberoid Building Products (Figs. 6.3 and 6.4); Scandinavian Colour Institute AB www.sci-sweden.se (Fig. 17.2); Securiglass Company Ltd (Fig. 7.11); Smith of Derby (Fig. 11.2); Solar Century – www.solarcentury.com (Figs. 14.2 and 14.3); The Steel Construction Institute (Figs. 5.7 and 5.12); Stone Federation of Great Britain (Fig. 9.3); TRADA Technology Ltd (Figs. 4.14 and 4.17); Trent Concrete Ltd (Figs. 1.17, 3.18, 3.19, 9.15, 11.5 and 11.6) and Zinc Development Association (Fig. 5.29).

The text uses the generic names for building materials and components wherever possible. However, in a few cases, products are so specific that registered trade names are required. In these cases the trade names are italicised in the text.

INFORMATION SOURCES

———

Specific information relating to the materials described in each chapter is given at the end of the appropriate section; however, the following are sources of general information relating to construction materials.

- Building Regulations 2000, Amendments and Approved Documents
- Specification
- RIBA Office Library and Barbour Index
- Building Research Establishment (BRE) publications
- Trade association publications
- Trade literature
- Architecture and built environment journals
- British Board of Agrément certificates
- British Standards
- European Standards
- Eurocodes.

European Standards (EN) have been published for a wide range of materials. A full European Standard, known in the UK as BS EN, is mandatory and overrules any conflicting previous British Standard which must be withdrawn. Prior to full publication, the draft European Standards are coded pr EN and are available for comment, but not implementation. Prospective standards, where documentation is in preparation, are published as European pre-standards (ENV). These are similar to the previous British Drafts for Development (DD) and would normally be converted to full European Standards (EN) after the three-year experimental period, when any conflicting national standards would have to be withdrawn. BRE Information Paper IP 3/99 (1999) identifies the issues relating to the adoption in the UK of the structural Eurocodes.

The Building Research Establishment (BRE) publishes informative and authoritative material on a wide range of subjects relating to construction. Trade associations produce advisory and promotional literature relating to their particular area of interest within the building industry. Architecture and building journals give news of innovations and illustrate their realisation in quality construction.

Information for this text has been obtained from a wide selection of sources to produce a student text with an overview of the production, nature and properties of a diverse range of building materials. New individual products and modifications to existing products frequently enter the market; some materials become unavailable. Detailed information and particularly current technical data relating to any specific product for specification purposes must therefore be obtained directly from the manufacturers or suppliers and cross-checked against current standards and regulations.

ABBREVIATIONS

——

General

AAC	autoclaved aerated concrete
ABS	acrylonitrile butadiene styrene
AC	aggressive chemical (environment)
ACEC	aggressive chemical environment for concrete
APM	additional protective measures
APP	atactic polypropylene
AR	alkali-resistant
ASR	alkali-silica reaction
BER	building emission rate
BRE	Building Research Establishment
BS	British Standard
CAD	computer-aided design
CEN	European committee for standardisation
CFCs	chlorofluorocarbons
CG	cellular glass
CPE	chlorinated polyethylene
CPVC	chlorinated polyvinyl chloride
CS	calcium silicate
CSPE	chlorosulfonated polyethylene
DC	design chemical (class)
DC	direct current
DD	draft for development
DPC	damp-proof course
DPM	damp-proof membrane
DR	dezincification-resistant
DS	design sulfate (class)
EN	Euronorm
ENV	Euronorm pre-standard
EP	expanded perlite
EPDM	ethylene propylene diene monomer
EPR	ethylene propylene rubber
EPS	expanded polystyrene
ETFE	ethylene tetrafluorethylene copolymer
EV	exfoliated vermiculite
EVA	ethylene vinyl acetate
FEF	flexible elastomeric foam
FPA	flexible polypropylene alloy
FRP	fibre reinforced polymer
GGBS	ground granulated blastfurnace slag
GRC	glass-fibre reinforced cement
GRG	glass-fibre reinforced gypsum
GRP	glass-fibre reinforced plastic or polyester
GS	general structural (timber)
HAC	high alumina cement
HCFCs	hydrochlorofluorocarbons
HD	high density
HDPE	high density polythene
HL	hydraulic lime
HLS	hue lightness saturation
ICB	expanded corkboard
ISO	international organisation for standardisation
LD	low density
LDPE	low density polythene
MAF	movement accommodation factor
MDF	medium density fibreboard
MF	melamine formaldehyde
MPa	mega pascal
MW	mineral wool
NCS	natural color system®
NHL	non-hydraulic lime
ODP	ozone depletion potential
OPC	ordinary Portland cement
OSB	oriented strand board
PAS	publicly available specification
PBAC	polystyrene-bead aggregate cement
PC	polycarbonate
PE	polyethylene
PEF	polyethylene foam
PEX	crosslinked polyethylene
PF	phenolic foam
PFA	pulverised fuel ash
PIB	polisobutylene
PIR	polyisocyanurate foam
PMMA	polymethyl methacrylate
PP	polypropylene

pr EN	draft Euronorm
PTFE	polytetrafluroethylene
PUR	rigid polyurethane foam
PV	photovoltaic
PVA	polyvinyl acetate
PVB	polyvinyl butyral
PVC	polyvinyl chloride (plasticised)
PVC-U	polyvinyl chloride (unplasticised)
PVC-UE	extruded polyvinyl chloride
RGB	red green blue
SAP	standard assessment procedure
SBEM	simplified building energy model
SBS	styrene butadiene styrene
Sg	specific gravity
SIP	structural insulated panel
SS	special structural (timber)
ST	standard (concrete mix)
T	tolerance (class)
TER	target emission rate
TFS	thin film silicon
THF	tetrahydro furan
UF	urea formaldehyde
UHPC	ultra high performance concrete
VET	vinyl ethylene terpolymer
VOC	volatile organic compounds
WF	wood fibre
WW	wood wool
XPS	extruded polystyrene

Units

dB	decibel
MPa	mega pascal
μm	micron (10^{-6}m)
nm	nanometre (10^{-9}m)

Chemical symbols

Al	aluminium
C	carbon
Ca	calcium
Cr	chromium
Cl	chlorine
Cu	copper
F	fluorine
Fe	iron
Mn	manganese
Mo	molybdenum
N	nitrogen
Ni	nickel
O	oxygen
S	sulfur
Si	silicon
Sn	tin
Ti	titanium
Zn	zinc

Cement notation

C_2S	dicalcium sicilate
C_3S	tricalcium silicate
C_3A	tricalcium aluminate
C_4AF	tetracalciumaluminoferrite

BRICKS AND BRICKWORK

—

Introduction

Originally, bricks were hand-moulded from moist clay and then sun-baked, as is still the current practice in certain arid climates. The firing of clay bricks dates back well over 5000 years, and is now a sophisticated and highly controlled manufacturing process; yet the principle of burning clay, to convert it from its natural plastic state into a dimensionally stable, durable, low-maintenance ceramic material, remains unchanged.

The quarrying of clay and brick manufacture are high-energy processes, which involve the emission of considerable quantities of carbon dioxide and other pollutants including sulfur dioxide. The extraction of clay also has long-term environmental effects, although in some areas former clay pits have now been converted to bird sanctuaries or to recreational use. However, well-constructed brickwork has a long life with low maintenance and although the use of Portland cement mortar prevents the recycling of individual bricks, the crushed material is frequently recycled as aggregate in further construction.

Clay bricks

The wide range of clays suitable for brick making in the UK gives a diversity to the products available. The effects of blending clays, the various forming processes, the application of surface finishes, and the adjustment of firing conditions further increase this variety. Earlier this century most areas had their own brickworks with characteristic products; however, ease of road transportation and continuing amalgamations within the industry have left a reduced number of major producers

and only a few small independent works. Most UK bricks are defined as high density (HD) fired-clay masonry units with a gross dry density greater than 1000 kg/m^3. The European standard (BE EN 771–1: 2003) refers also to low density (LD) fired-clay masonry units and these blocks are described in Chapter 2.

The main constituents of brick-making clays are silica (sand) and alumina, but with varying quantities of chalk, lime, iron oxide and other minor constituents, e.g. fireclay, according to their source. The largest UK manufacturer uses the Lower Oxford clays of Bedfordshire, Buckinghamshire and Cambridgeshire to produce the *Fletton* brick. This clay contains some carbonaceous content that reduces the amount of fuel required to burn the bricks, lowering cost and producing a rather porous structure. Other particularly characteristic bricks are the strongly coloured *Staffordshire Blues* and *Accrington Reds* from clays containing high iron content and the yellow *London Stocks* from the Essex and Kent chalky clays with lower iron content.

SIZE

The standard metric brick is 215 × 102.5 × 65 mm, weighing between 2 and 4 kg, and is easily held in one hand. The length of a brick (215 mm) is equal to twice its width (102.5 mm) plus one standard 10 mm joint and three times its height (65 mm) plus two standard joints (Fig. 1.1).

The building industry modular co-ordination system (BS 6750: 1986) is based on the module (M) of 100 mm and multimodules of 3M, 6M, 12M, 15M, 30M and 60M. For metric brickwork, the base unit is 3M or 300 mm. Thus four courses of 65 mm brickwork with joints give a vertical height of 300 mm, and four stretchers with joints co-ordinate to 900 mm.

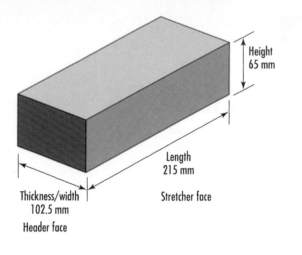

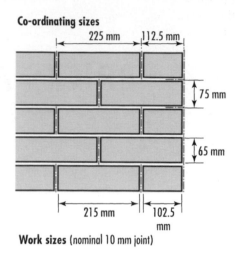

Work sizes (nominal 10 mm joint)

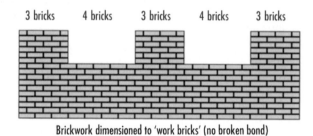

Brickwork dimensioned to 'work bricks' (no broken bond)

Fig. 1.1 Brick and co-ordinating sizes

Table 1.1 illustrates the two types of dimensional tolerance limits set for clay masonry units including the metric brick, which relate to the square root of the work size dimension. Measurements are based on a random sample of ten bricks. The calculation based on the use of the square root of work size ensures that the dimensional tolerance limits are appropriate for the wide range in size of clay masonry units used within the European Union (BS EN 771–1: 2003).

Tolerances

Mean value

Tolerance limits are set for the difference between the stated work size (e.g. 215, 102.5 and 65 mm) and the

Table 1.1   Tolerances on brick sizes

	Brick (work) dimensions (mm)	Maximum deviation ($\pm$) of mean from declared work dimension (mm)		Maximum range of size within sample of ten bricks (mm)	
		T1	T2	R1	R2
Length	215	6	4	9	4
Width	102.5	4	3	6	3
Height	65	3	2	5	2

Limits for Tm and Rm are as declared by the manufacturer.

measured mean from the samples, for each of the three brick dimensions (length, width and height). These are categorised as T1, T2 and Tm where Tm is a tolerance quoted by the manufacturer.

T1 $\pm 0.40 \sqrt{}$(work size dimension) mm or 3 mm if greater

T2 $\pm 0.25 \sqrt{}$(work size dimension) mm or 2 mm if greater

Tm   deviation in mm declared by the manufacturer

Range

The maximum range of size for any dimension is designated by categories R1, R2 and Rm.

R1 $0.6 \sqrt{}$(work size dimension) mm

R2 $0.3 \sqrt{}$(work size dimension) mm

Rm   range in mm declared by the manufacturer

There is no direct correlation between the limits on mean value (T) and those for the range (R), thus a brick conforming to category T2 may be within the wider range R1. Category R2 bricks may only be required for very tight dimensional control, as in short runs of brickwork.

Alternative sizes

The metric standard evolved from the slightly larger Imperial sizes, typically $9 \times 4^{7}/_{8} \times 2^{5}/_{8}$ in (229 × 111 × 66 mm). Some manufacturers offer a limited range of bricks to full Imperial dimensions, alternatively to a depth of 66 mm for bonding in to Imperial brickwork for restoration and conservation work.

The 1970s also saw the introduction of metric modular bricks with co-ordination sizes of either 200 or 300 mm in length, 100 mm wide and either 75 or 100 mm in height. The popularity of these bricks has now declined but they did give the architect opportunities for increasing or reducing horizontal emphasis and scale within the context of traditional brickwork. The British Standard BS 6649: 1985 now only refers to the 200 × 100 × 75 mm modular co-ordinating format.

MANUFACTURE OF CLAY BRICKS

There are five main processes in the manufacture of clay bricks:

- extraction of the raw material;
- forming processes;
- drying;
- firing;
- packaging and distribution.

EXTRACTION OF THE RAW MATERIAL

The process begins with the extraction of the raw material from the quarry and its transportation to the works, by conveyor belt or road transport. Topsoil and unsuitable overburden is removed first and used for site reclamation after the usable clay is removed.

The raw material is screened to remove any rocks, then ground into fine powder by a series of crushers and rollers with further screening to remove any over-size particles. Small quantities of pigments or other clays may be blended in at this stage to produce various colour effects; for example, manganese dioxide will produce an almost black brick and fireclay gives a teak brown effect. Occasionally, coke breeze is added into the clay as a source of fuel for the firing process. Finally, depending on the subsequent brick forming process, up to 25% water may be added to give the required plasticity.

Forming processes

Handmade bricks

The *handmade* process involves the throwing of a suitably sized clot of wet clay into a wooden mould on a bench. The surplus clay is struck off with a framed wire and the green brick removed. The bricks produced are irregular in shape with soft arrises and interestingly folded surfaces. Two variations of the process are pallet moulding and slop moulding.

In pallet moulding, a stock board, the size of the bed face of the brick, is fixed to the bench. The mould fits loosely over the stock board, and is adjusted in height to give the appropriate thickness to the green brick. The mould and board are sanded to ease removal of the green brick, which is produced with a *frog* or depression on one face. In the case of slop moulding, the stock mould is placed directly on the bench, and is usually wetted rather than sanded to allow removal of the green brick, which unlike the pallet-moulded brick is smooth on both bed faces (Fig. 1.2).

Soft mud process

The handmade process has now been largely automated, with the clay being mechanically thrown into pre-sanded moulds; the excess clay is then removed and the bricks released from the mould. These *soft mud*

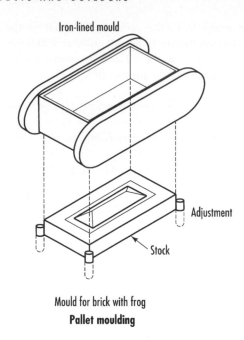

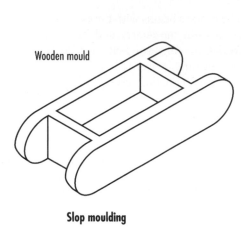

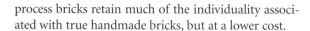

Slop moulding

Pallet moulding

Mould for brick with frog

Fig. 1.2 Moulds for handmade bricks

process bricks retain much of the individuality associated with true handmade bricks, but at a lower cost.

Pressed bricks

In the *semi-dry* process used for *Fletton* bricks the appropriate quantity of clay is subjected to a sequence of four pressings within steel moulds to produce the green brick. These bricks usually have a deep frog on one bed face. For facing bricks, texturing on both headers and one stretcher may be applied by a series of rollers. A water spray to moisten the surface, followed by a blast of a sand/pigment mixture produces the sand-faced finish.

With clays that require a slightly higher water content for moulding, the *stiff plastic* process is used in which brick-size clots of clay are forced into the moulds. A single press is then required to form the brick. Engineering bricks made by this process often have shallow frogs on both bed faces. In all cases the size of the mould is calculated to allow for the anticipated drying and firing shrinkage.

Extruded wire-cut bricks

In this process clay with a water content of up to 25% is fed into a screw extruder which consolidates the clay and extracts the air. The clay is forced through a die and forms a continuous column with dimensions equal to the length and width of a green brick (Fig. 1.3). The

Fig. 1.3 Extruding wire-cut bricks

surface may then be textured or sanded, before the clay column is cut into brick units by a series of wires. The bed faces of wire-cut bricks often show the drag marks where the wires have cut through the extruded clay. Perforated wire-cut bricks are produced by the incorporation of rods or tines between the screw extruder and the die. The perforations save clay and allow for a more uniform drying and firing of the bricks without significant loss of strength. Thermal performance is not significantly improved by the incorporation of voids.

Drying

To prevent cracking and distortion during the firing process, green bricks produced from wet clays must be allowed to dry out and shrink. Shrinkage is typically 10% on each dimension depending upon the moisture content. The green bricks, laid in an open chequerwork pattern to ensure a uniform loss of moisture, are stacked in, or passed through, drying chambers which are warmed with the waste heat from the firing process. Drying temperatures and humidity levels are carefully controlled to ensure shrinkage without distortion.

Firing

Both intermittent and continuous kilns are used for firing bricks. The former is a batch process in which the single kiln is loaded, fired, cooled and unloaded. In continuous kilns, the firing process is always active; either the green bricks are moved through a fixed firing zone, or the fire is gradually moved around a series of interconnecting chambers to the unfired bricks. Both continuous systems are more energy efficient than the intermittent processes. Generally, for large-scale production, the continuous tunnel kiln (Fig. 1.4) and the Hoffman kiln (Fig. 1.5) are used. Down-draught kilns, clamps and intermittent gas-fired kilns are used for the more specialised products. Dependent on the composition of the clay and the nature of the desired product, firing temperatures are set to sinter or vitrify the clay. Colour variations called *kiss-marks* occur where bricks were in contact with each other within the kiln and are particularly noticeable on *Flettons*.

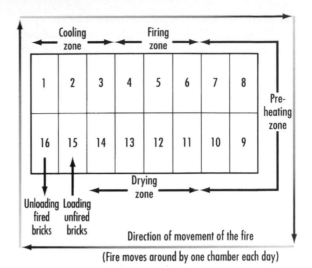

Fig. 1.5 Hoffman kiln plan

Tunnel kiln

In the *tunnel kiln* process the bricks are loaded 10 to 14 high on kiln cars which are moved progressively through the preheating, firing and cooling zones. A carefully controlled temperature profile within the kiln and an appropriate kiln car speed ensures that the green bricks are correctly fired with the minimum use of fuel, usually natural gas. The maximum firing temperature within the range 940°C and 1200°C depends upon the clay, but is normally around 1050°C, with an average kiln time of three days. The oxygen content within the atmosphere of the kiln will affect the colour of the brick products. Typically a high temperature and low oxygen content are used in the manufacture of blue bricks. A higher oxygen content will turn any iron oxide within the clay red.

Hoffman kiln

Introduced in 1858, the Hoffman kiln is a continuous kiln in which the fire is transferred around a series of

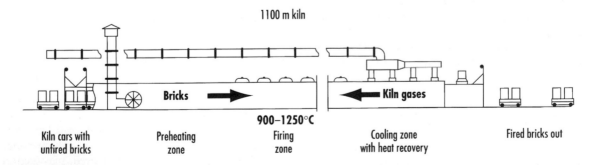

Fig. 1.4 Tunnel kiln

chambers which can be interconnected by the opening of dampers. There may be 12, 16 or 24 chambers, although 16 is usual. The chambers are filled with typically 100 000 green bricks. The chambers in front of the fire, as it moves around, are preheated, then firing takes place (960°–1000°C), followed by cooling, unloading and resetting of the next load. The sequence moves on one chamber per day, with three days of burning. The usual fuel is natural gas, although low-grade coal and landfill methane are used by some manufacturers.

Intermittent gas-fired kilns

Intermittent gas-fired kilns are frequently used for firing smaller loads, particularly *specials*. In one system, green bricks are stacked onto a concrete base and a mobile kiln is lowered over the bricks for the firing process. The firing conditions can be accurately controlled to match those within continuous kilns.

Clamps

The basis of clamp firing is the inclusion of coke breeze into the clay, which then acts as the major source of energy during the firing process. In the traditional process alternate layers of unfired bricks and additional coke breeze are stacked up and then sealed over with waste bricks and clay. The clamp is then ignited with kindling material and allowed to burn for two to five weeks. After firing, the bricks are hand selected because of their variability from under- to over-fired. More recently, gas-fired clamps have been developed which give a fully controlled firing process but still produce bricks with the characteristic dark patches on their surfaces due to the burnt breeze content.

Down-draught kilns

Down-draught kilns are used for the high-temperature firing, especially of engineering bricks, in an intermittent process. Fuel is burnt around the perimeter of the kiln, which is stacked with green bricks. The hot gases rise towards the domed roof, forcing down the cooler gases through a perforated floor and out to the chimney. Thus the heat is retained and the very high temperatures for hard-burnt bricks are achieved.

Packaging and distribution

Damaged or cracked bricks are removed prior to packing. Most bricks are now banded and shrink-wrapped into packs of between 300 and 500, for easy transportation by fork-lift truck and specialist road vehicles. Special shapes are frequently shrink-wrapped onto wooden pallets.

SPECIFICATION OF CLAY BRICKS

To specify a particular brick it is necessary to define certain key criteria, which relate to form, durability and appearance. The British Standard BS 3921: 1985 gives a performance specification based on size, frost resistance, soluble-salt content, compressive strength and visual appearance. The European Standard BS EN 771–1: 2003 requires an extensive minimum description for masonry units including, the European Standard number and date (e.g. BS EN 771–1: 2003), the type of unit (e.g. high density – HD), dimensions and tolerances from mean value, configuration (e.g. a solid or frogged brick), compressive strength and freeze/thaw resistance. Also, depending upon the particular end use, additional description may be required. This may, as appropriate, include dry density, dimensional tolerance range, water absorption, thermal properties, active soluble salts content, moisture movement, reaction to fire and vapour permeability.

Within the building industry the classification usually also includes some traditional descriptions:

- place of origin and particular name (e.g. *Staffordshire smooth blue*);
- clay composition (e.g. Gault, Weald or Lower Oxford Clay, Etruria Marl, Keuper Marl [Mercian Mudstones] or shale);
- variety – typical use (e.g. Class A engineering, common or facing);
- type – form and manufacturing process (e.g. solid, frogged, wire cut);
- appearance – colour and surface texture (e.g. coral red rustic).

Variety

Bricks may be described as common, facing or engineering.

Common bricks

Common bricks have no visual finish, and are therefore usually used for general building work especially where the brickwork is to be rendered, plastered or will be unseen in the finished work.

Facing bricks

Facing bricks are manufactured and selected to give an attractive finish. The particular colour, which may be uniform or multicoloured, results from the blend of clay used, and the firing conditions. Additionally, the surface may be smooth, textured or sand-faced as required. Facing bricks are used for most visual brickwork where a pleasing and durable finish is required.

Engineering bricks

Engineering bricks are dense and vitreous, with specific load bearing characteristics and low water absorption. The two classes (A and B) are defined specifically according to their minimum crushing strengths and maximum water absorption (Table 1.2), but in addition most engineering bricks have high density, good frost resistance and low soluble-salt content. Engineering bricks are used to support heavy loads, and also in positions where the effects of impact damage, water absorption or chemical attack need to be minimised. They are generally *reds* or *blues* and more expensive than other machine-made facing bricks because of their higher firing temperature.

Type

Type refers to the form of the brick and defines whether it is solid, frogged, cellular, perforated, or of a special shape (Figs. 1.6 and 1.7). Bricks may be frogged on one or both bed faces; perforations may be few and large or many and small. Cellular bricks have

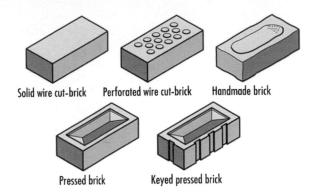

Fig. 1.6 Brick types

cavities closed at one end. Keyed bricks are used to give a good bond to plaster or cement rendering. Because of the wide range of variation within brick types, the manufacturer is required to give details of the orientation and percentage of perforations in all cases.

For maximum strength, weather resistance and sound insulation, bricks should be laid with the frogs uppermost so that they are completely filled with mortar; with double-frogged bricks the deeper frog should be uppermost. However, for cheapness, speed and possibly to minimise the dead weight of construction, frogged bricks are frequently laid frog-down. Inevitably this leads to a resultant reduction in their load-bearing capacity.

Standard specials

Increasingly, *specials* (special shapes) are being used to enhance the architectural quality of brickwork. British Standard BS 4729: 2005 illustrates the range of standard specials, which normally can be made to order to match standard bricks (Fig. 1.7).

Designation of standard specials:

Angle and cant bricks
Bullnose bricks
Copings and cappings
Plinth bricks
Arch bricks
Radial bricks
Soldier bricks
Cuboid bricks
Bonding bricks
Brick slips

Table 1.2   Properties of clay engineering bricks

Physical property	Clay engineering bricks	
	Class A	Class B
Defined properties		
Minimum compressive strength (MPa)	≥ 70	≥ 50
Maximum water absorption (% by mass)	< 4.5 (and DPC1)	< 7.0 (and DPC2)
Typical additional properties		
Net dry density (kg/m³)	≥ 2200	≥ 2100
Freeze/thaw resistance class	F2	F2
Active soluble salts content class	S2	S2

Note: The water absorption limits for all clay bricks used for damp-proof courses for buildings (DPC1) and external works (DPC2) are included in the table.

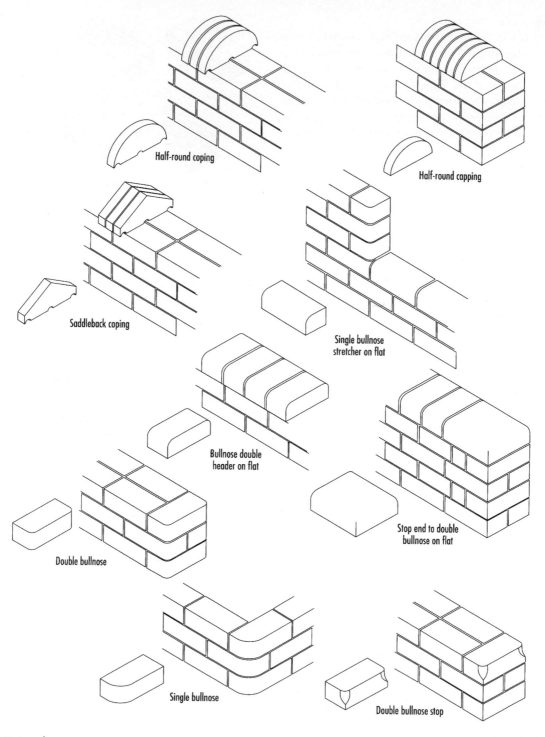

Fig. 1.7 Specials.

Manufacturers also frequently make purpose-made specials (*special specials*) to the particular requirements of the architect or builder. Inevitably, delivery on specials takes longer than for ordinary bricks, and their separate firing frequently leads to some colour variation between the specials and the standard bricks, even where the clay used is identical. The more complex specials are handmade, usually in specially shaped stock

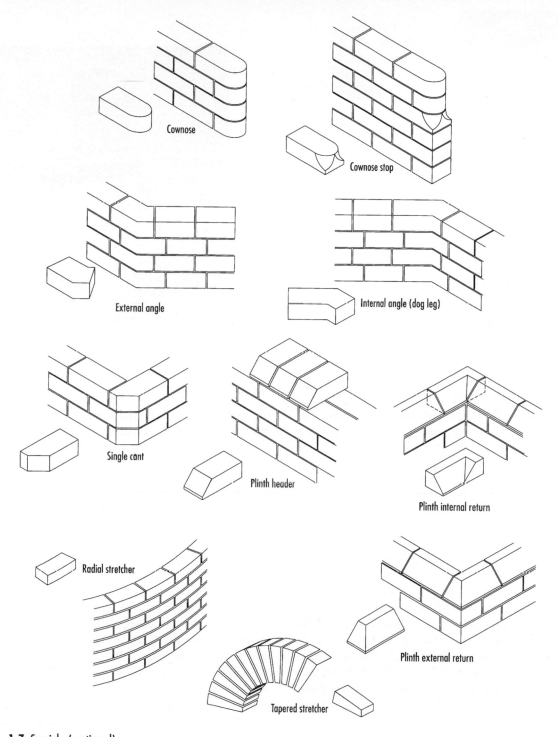

Fig. 1.7 Specials. (continued)

moulds, although some can be made by modifying standard bricks before firing. The range of shapes includes copings and cappings (for parapets and freestanding walls), bullnose (for corner details, e.g. window and door reveals), plinths (for corbelling details and cills), cants (for turning angles), arches and brick slips (to mask reinforced concrete lintels, etc.). Special bricks are also manufactured by cutting standard bricks, then, if

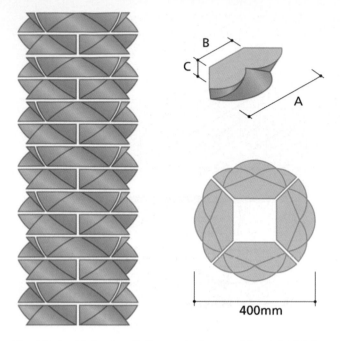

Fig. 1.7 Specials (continued). Photograph: Courtesy of Ibstock Brick Ltd

necessary, bonding the pieces with epoxy resins. This has the advantage of ensuring an exact colour match to the standard bricks. Many brick slips and arch voussoir sets (bricks to create an arch) are produced by this method.

APPEARANCE

The colour range of bricks manufactured in the UK is extensive. The colours range from the light buffs, greys and yellows through pastel pink to strong reds, blues, browns and deep blue/black, depending mainly upon the clay and the firing conditions, but also on the addition of pigments to the clay or the application of a sand facing. Colours may be uniform, varied over the surface of individual bricks or varied from brick to brick. The brick forms vary from precise to those with rounded arrises; textures range from smooth and sanded to textured and deeply folded, depending upon the forming process (Fig. 1.8).

In view of the variability of bricks from batch to batch it is essential that they should be well mixed, preferably at the factory before palleting, or failing this, on site. If this is not done sufficiently, accidental colour banding will appear as the brickwork

proceeds. Sand-faced bricks are liable to surface damage on handling, which exposes the underlying colour of the brick. Chipping of the arrises on bricks with *through colour* is visually less detrimental. Where rainwater run-off is an important factor, e.g. on cills and copings, smooth rather than heavily rusticated bricks should be used, as the latter would saturate and stain. Handmade bricks with deep surface folds should be laid frog-up so that the creases or *smiles* tend to shed the rainwater from the face of the brickwork.

Glazed bricks, available in a wide range of intense colours, are sometimes used for their strong aesthetic effect (Fig. 1.9) or resistance to graffiti. They are commonly manufactured in a two-stage process, which involves the initial firing of the green brick to the *biscuit* stage, followed by the application of a *slip glaze* and a second firing. In an alternative one-stage process, a clear slip glaze is applied before firing to allow the natural colour of the brick to show through.

The visual acceptability of facing bricks and the quality of the bricklaying would normally be assessed on site by the construction of a reference panel to an agreed standard, using at least 100 randomly selected bricks with examples of any colour banding, the proposed bonding, mortar and jointing. All subsequent

Handmade multi orange

Staffordshire blue smooth brindle

Multi yellow stock

Dark red stock

Buff rusticated wire-cut

Red creased wire cut

Silver grey combed wire-cut

Red mixture wire-cut

Fig. 1.8 Typical range of clay brick textures and colours. Photographs: Courtesy of Ibstock Brick Ltd

Fig. 1.9 Glazed bricks – Atlantic House, London. Architects: Proun Architects. Photograph: Courtesy of Ibstock Brick Ltd

Table 1.3 Designation of freeze/thaw resistance and active soluble salts content for clay bricks

Durability designation	Freeze/thaw resistance
F2	masonry subjected to severe exposure
F1	masonry subjected to moderate exposure
F0	masonry subjected to passive exposure
	Active soluble salts content
S2	sodium/potassium 0.06%, magnesium 0.03%
S1	sodium/potassium 0.17%, magnesium 0.08%
S0	no requirement

F2 bricks are totally resistant to repeated freezing and thawing when in a saturated condition. Category F1 bricks are durable, except when subjected to repeated freezing and thawing under saturated conditions. Therefore, category F1 bricks should not be used in highly exposed situations such as below damp-proof courses, for parapets or brick-on-edge copings, but they are suitable for external walls which are protected from saturation by appropriate detailing. Category F0 bricks must only be used where they are subject to passive exposure, as when protected by cladding or used internally.

Soluble-salt content

The soluble-salt content of bricks is defined by three categories: low (S2), normal (S1) and no limits (S0) (Table 1.3). Both the S2 and S1 categories have defined maximum limits for sodium/potassium and magnesium salt contents. The soluble salts derive from the original clay or from the products of combustion during the firing process. Soluble salts can cause efflorescence and soluble sulfates may migrate from the bricks into the mortar or any rendering, causing it to expand and deteriorate by sulfate attack. If used in an exposed situation S1 and S0 category bricks should be bonded with sulfate-resisting cement mortar.

Efflorescence

Efflorescence sometimes appears as a white deposit on the surface of new brickwork. It is caused by moisture carrying salts from inside the bricks and mortar to the surface where the water evaporates leaving the crystalline salts. Under most conditions it disappears without deleterious effect within one year. In exposed

brick deliveries and constructed brickwork should then be checked against the reference panel.

DURABILITY

Frost resistance

Bricks are classified into one of the three categories, F2, F1 and F0 according to their frost resistance within a standardised freezing test (Table 1.3). Only category

brickwork that is constantly subjected to a cycle of wetting and drying, efflorescence can occur at any time; further, a build-up and expansion of crystalline salts under the surface (*crypto-efflorescence*) may cause the face of the brickwork to crumble or spall.

Staining

The surface of brickwork may be stained by cement during the building process, or by lime leaching out of the fresh mortar (Fig. 1.10). In either case the excess should be brushed and washed off, without saturating the brickwork.

PHYSICAL PROPERTIES

Compressive strength

High density (HD) clay bricks are available with a range of compressive strengths from around 5 MPa to well over 100 MPa. The criteria for general use, damp-proof courses and engineering use are set out in Table 1.2 (p. 7).

To determine the crushing strength of bricks, both bed faces are ground down until flat and parallel. The bricks are then crushed without filling the voids or frogs. Where frogs are to be laid upwards and filled in the construction, the crushing strength (MPa) is based on the net bearing area. Where frogs or voids are not to be filled, the crushing strength is based on the full gross area of the bed face.

Water absorption and suction

The level of water absorption is critical when bricks are to be used for damp-proof courses, or as engineering bricks. Appropriate limits are shown in Table 1.2, although generally absorption ranges from 1 to 35%. Suction rates are now quoted by most brick manufacturers, as high values can adversely affect the bricklaying process. Bricks with high suction rates absorb water rapidly from the mortar, making it insufficiently plastic to allow for repositioning of the bricks as the work proceeds. Generally, low or medium suction rates (1.0–2.0 kg/m^2 per min) are advantageous. In warm weather, high-suction-rate bricks may be wetted in clean water before laying, but any excess water will cause the brick to float on the mortar bed and will also increase the risk of subsequent efflorescence and staining.

Fig. 1.10 Lime leaching on brickwork

Moisture and thermal movement

After the firing process bricks absorb moisture from the atmosphere and expand irreversibly, up to a maximum of 0.1%. It is therefore recommended that bricks should not be used for at least two weeks after firing, (although it is now recognised that this irreversible process may continue at a decreasing rate for 20 years). Subsequent moisture and thermal movements are largely reversible and movement joints allowing for a 1 mm movement per 1 m of brickwork should be allowed, typically at 10–12 m centres and at a maximum of 15 m, in restrained walls. Unrestrained or lightly restrained walls should have movement joints at 7–8 m centres. Horizontal movement joints should be at approximately 12 m intervals, as the vertical movement is of the same order as movement in the horizontal direction.

For many buildings the necessary movement joints can be made inconspicuous by careful detailing or featured as part of the design. Appropriate locations for movement joints would be where differing structural forms adjoin, such as abutments between walls and columns or where the height or thickness of a wall changes; alternatively, at design details such as brickwork returns, re-entrant corners, or the recesses for downpipes. In expansion joints, fillers such as cellular polythene, polyurethane or foam rubber should be used, as these are easily compressible. Pointing should be with a flexible sealing compound such as two-part polysulfide.

Typical reversible moisture movement = 0.02%
Typical reversible thermal movement = 0.03%
Thermal movement = $5–8 \times 10^{-6}$ deg C^{-1}

Thermal conductivity

The thermal conductivity of brickwork is dependent upon its density and moisture content but generally clay bricks are poor thermal insulators. Brick manufacturers quote thermal conductivities at a standard 5% moisture content for exposed brickwork, and may also give the 1% moisture content figure for protected brickwork.

Using bricks with an average thermal conductivity of 0.96 W/m K, a typical partially filled cavity system is:

102.5 mm fairfaced brickwork
50 mm clear cavity
50 mm foil-faced rigid polyurethane insulation (λ = 0.023 W/m K)

100 mm lightweight blockwork (λ = 0.15 W/m K)
12.5 mm plasterboard on dabs

giving a U-value of 0.27 W/m^2 K

The thermal conductivity of clay bricks at 5% moisture content typically ranges between 0.65 and 1.95 W/m K.

Fire resistance

Clay brickwork generally offers excellent fire resistance by retaining its stability, integrity and insulating properties. The standard (BS 5628–3: 2001) indicates that 100 mm and 200 mm of load-bearing solid clay brick masonry will give 120 minutes and 360 minutes of fire resistance, respectively. Bricks with less than 1% organic material are automatically categorised as Euroclass A1 with respect to reaction to fire.

Acoustic properties

Good-quality brickwork is an effective barrier to airborne sound, provided that there are no voids through the mortar for the passage of sound. All masonry joints should be sealed and bricks laid with filled frogs to achieve the necessary mass per unit area and avoid air pathways.

At the junction between a cavity blockwork separating wall and an external brick and blockwork wall, if the external cavity is not fully filled with thermal insulation, then the separating wall cavity must be closed with a flexible cavity stop to reduce sound transmission sufficiently to comply with the Building Regulations Part E performance requirements.

Impact sound absorption by brickwork over the normal frequency range is fairly low and further decreased by the application of dense plaster or paint. However, the application of acoustic plasters or the addition of an independent panel of plasterboard backed by absorbent material improves impact sound insulation.

QUALITY CONTROL

To meet the consistent standards of quality required by clients, many brick manufacturers are now operating quality-assurance systems. These require manufacturers to document all their operational procedures and set out standards to which products must adhere. Quality is controlled by a combination of an internal self-monitoring system and two to four independent spot-check reviews per year. Both the content of the technical literature and the products themselves are subjected to this scrutiny.

Brickwork

CLAY BRICKWORK

The bonding, mortar colour and joint profile have a significant visual effect on brickwork. The overall effect can be to emphasise as a feature, or reduce to a minimum, the impact of the bonding mortar on the bricks. Additionally the use of polychromatic brickwork with complementary or contrasting colours for quoins, reveals, banding and even graphic designs can have a dramatic effect on the appearance of a building. The three-dimensional effects of decorative dentil courses and projecting corbelled features offer the designer further opportunities to exploit the effects of light and shade. Normally, a projection of 10–15 mm is sufficient for the visual effect without causing increased susceptibility to staining or frost damage. Curved brickwork constructed in stretcher bond shows faceting and the overhang effect, which is particularly accentuated in oblique light. With small-radii curvatures, the necessary change of bonding pattern to header bond can also be a visual feature, as an alternative to the use of curved-radius bricks.

The *Gothic Revival* exterior of the Queens Building, De Montfort University, Leicester (Fig. 1.11), illustrates the visual effects of polychromatic brickwork and voussoir specials. The energy-efficient building maximises use of natural lighting, heating and ventilation, using massive masonry walls to reduce peak temperatures. The mortar, which matches the external coral-red brickwork, reduces the visual impact of the individual bricks, giving the effect of planes rather than walls. This is relieved by the colour and shadow effects of the polychromatic and corbelled features, which are incorporated in the ventilation grilles and towers. The special bricks, cill details and banding are picked out in a deeper cadmium red and silver buff to contrast with the characteristic Leicestershire red-brick colouring.

Mortars

The mortar in brickwork is required to give a bearing for the bricks and to act as a sealant between them. Mortars should be weaker than the individual bricks, to ensure that any subsequent movement does not cause visible cracking of the bricks, although too weak a mix would adversely affect durability of the brickwork. Mortar mixes are based on blends of either cement/lime/sand, masonry cement/sand or cement/sand with plasticiser. When the mix is gauged by volume an allowance has to be made for bulking of damp sand. The five mix designations are shown in Table 1.4. A typical 1 : 1 : 6 (cement : lime : sand) mix (designation (iii)) would generally be appropriate and durable for low-rise construction, but for calculated structural brickwork or for increased resistance to frost in exposed situations a greater-strength mortar (designation (i) or (ii)) may be required. In the repointing of old brickwork it is particularly important to match the porosity of the brick to the water-retention characteristics of the mortar. This prevents excessive loss of water from the mortar before hydration occurs, which may then cause the pointing to crumble.

The use of lime mortar, as in the Building Research Establishment environmental building in Garston, Watford, will allow for the ultimate reuse of the bricks at the end of the building's life-cycle. The recycling of bricks is not possible, except as rubble, when strong Portland cement mortar is used.

Sands for mortars are normally graded to BS EN 13139: 2002 into categories designated by a pair of sieve sizes d/D which define the lower and upper size limits in mm respectively. The majority of the particle size distribution should lie between the stated limits. The preferred grades are 0/1 mm, 0/2 mm, 0/4 mm, 0/8 mm, 2/4 mm and 2/8 mm. Typically, between 85 and 99% of the sand should pass through the larger sieve limit, and between 0 and 20% should pass through the smaller sieve size limit. The grades with more fines (63 micron or less) require more cement to achieve the same strength and durability as the equivalent mortars mixed with a lower fines content.

Ideally, brickwork should be designed to ensure the minimal cutting of bricks, and should be built with a uniform joint width and vertical alignment of the joints (perpends). During construction, brickwork should be kept clean and protected from rain and frost. This reduces the risk of frost damage, patchiness and efflorescence. Brickwork may be rendered externally or plastered internally if sufficient mechanical key is provided by appropriate jointing or the use of keyed bricks. For repointing existing brickwork, it is necessary to match carefully the mortar sand, and to use lime mortar where it was used in the original construction.

Fig. 1.11 Decorative brickwork – Queens Building, De Montfort University, Leicester. Architects: Short Ford & Associates. Photograph: Arthur Lyons

Bonding

Figure 1.12 illustrates the effects of bonding. The stretcher bond is standard for cavity walls and normally a half-lap bond is used, but an increase in horizontal emphasis can be achieved by the less standard quarter or third bond. In conservation work it may be necessary to use half bricks (snap headers) to match the appearance of bonding in solid brick walls. For one-brick-thick walls more variations are possible; most typical are the English and Flemish bonds. The equivalent English and Flemish garden wall bonds, which have more stretchers, are primarily used for one-brick-thick walls where the reduced number of headers makes it easier to build both sides fairfaced. Panels of herringbone brickwork (raking bond), or dog tooth and dentil courses as in Victorian brickwork, can generate interesting features.

In all cavity brickwork, wall ties manufactured from galvanised steel, stainless steel or polypropylene to BS EN 845–1: 2003 should be incorporated (Fig. 1.13). They should be laid drip down and level or sloping down towards the outer leaf. Where mortar bed-joints do not co-ordinate between masonry leaves, slope-tolerant cavity wall ties must be used. In

Table 1.4 Mortar mix designations

Designation	Cement:lime:sand	Masonry cement:sand	Cement:sand with plasticiser
(i)	1:0:3 - 1:¼:3		
(ii)	1:½:4 - 1:½:4 ½	1:2 ½ - 1:3 ½	1:3 - 1:4
(iii)	1:1:5 - 1:1:6	1:4 - 1:5	1:5 - 1:6
(iv)	1:2:8 - 1:2:9	1:5 ½ - 1:6 ½	1:7 - 1:8
(v)	1:3:10 - 1:3:12	1:6 ½ - 1:7	1:8

partially filled cavities, the wall ties should clip the insulation cavity batts to the inner leaf. In all cases the cavity, insulation and ties should be kept clear of mortar droppings and other residues by using a protective board. With the widening of cavities associated with increased insulation, the use of the traditional butterfly, double triangle and vertical twist ties in galvanised steel will be increasingly replaced by longer stainless steel ties, which do not suffer from corrosion in the more aggressive environments. Asymmetric wall ties are used for fixing masonry to timber or thin-joint aircrete blockwork. Movement-tolerant wall ties bend, or slide within a slot system fixed to one leaf of the masonry.

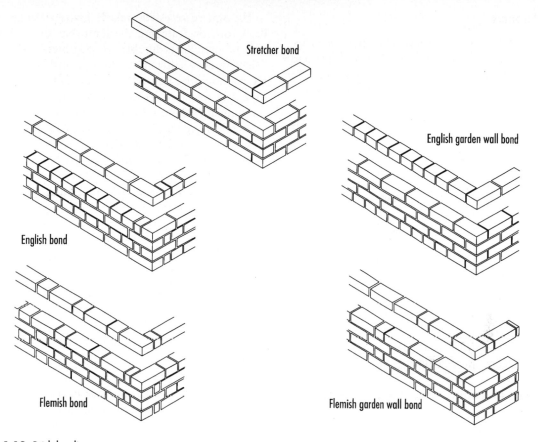

Stretcher bond

English garden wall bond

English bond

Flemish bond

Flemish garden wall bond

Fig. 1.12 Brick bonding

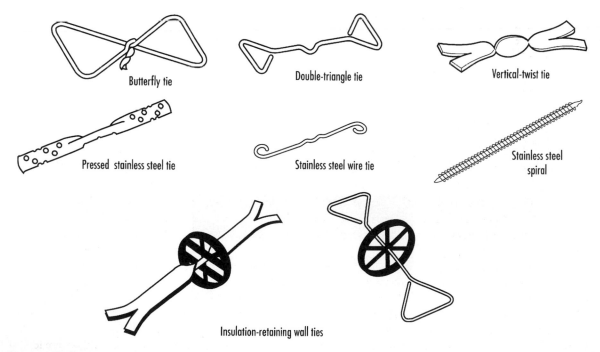

Butterfly tie

Double-triangle tie

Vertical-twist tie

Pressed stainless steel tie

Stainless steel wire tie

Stainless steel spiral

Insulation-retaining wall ties

Fig. 1.13 Wall ties

Coloured mortars

Mortar colour has a profound effect on the overall appearance of the brickwork as, with stretcher bond and a standard 10 mm joint, the mortar accounts for 17% of the brickwork surface area. A wide range of light-fast coloured mortars is available which can be used to match or contrast with the bricks, thus highlighting the bricks as units or creating a unity within the brickwork. The coloured mortars contain inert pigments, which are factory-blended to a tight specification to ensure close colour matching between batches. Occasionally, black mortars may bloom due to lime migration to the surface. Coloured mortars can be used creatively to enhance the visual impact of the brickwork and even create designs on sections of otherwise monochromatic brickwork. The quantity of pigment should not exceed 10% by weight of the cement.

Mortar colours may also be modified by the use of stains after curing; however, such applications only penetrate 2 mm into the surface, and therefore tend to be used more for remedial work. Through-body colours are generally more durable than surface applications.

Joint profiles

The standard range of joint profiles is illustrated in Figure 1.14. It is important that the main criteria should be the shedding of water to prevent excessive saturation of the masonry, which could then deteriorate. Normally the brickwork is jointed as the construction proceeds. This is the cheapest and best method as it gives the least disturbance to the mortar bed. Pointing involves the raking out of the *green* mortar to a depth of 13–20 mm, followed by refilling the joint with fresh mortar. This is only appropriate when the desired visual effect cannot be obtained directly by jointing; for example, when a complex pattern of coloured mortar joints is required for aesthetic reasons.

The square recessed (raked) joints articulate the brickwork by featuring the joint, but these should only be used with durable (F2, S2) high-absorption bricks under sheltered conditions; furthermore, the recess should be limited to a maximum depth of 6 mm. The struck or weathered joint also accentuates the light and shade of the brickwork while, as a tooled joint, offering good weather resistance in all grades of exposure. If the visual effect of the joint is to be diminished, the flush joint may be used, but the curved recessed (bucket-handle) joint, which is compressed by tooling, offers better appearance and weathering properties. No mortar should be allowed to smear the brickwork, as it is difficult to remove subsequently without the use of dilute acid or pressure jets of water.

Reinforced brickwork

Reinforcement may be introduced vertically or horizontally into brickwork (Fig. 1.15). Bed-joint reinforcement, usually austenitic stainless steel, should be completely surrounded by mortar with a minimum cover of 15 mm. For continuity in long walls, sections of reinforcement should be sufficiently end lapped. Vertical reinforcement is possible in the cavity or in pocket-type walls, where the void spaces are formed in the brickwork, then reinforcement and concrete are introduced after the masonry is completed. Care should be taken in the use of vibrators to compact the concrete within new masonry.

Decorative brickwork

Plaques, motifs, murals and sophisticated sculptures (Fig. 1.16) can be manufactured to individual designs both for new buildings and for the renovation or refurbishment of Victorian *terracotta*. The designs are carved as a bas-relief in soft solid through-colour brickwork or moulded in the unfired clay in relatively small units and joined on site with a matching mortar.

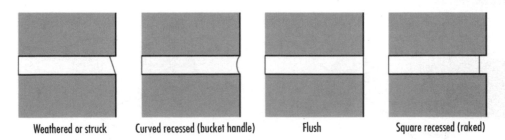

Weathered or struck   Curved recessed (bucket handle)   Flush   Square recessed (raked)

Fig. 1.14 Joint profiles

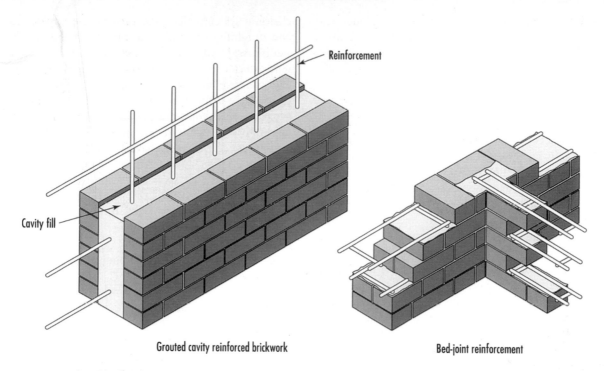

Grouted cavity reinforced brickwork

Bed-joint reinforcement

Fig. 1.15 Reinforced brickwork

Fig. 1.16 Decorative carved brickwork. Photograph: Courtesy of Baggeridge Brick plc

For repetitive units, the clay is shaped in an appropriate wooden mould. Relief depths of 10–30 mm give shadow and contrast sufficient for most sculptural effects to be seen, although the viewing distances and angles must be considered. For large brickwork sculptures, the whole unit may be built in green bricks, with allowances made for the mortar joints and drying contraction. The design is then carved, numbered, dismantled, fired and reassembled on site.

Thin-bed masonry

The use of thin-bed masonry, with joints of between 2 and 6 mm, significantly reduces the visual effect of the mortar joints from 17% in 10-mm-joint standard brickwork, to only 8% in 4 mm joints. This effect is further enhanced by the use of glue-mortars which are applied to create a recessed joint. Thus the joint becomes only a shade line and the visual effect of the wall is totally determined by the colour and texture of the bricks. Because the glue-mortar is stronger than traditional mortar and has tensile properties, the brickwork patterns are not constrained to standard stretcher bonding. The glue-mortar is applied in two lines to both the horizontal and vertical joints, and therefore solid or perforated bricks rather than frogged bricks are most appropriate. Thin-bed masonry wall ties and special aramid bed-joint reinforcement are used as appropriate. The system offers the creative designer significant alternative aesthetic opportunities.

Preassembled brickwork

The use of pre-assembled brickwork supported on reinforced concrete or steel frames offers the builder a potentially higher level of quality control and increased speed of construction on site (Fig. 1.17). It also offers the scope to create complex details, and forms such as long low arches, that would be expensive or impossible in traditional brick construction. Specialist manufacturers produce large complete brick-clad precast-concrete panels either with whole bricks or brick slips. Typically, the rear faces of brick slips are drilled at an angle, then stainless steel rods inserted and fixed with resin adhesive. The brick slips are laid out with spacers within the panel mould, prior to the addition of steel reinforcement and concrete. Finally, the brick slips are pointed up giving the appearance of normal brickwork.

Brick cladding systems

A significant revolution for brick-faced building has been the development of brick slip and brick tile cladding systems, designed to have the appearance and durability of traditional brickwork, but with a significantly reduced construction time. In one system, external walls are constructed with 215 mm aerated concrete blockwork and faced with an extruded polystyrene insulation panel to which 16 mm brick slips are applied onto the pre-formed grid, giving the appearance of standard external leaf brickwork. The polystyrene grid panels have an overlap to ensure horizontal joints are watertight and are tongued and grooved to interlock vertically. Adhesive is applied to the polystyrene and the brick slips are pushed into place with the appropriate horizontal spacing. Mortar is applied either with a pointing gun or a mortar bag and tooled to the required joint profile. With the use of highly insulating blocks, this type of construction can achieve U-values as low as 0.27 W/m^2 K.

An alternative system uses a plastic-coated galvanised steel profile fixed to the structural wall (Fig. 1.18). The specially shaped brick tiles then clip into the steel system with appropriate vertical joint spacing. Mortar (typically a 1 : 1 : 6 mix) is applied with a pointing gun and smoothed off to the required profile, usually bucket-handle. A range of special tiles is manufactured to produce dados,

Fig. 1.17 Preassembled brickwork. Photograph: Courtesy of Trent Concrete Ltd

Fig. 1.18 Brick cladding system. Photograph: Courtesy of Corium, a division of Baggeridge Brick plc

plinths, cills and external returns, giving the appearance of traditional brickwork. Because the brickwork is non-structural, a range of bond patterns including stack, quarter and diagonal is optional. This type of pre-fabrication offers the potential for increased off-site construction work, and some manufacturers supply pre-formed brick-tile panels ready for fixing on site.

CLAY BRICK PAVING

Many clay brick manufacturers produce a range of plain and chamfered paving bricks together with a matching range of paver accessories. Bricks for flexible paving are usually nibbed to set the spacing correctly. The material offers a human scale to large areas of hard landscape, especially if creative use is made of pattern and colour. Typical patterns (Fig. 1.19) include herringbone, running bond, stack bond, basket-weave and the use of borders and bands. Profiled brick designs include decorative diamond and chocolate-bar patterns, and pedestrian-management texturing. The paving bricks may be laid on a hard base with mortar joints or alternatively on a flexible base with fine sand brushed between the pavers. Edge restraint is necessary to prevent lateral spread of the units.

The British Standard (BS EN 1344: 2002) stipulates minimum paver thicknesses of 40 mm and 30 mm for flexible and rigid construction respectively. However, 50 mm pavers are generally used for flexible laying and 60 mm pavers are necessary when subjected to substantial vehicular traffic (BS 7533–1: 2001). Table 1.5 shows the standard sizes. Clay pavers are classified by freeze/thaw resistance. Pavers with designation FP0 are unsuitable for saturated freezing conditions, while pavers designated FP100 may be used under freeze/thaw conditions. The Standard BS EN 1344: 2002 classifies five categories (T0 to T4) of transverse breaking strength, with the lowest category T0 being only appropriate for rigid construction. Slip resistance for the unpolished pavers is categorised as high, moderate, low or extremely low. This factor needs to be considered particularly for potentially wet conditions to ensure safe pedestrian and traffic usage.

Fig. 1.19 Typical range of clay pavers. Photographs: Courtesy of Baggeridge Brick plc

Table 1.5 Standard work sizes for pavers

Length (mm)	Width (mm)	Thickness (mm)
215	102.5	50
215	102.5	65
210	105	50
210	105	65
200	100	50
200	100	65

Calcium silicate bricks

Calcium silicate bricks, also known as sandlime or flintlime bricks, were first produced commercially in Germany in 1894, and then in the UK in 1905. Initially their use was confined to common brick applications, but in the 1950s, their durability for foundations was exploited. Research into mix design and the development of improved manufacturing processes subsequently led to the production of a full range of load-bearing-strength classes and attractive facings. Calcium silicate bricks are competitively priced and account for about 3% of the UK brick market.

SIZE

The work size for calcium silicate bricks is $215 \times 102.5 \times 65$ mm, the same as for clay bricks, with a co-ordinating size of $225 \times 112.5 \times 75$ mm, allowing

Stack bond

Herringbone

Running bond

Basket-weave

Fig. 1.19 Continued. Typical range of clay pavers and hard land-scape at Birmingham. Photographs: Courtesy of Baggeridge Brick Plc

for 10 mm mortar joints. Generally, calcium silicate bricks are more accurate in form and size than fired clay bricks, which inevitably distort in the manufacturing process. The dimensional tolerances for calcium silicate bricks defined in the standard (BS EN 771–2: 2003) are generally ±2 mm on each dimension, except for thin layer mortar when a maximum of only ±1 mm tolerance is permitted on the height.

MANUFACTURE OF CALCIUM SILICATE BRICKS (SANDLIME AND FLINTLIME BRICKS)

The raw materials are silica sand (approximately 90%), hydrated lime, crushed flint, colouring pigments and water. (If quicklime is used, it is fully hydrated before the bricks are pressed, to prevent expansion under the steam treatment.) A mixture of sand, lime and water is used to manufacture the natural white sandlime brick. The addition of colouring pigments or crushed-flint aggregate to the standard components, or the application of texturing to the brick surface, gives the wider product range.

The appropriately proportioned blend is pressed into brick units, stacked on bogies, moved into the autoclave and subjected to steam pressure (0.8–1.3 MPa) for 4 to 15 hours at 180°C (Fig. 1.20). This causes the hydrated lime to react chemically with the surface of the sand particles, enveloping them with hydrated calcium silicates which fill much of the void spaces between the sand particles. Subsequently the calcium silicates react slowly with carbon dioxide from the atmosphere to produce calcium carbonate, with a gradual increase in the strength of the bricks.

APPEARANCE

The manufacturing process results in accurate shapes and dimensions, and with the untextured calcium silicate bricks, a smooth finish. The colour range is extensive, from white and pastel shades through to deep reds, blues, browns, greens and yellows. The visual effect on the brickwork tends to be that of precision. The bricks tend to be more brittle than clay bricks and are therefore more susceptible to damage on their arrises.

SPECIFICATION OF CALCIUM SILICATE BRICKS

Types

Both solid and frogged calcium silicate bricks are available. Manufacturers produce a wide range of matching specials to BS 4729: 2005; *special specials* to clients' requirements; and brick slips for facing reinforced concrete.

Durability

Calcium silicate bricks have good frost resistance, but should not be exposed repeatedly to either strong salt solutions, acids or industrial effluent containing magnesium or ammonium sulfates. The bricks have a negligible salt content and therefore efflorescence, and sulfate attack on the mortar, cannot arise from within the bricks. The bricks are themselves resistant to sulfate attack and can therefore be used below ground with a suitable sulfate-resisting cement mortar. However, calcium silicate bricks should not be used as pavers where winter salting can be expected.

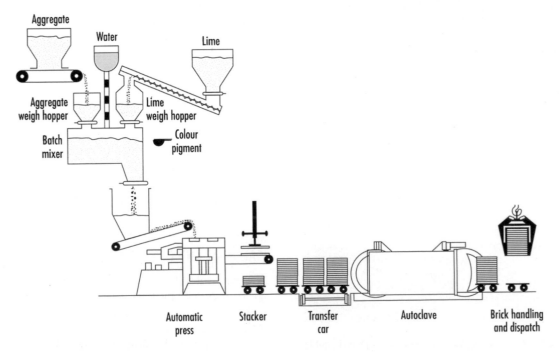

Fig. 1.20 Manufacture of calcium silicate bricks

PHYSICAL PROPERTIES

Compressive strength

The British Standard BS EN 771–2: 2003 defines the range of compressive strength classes, as shown in Table 1.6.

Weight

Most standard calcium silicate bricks weigh between 2.4 and 3.0 kg, but densities can range from below 500 to above 2200 kg/m^3.

Water absorption

Water absorption is usually in the range 8–15% by weight.

Moisture and thermal movement

Unlike clay bricks, which expand after firing, calcium silicate bricks contract. This shrinkage is increased if the bricks become wet before use, therefore site protection of brick stacks from saturation is essential. Similarly, unfinished brickwork should be protected from both saturation and freezing during construction. Reversible moisture movement for calcium silicate bricks is greater than for clay bricks, so expansion joints must be provided at intervals between 7.5 and 9.0 m. Such movement joints should not be bridged by rigid materials. Generally, a weak mortar mix should be used (e.g. 1 : 2 : 9 cement : lime : sand), except below damp-proof course level (DPC) and for copings, to prevent visible cracking of either the mortar or the bricks.

Typical reversible moisture movement = $\pm 0.05\%$
Typical reversible thermal movement = $\pm 0.05\%$
Thermal movement = $8 - 14 \times 10^{-6}$ deg C^{-1}

Thermal conductivity

The thermal conductivities are equivalent to those of clay bricks of similar densities.

The thermal conductivity of calcium silicate brick ranges from 0.6 W/m K (Class 20) to 1.3 W/m K (Class 40).

Fire resistance

The fire resistance of calcium silicate bricks is similar to that of clay bricks, with solid 100 mm calcium silicate brickwork giving 120 minutes and 200 mm giving 360 minutes' fire resistance, according to BS 5628–3: 2001. The standard illustrates only marginal differences in fire resistance between calcium silicate and clay bricks. Calcium silicate bricks (with less than 1% organic material) are designated Euroclass A1 with respect to reaction to fire.

Acoustic properties

Acoustic properties are related to mass and are therefore the same as for clay bricks of equivalent density.

CALCIUM SILICATE BRICKWORK

Most design considerations are the same for either clay or calcium silicate brick. However, calcium silicate bricks are particularly popular for their light reflecting properties, for example in light wells or atria. Their smooth crisp appearance with a non-abrasive surface is particularly appropriate for some interior finishes and also forms an appropriate base for painted finishes. The use of complementary coloured mortars enhances the aesthetic effect when using strongly coloured bricks. Their dimensional accuracy gives some advantage in the bricklaying process, and cost is comparable to that of the equivalent clay bricks.

The interior of the Queens Building of De Montfort University, Leicester (Fig. 1.21) illustrates

Table 1.6 Minimum compressive strength for calcium silicate bricks

Compressive strength class	Normalised compressive strength (MPa)
5	5.0
7.5	7.5
10	10.0
15	15.0
20	20.0
25	25.0
30	30.0
35	35.0
40	40.0
45	45.0
50	50.0
60	60.0
75	75.0

Fig. 1.21 Polychromatic calcium silicate brickwork – Queens Building, De Montfort University, Leicester. Architects: Short Ford & Associates. Photographs: Lens-based media, De Montfort University

the effective use of calcium silicate brickwork in creating a light internal space. Incorporated within the ivory Flemish-bond brickwork are restrained bands of polychromatic features and robust articulation of obtuse-angle quoins. The accuracy of the brickwork emphasises the clarity of the internal form, reflecting the disciplines of engineering that the building houses.

Concrete bricks

Developments in the use of iron oxide pigments have produced a wide range of colour-stable quality concrete-brick products. Currently concrete bricks are competitively priced and hold approximately 10% of the total brick market share.

SIZE

The standard size of concrete bricks is $215 \times 103 \times 65$ mm as for clay bricks, but due to their manufacturing process, concrete bricks can be made to close tolerances, so accurate alignment is easy to achieve on site. Half-brick walls can readily be built fairfaced on both sides. Other sizes, as shown in Table 1.7, are listed in BS 6073–2: 1981.

MANUFACTURE OF CONCRETE BRICKS

Concrete bricks are manufactured from blended dense aggregates (e.g. crushed limestone and sand) together with cement under high pressure in steel moulds. Up to 8% of appropriately blended iron oxide pigments, depending on the tone and depth of colour required, is added to coat the cement particles which will then form the solid matrix with the aggregate. The use of coloured aggregates also increases the colour range. The accurate manufacturing process produces bricks that have clean arrises.

APPEARANCE

A wide range of colours, including multicolours, is available, from red, buff and yellow to green and black. Surfaces range from smooth to simulated natural stone, including those characteristic of handmade and textured clay bricks. Because of the wide range of pigments used in the manufacturing process, it is

Table 1.7 Standard and modular sizes for concrete bricks

	Length (mm)	Width (mm)	Height (mm)
standard	215	103	65
modular	290	90	90
	190	90	90
	190	90	65

possible to match effectively new concrete bricks to old and weathered clay bricks for the refurbishment or extension of old buildings.

SPECIFICATION OF CONCRETE BRICKS

Types

Concrete bricks may be solid, perforated, or frogged, according to the manufacturer.

Three categories are defined: common, facing and engineering. The latter can be manufactured with a range of strengths and densities to specific requirements. A normal range of specials to BS 4729: 2005 is produced, although as with clay and calcium silicate bricks, a longer delivery time must be anticipated. The manufacturer's reference, the crushing strength, the dimensions and the brick type must be clearly identified with each package of concrete bricks. Engineering quality concrete bricks should be used below ground where significant sulfate levels are present according to the classification given in the BRE Special Digest 1, *Concrete in aggressive ground* (2005).

DURABILITY

Concrete bricks are resistant to frost and are therefore usable in all normal levels of exposure. Like all concrete products, they harden and increase in strength with age. As with calcium silicate bricks, they can be made free of soluble salts and thus free from efflorescence. Concrete bricks should not be used where industrial effluents or acids are present.

PHYSICAL PROPERTIES

Weight and compressive strength

The standard brick weighs approximately 3.2 kg and has minimum crushing strength of 7.0 MPa, although

20–40 MPa is the typical range. Engineering bricks have a strength of 40 MPa with a sulfate-resisting Portland or equivalent cement content of 350 kg/m³ (BS 6073–2: 1981).

Water absorption

Water absorption is typically 8%, but engineering-quality bricks average less than 7% after 24 hours cold immersion, and are suitable for aggressive conditions such as retaining walls, below damp-proof course level and for inspection chambers.

Moisture and thermal movement

Concrete bricks have a typical drying shrinkage of 0.04%, with a maximum of 0.06%. Moisture and thermal movements are greater than for calcium silicate bricks and movement joints should be at 5–6 m centres. Because of their moisture movement, prior to laying, concrete bricks should not be wetted to overcome excessive suction, but the water retentivity of the mortar should be adjusted accordingly. Brick stacks should be protected on site from rain, frost and snow.

Thermal conductivity

The thermal conductivities of concrete bricks are equivalent to those of clay and calcium silicate bricks of similar densities. Partially filled cavities, maintaining a clear cavity, are recommended to prevent water penetration to the inner leaf.

The thermal conductivity of concrete bricks ranges between 1.4 and 1.8 W/m K.

An appropriate level of thermal insulation for external walls can be achieved using concrete brickwork. A typical partial cavity fill system is:

102.5 mm concrete-facing brick
50 mm clear cavity airspace
45 mm foil-faced rigid polyurethane insulation ($\lambda = 0.023$ W/m K)
115 mm high performance lightweight blockwork ($\lambda = 0.11$ W/m K)
12.5 mm plasterboard on dabs

giving a U-value of approximately 0.27 W/m² K depending on the thermal conductivity of the concrete bricks used.

Fire resistance

The fire resistance of concrete bricks is of the same order as clay and calcium silicate bricks. Concrete bricks (with less than 1% organic material) are designated Euroclass A1 with respect to reaction to fire.

Acoustic properties

Dense concrete bricks are suitable for the reduction of airborne sound transmission. On a weight basis, they are equivalent to clay and calcium silicate bricks.

CONCRETE BRICKWORK

With the wide range of colour and texture options now offered by concrete-brick manufacturers, it is frequently difficult to distinguish visually, except at close quarters, between concrete and clay brickwork. The visual effects of using coloured mortars and various jointing details are as for clay bricks, but for exposed situations the use of raked joints is not recommended.

References

FURTHER READING

Brick Development Association. 2005: *The BDA guide to successful brickwork*. 3rd ed. London: Butterworth-Heinemann.

Brunskill, R.W. 1997: *Brick building in Britain*. London: Weidenfeld Nicolson Illustrated.

Campbell, J.W.P. 2003: *Brick: A world history*. London: Thames and Hudson.

Hammett, M. 1997: *Resisting rain penetration with facing brickwork*, Windsor: Brick Development Association.

Hammett, M. 2003: *Brickwork and paving for house and garden*. Marlborough: Crowood.

Hendry, A.W., Sinha, B.P. and Davies, S.R. 1997: *Design of masonry structures*. 3rd ed. London: E. & F.N. Spon.

Knight, T.L. 1997: *Creative brickwork*. London: Arnold.

Kreh, D. 1999: *Building with masonry: Brick, block and concrete*. Newtown, CT, USA: Taunton Press.

Lynch, G. 1994: *Brickwork: History, technology and practice*. Vol. 1, London: Donhead Publishing.

Lynch, G. 1994: *Brickwork: History, technology and practice*. Vol. 2, London: Donhead Publishing.

McKenzie, W.M.C. 2001: *Design of structural masonry*. London: Palgrave.

Nash, W.G. 1990: *Brickwork 1*, 3rd ed. Cheltenham: Stanley Thornes.

Sovinski, R.W. 1999: *Brick in the landscape: a practical guide to specification and design*. Bognor Regis: Wiley.

Thomas, K. 1996: *Masonry Walls – Specification and design*. Oxford: Butterworth-Heinemann.

Warren, J. 1997: *Conservation of brick*. Oxford: Butterworth-Heinemann.

Weston, R. 2003: *Materials, Form and Architecture*. London: Laurence King Publishing.

STANDARDS

BS 187: 1978 Specification for calcium silicate (sandlime and flintlime) bricks.

BS 743: 1970 Materials for damp-proof courses.

BS 1243: 1978 Specification for metal ties for cavity wall construction.

BS 3921: 1985 Specification for clay bricks.

BS 4729: 2005 Clay and calcium silicate bricks of special shapes and sizes – Recommendations.

BS 5628 Code of practice for use of masonry:
   Part 1: 1992 Structural use of unreinforced masonry.
   Part 2: 2000 Structural use of reinforced and prestressed masonry.
   Part 3: 2001 Materials and components, design and workmanship.

BS 6073 Precast concrete masonry units:
   Part 1: 1981 Specification for precast concrete masonry units.
   Part 2: 1981 Method for specifying precast concrete masonry units.

BS 6100 Glossary of building and civil engineering terms:
   Part 0: 2002 Introduction.
   Part 5 Masonry.
   Sec. 5.1: 1992 Terms common to masonry.
   Sec. 5.3: 1984 Bricks and blocks.

BS 6477:1992 Specification for water repellents for masonry surfaces.

BS 6649:1985 Specification for clay and calcium silicate modular bricks.

BS 6676: Thermal insulation of cavity walls using man-made mineral fibre batts (slabs):
   Part 1: 1986 Specification for man-made mineral fibre batts (slabs).
   Part 2: 1986 Code of practice for installation of batts (slabs) filling the cavity.

BS 6677 Clay and calcium silicate pavers for flexible pavements:
   Part 1: 1986 Specification for pavers.

BS 6717: 2001 Precast, unreinforced concrete paving blocks.

BS 6750: 1986 Specification for modular co-ordination in building.

BS 7533 Pavements constructed of clay, natural stone or concrete pavers.

BS 8000 Workmanship on building sites:
   Part 3: 2001 Code of practice for masonry.

BS 8208 Guide to assessment of suitability of external cavity walls for filling with thermal insulation:
   Part 1: 1985 Existing traditional cavity construction.

BS 8215: 1991 Code of practice for design and installation of damp-proof courses in masonry construction.

BS EN 771 Specification for masonry units:
   Part 1: 2003 Clay masonry units.
   Part 2: 2003 Calcium silicate masonry units.
   Part 3: 2003 Aggregate concrete masonry units.
   Part 4: 2003 Autoclaved aerated concrete masonry units.

BS EN 772 Methods of test for masonry units.

BS EN 845 Specification for ancillary components for masonry:
   Part 1: 2003 Ties, tension straps, hangers and brackets.
   Part 2: 2003 Lintels.
   Part 3: 2003 Bed-joint reinforcement of steel meshwork.

BS EN 934–3: 2003 Admixtures for masonry mortar.

BS EN 998–2: 2003 Specification for mortar for masonry. Masonry mortar.

BS EN 1015 Methods of test for mortar for masonry.

BS EN 1052 Methods of test of masonry.

BS EN 1344: 2002 Clay pavers – requirements and test methods.

BS EN 1365–1: 1999 Fire resistance tests for load-bearing elements – walls.

BS EN 1996 Eurocode 6: Design of masonry structures:
   Part 1.1: 2005 Rules for reinforced and unreinforced masonry.
   Part 1.2: 2005 Structural fire design.
   Part 2: 2006 Design considerations, selection of materials and execution of masonry.
   Part 3: 2006 Simplified calculation methods for unreinforced masonry structures.

PAS 70: 2003 HD clay bricks – guide to appearance and site measured dimensions and tolerance.

BUILDING RESEARCH ESTABLISHMENT PUBLICATIONS

BRE Digests

BRE Digest 245: 1986 Rising damp in walls: diagnosis and treatment.

BRE Digest 273: 1983 Perforated clay bricks.

BRE Digest 329: 2000 Installing wall ties in existing construction.

BRE Digest 359: 1991 Repairing brick and block masonry.

BRE Digest 360: 1991 Testing bond strength of masonry.

BRE Digest 361: 1991 Why do buildings crack?

BRE Digest 362: 1991 Building mortars.

BRE Digest 380: 1993 Damp-proof courses.

BRE Digest 401: 1995 Replacing wall ties.

BRE Digest 441: 1999 Clay bricks and clay brick masonry (Parts 1 and 2).

BRE Digest 460: 2001 Bricks, blocks and masonry made from aggregate concrete (Parts 1 and 2).

BRE Digest 461: 2001 Corrosion of metal components in walls.

BRE Digest 487: 2004 Structural fire engineering design: materials and behaviour – masonry (Part 3).

BRE Special Digest

BRE SD4: 2003 Masonry walls and beam and block floors: U-values and building regulations.

BRE Defect action sheets

BRE DAS 115: 1989 External masonry cavity walls – selection and specification.

BRE DAS 116: 1989 External masonry cavity walls: wall ties – installation.

BRE DAS 128: 1989 Brickwork: prevention of sulphate attack.

BRE DAS 129: 1989 Freestanding masonry boundary walls: stability and movement.

BRE DAS 130: 1989 Freestanding masonry boundary walls: materials and construction.

BRE Good building guides

BRE GBG 14: 1994 Building simple plan brick or blockwork freestanding walls.

BRE GBG 17: 1993 Freestanding brick walls: repairs to copings and cappings.

BRE GBG 19: 1994 Building reinforced, diaphragm and wide plan freestanding walls.

BRE GBG 62: 2004 Retro-installation of bed joint reinforcement in masonry.

BRE GBG 66: 2005 Building masonry with lime-based bedding mortars.

BRE Information papers

BRE IP 6/86 The spacing of wall ties in cavity walls.

BRE IP 16/88 Ties for cavity walls: new developments.

BRE IP 12/90 Corrosion of steel wall ties: history of occurrence, background and treatment.

BRE IP 13/90 Corrosion of steel wall ties: recognition and inspection.

BRE IP 10/93 Avoiding latent mortar defects in masonry.

BRE IP 10/99 Cleaning exterior masonry.

BRE IP 11/00 Ties for masonry walls: a decade of development.

BRE Report

BR117: 1988 Rain penetration through masonry walls: diagnosis and remedial details.

BRICK DEVELOPMENT ASSOCIATION PUBLICATIONS

Design Notes

DN 7: 1986 Brickwork durability. J.R. Harding and R.A. Smith.

DN 8: 1995 Rigid paving with clay pavers. M. Hammett and R.A. Smith.

DN 9: 1988 Flexible paving with clay pavers. R.A. Smith.

DN 11: 1990 Improved standards of insulation in cavity walls with outer leaf of facing brickwork. R.W. Ford and W.A. Durose.

DN 12: 1991 The design of curved brickwork. M. Hammett and J. Morton.

DN 13: 1993 The use of bricks of special shape. M. Hammett.

DN 15: 1992 Brick cladding to timber frame construction. B. Keyworth.

Building notes

BN 1: 1991 Brickwork – Good site practice. T. Knight.

BN 2: 1986 Cleaning of brickwork. J. Harding and R.A. Smith.

Technical information papers

TIP 8: 1988 A basic guide to brickwork mortars.
TIP 10: 1988 Brickwork dimension tables.

Technical papers

Hammett, M. (1988) The repair & maintenance of brickwork. *Building Technical Note* 20.

Lilley, A.A. (1990) Flexible brick paving: application & design. *Highways & Transportation* **10** (37).

ADVISORY ORGANISATIONS

Brick Development Association, Woodside House, Winkfield, Windsor, Berks. SL4 2DX (01344 885651). Mortar Industry Association, 156 Buckingham Palace Road, London SW1W 9TR (020 7730 8194).

2

BLOCKS AND BLOCKWORK

Introduction

The variety of commercially available concrete blocks is extensive, from dense through to lightweight, offering a range of load-bearing strength, sound and thermal insulation properties. Where visual blockwork is required, either internally or externally, fairfaced blocks offer a selection of textures and colours at a different visual scale compared to that associated with traditional brickwork. Externally, visual concrete blockwork weathers well, providing adequate attention is given to the quality of the material and rainwater run-off detailing. Blockwork has considerable economic advantages over brickwork in respect of speed of construction, particularly as the lightweight blocks can be lifted in one hand.

Whilst clay blocks are used extensively for masonry construction on the continent of Europe, until recently there had been little demand from the building industry within the UK. However, both fired and unfired clay blocks are now commercially available within the UK. The use of clay blocks for floor construction has been superseded by the use of reinforced concrete inverted T-beams with concrete infill blocks.

Concrete paving blocks, which offer opportunities for creative hard landscaping with their diversity of form and colour, are widely used for town pedestrian precincts and individual house driveways. Concrete interlocking blocks with planting are used to create environmental walls.

Concrete blocks

TYPES AND SIZES

Concrete blocks are defined as solid, cellular or hollow, as illustrated in Figure 2.1.

Concrete blocks are manufactured to various workface dimensions in an extensive range of thicknesses, offering a wide choice of load-bearing capacity and level of insulation. The standard workface size, which co-ordinates to three courses of metric brickwork allowing for 10 mm mortar joints, is 440×215 mm (Fig. 2.2), but the other sizes in Table 2.1 are marketed for aesthetic and constructional reasons. For example, narrow bands of a different colour may be used as visual features within fairfaced blockwork, and foundation or party wall blocks are normally laid flat. The use of thin-joint masonry offers speedier construction, especially when using large format blocks (Fig. 2.3), which are approximately equivalent in size to two standard units. However, blocks heavier than 20 kg should not be lifted repeatedly by a single person as this potentially can lead to injury.

The European Standard (BS EN 771–3: 2003) describes a wider range of aggregate concrete masonry units incorporating either dense or lightweight aggregates. Under the European Standard, the minimum description for concrete blocks includes the European Standard number and date (e.g. BS EN 771–3: 2003), the type of unit (e.g. common or facing), work size dimensions and tolerance category, configuration (e.g. solid or with voids) and compressive strength. Also, depending upon the particular end use, additional description may be required. This may, as appropriate, include surface finish, net and gross dry density, co-ordinating size, thermal properties and moisture movement. Tolerance limits for regular-shaped blocks are defined at three levels in Table 2.2. Compressive strengths of concrete masonry units are classified to Category I or Category II. Category I units have the tighter control with only a 5% risk of the units not achieving the declared compressive strength.

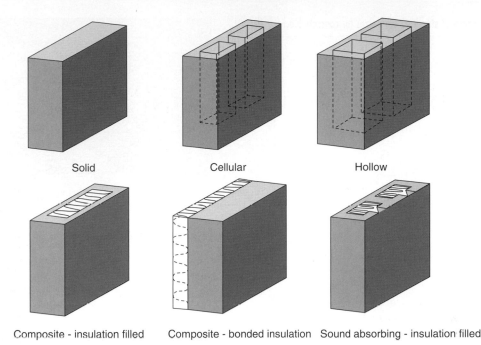

Solid   Cellular   Hollow

Composite - insulation filled   Composite - bonded insulation   Sound absorbing - insulation filled

Fig. 2.1 Types of concrete blocks

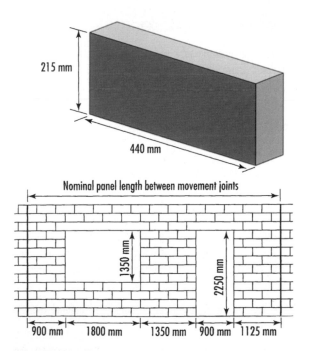

Fig. 2.2 Co-ordinating sizes for blockwork

The European Standard (BS EN 771–4: 2003) gives the specification for autoclaved aerated concrete (AAC) masonry units. The maximum size of units within the standard is 1500 mm length × 1000 mm

height – 600 mm width. The tolerance limits on the dimensions are defined in Table 2.3, and are dependent on whether the units are to be erected with standard or thin layer mortar joints. The standard manufacturer's description for AAC masonry units must include the European Standard number and date (e.g. BS EN 771–4: 2003), dimensions and tolerances, compressive strength (Category I or II, as for concrete units) and dry density. Further description for specific purposes may include durability, configuration (e.g. perforations or tongued and grooved jointing system) and intended use.

MANUFACTURE

Dense concrete blocks, which may be hollow, cellular or solid in form are manufactured from natural dense aggregates including crushed granite, limestone and gravel. Medium and lightweight concrete blocks are manufactured incorporating a wide range of aggregates including expanded clay, expanded blast furnace slag, sintered ash and pumice. Concrete is cast into moulds, vibrated and cured. Most aerated (aircrete or autoclaved aerated concrete) blocks are formed by the addition of aluminium powder to a fine mix of sand, lime, fly ash (pulverised-fuel ash) and Portland cement. The hydrogen gas generated by

Table 2.1 A range of standard work sizes for concrete blocks

	Work size	
	Length (mm)	Height (mm)
Coursing blocks		
	215	40
	215	65
	215	70
	215	100
Wall blocks		
	440	140
	440	215
	440	430
	540	440
	610	140
	610	215
	610	270
	620	215
	620	300
	620	430
Floor blocks		
	440	215
	440	350
	540	440
	610	350
	620	215
	620	540
Foundation blocks		
	440	140
	440	215
	620	140
	620	215

Only a selection of thicknesses are produced by most manufacturers within the range 50 to 350 mm.
Generally available thicknesses are: 70, 75, 100, 115, 125, 130, 140, 150, 175, 190, 200, 215, 250, 255, 265, 275, 280 and 300 mm.

the dissolution of the metal powder produces a non-interconnecting cellular structure. The process is accelerated by pressure steam curing in an autoclave (Fig. 2.4). For some products, additional insulation is afforded by the filling of voids in the cellular blocks or by bonding on a layer of extruded polystyrene, polyurethane or foil-faced phenolic foam (Fig. 2.1). Standard blocks, typically natural grey or buff in colour, are usually shrink-wrapped for delivery. Different grades of blocks are usually identified by scratch marks or colour codes.

Fig. 2.3 Thin-joint masonry using large format blocks. Photograph reproduced from GBG 58 by permission of BRE and courtesy of Aircrete Products Association

Table 2.2 Limit of tolerances on block sizes

Tolerance category	D1	D2	D3
Length (mm)	+3	+1	+1
	−5	−3	−3
Width (mm)	+3	+1	+1
	−5	−3	−3
Height (mm)	+3	+2	+1.5
	−5	−2	−1.5

Table 2.3 Limit of tolerances on autoclaved aerated concrete block sizes

	Standard joints	Thin layer mortar joints	
		TLMA	TLMB
Length (mm)	−5 to +3	±3	±3
Width (mm)	−5 to +3	±2	±1
Height (mm)	±3	±2	±2

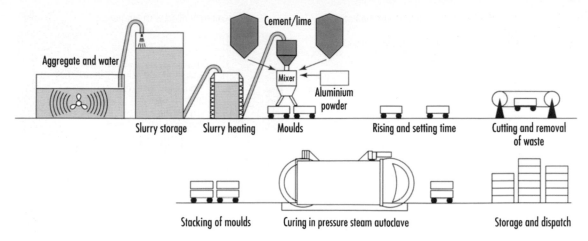

Fig. 2.4 Manufacture of aerated blocks

PROPERTIES

Density and strength

Concrete blocks range in compressive strength from 2.8 MPa to 30 MPa, with associated densities of 420 to 2200 kg/m³ and thermal conductivities from 0.10 to 1.5 W/m K at 3% moisture content (Table 2.4). Drying shrinkages are typically in the range 0.03 to 0.05%.

Durability

Dense concrete blocks and certain aerated lightweight blocks are resistant to freeze/thaw conditions below damp-proof course (DPC) level. However, some lightweight concrete blocks, with less than 7 MPa crushing strength, should not be used below DPC level, except for the inner skin of cavity construction.

Fixability

Aerated and lightweight concrete blocks offer a good background for fixings. For light loads, nails to a depth of 50 mm are sufficient. For heavier loads, wall plugs and proprietary fixings are necessary. Fixings should avoid the edges of the blocks.

Thermal insulation

The Building Regulations Approved Document Part L (2006 edition) requires new dwellings (Part L1A) and other new building types (Part L2A) to be compliant with an overall energy and carbon performance, the Target Emissions Rate (TER) based on the whole building (Chapter 7, page 223). Individual U-values for elements are therefore not set, except for extensions on existing dwellings (Part L1B) and other existing buildings (Part L2B) where an indicative U-value of 0.30 W/m² K is the standard for new exposed walls. The limiting area-weighted U-value standard for wall elements in new buildings is 0.35 W/m² K, but to achieve the Target Emission Rate overall, most buildings will require wall U-values within the range 0.27–0.30 W/m² K.

The following material cominations achieve a U-value of 0.27 W/m² K (Fig. 2.5).

Partially filled cavity

102.5 mm fairfaced brickwork outer leaf
50 mm clear cavity
50 mm foiled-faced polyurethane foam ($\lambda = 0.023$ W/m K)
100 mm lightweight blocks ($\lambda = 0.15$ W/m K)
12.5 mm plasterboard on dabs ($\lambda = 0.16$ W/m K)

Table 2.4 Typical relationship between density and thermal conductivity for concrete blocks

Nominal density (kg/m³)	2200	2000	1800	1600	1400	1200	1000	900	800	700	600	500	420
Typical thermal conductivity (W/m K)	1.5	1.10	0.83	0.63	0.47	0.36	0.27	0.24	0.20	0.17	0.15	0.12	0.10

(Blocks of differing compositions will vary from these average figures.)

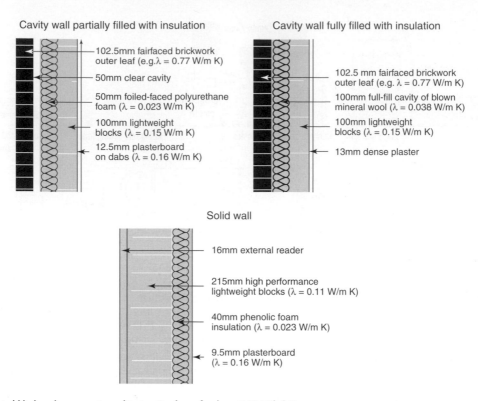

Fig. 2.5 Typical blockwork construction achieving U-values of at least 0.27 W/m² K

Fully filled cavity

102.5 mm fairfaced brickwork outer leaf
100 mm full-fill cavity of blown mineral wool (λ = 0.038 W/m K)
100 mm lightweight blocks (λ = 0.15 W/m K)
13 mm dense plaster

Similarly, a U-value of 0.27 W/m² K can be achieved with 100 mm external fairfaced blockwork as an alternative to fairfaced brickwork, providing that the necessary additional thermal resistance is provided by slightly increased cavity insulation. The thin-joint mortar system for inner leaf blockwork gives slightly enhanced U-values compared to the equivalent standard 10 mm joint blockwork construction. Rendered solid wall construction can also achieve a U-value of 0.27 W/m² K (Fig. 2.5).

Solid wall

16 mm external render
215 mm high performance lightweight blocks (λ = 0.11 W/m K)
50 mm lining of 9.5 mm plasterboard (λ = 0.16 W/m K) backed with 40 mm phenolic foam insulation (λ = 0.023 W/m K)

Fire resistance

Concrete block construction offers good fire resistance. Solid unplastered 90 mm blocks can give up to 60 minutes fire protection when used as load-bearing walls; certain 150 mm and most 215 mm solid blocks can achieve 360 minutes protection. Dense, lightweight and autoclaved aerated concrete blocks with less than 1% organic material are automatically categorised as Euroclass A1 with respect to reaction to fire.

Sound insulation

The Building Regulations 2000 Approved Document E (2003) provides guidance on minimum standards of acoustic insulation for internal and separating walls of new dwellings. The regulations require minimum sound insulation of 45 dB for separating walls and 40 dB for internal bedroom or WC walls. The passage of airborne sound depends upon the density and porosity of the material. The following alternative systems should perform to the required airborne insulation standard for separating walls of new build dwellings, subject to appropriate site testing.

    12.5 mm plasterboard on dabs
    8 mm render

100 mm dense (1600–2200 kg/m³) or lightweight (1350–1600 kg/m³) blockwork

75 mm clear cavity only linked by appropriate wall ties

100 mm dense (1600–2200 kg/m³) or lightweight (1350–1600 kg/m³) blockwork

8 mm render

12.5 mm plasterboard on dabs

These alternatives will only perform to the required standard if there are no air leaks within the construction, all joints are filled, the cavities are kept clear except for the approved wall ties, and any chasing out on opposite sides of the construction is staggered. Vertical chases should, in any case, not be deeper than one third of the block thickness. Horizontal chases should be restricted to not more than one sixth of the block thickness, due to the potential loss of structural strength.

Sound absorption

The majority of standard concrete blocks with hard surfaces are highly reflective to sound, thus creating long reverberation times within building enclosures. Acoustic-absorbing concrete blocks are manufactured with a slot on the exposed face which admits sound into the central cavity (Fig. 2.1). Since the void space is lined with sound-absorbing fibrous filler, incident sound is dissipated rather than reflected, significantly reducing reverberation effects. Acoustic control blocks in fairfaced concrete are suitable for use in swimming pools, sports halls, industrial buildings and auditoria.

SPECIALS

Most manufacturers of blocks produce a range of *specials* to match their standard ranges. Quoins, cavity closers, splayed cills, flush or projecting copings, lintel units, bullnose ends and radius blocks are generally available, and other specials can be made to order (Fig. 2.6). The use of specials in fairfaced blockwork can greatly enhance visual qualities. Matching full-length lintels may incorporate dummy joints and should bear on to full, not cut, blocks.

FAIRFACED BLOCKS

Fairfaced concrete blocks are available in a wide range of colours from white, through buff, sandstone, yellow, to pink, blue, green and black. Frequently the colour is *all through*, although some blocks have an

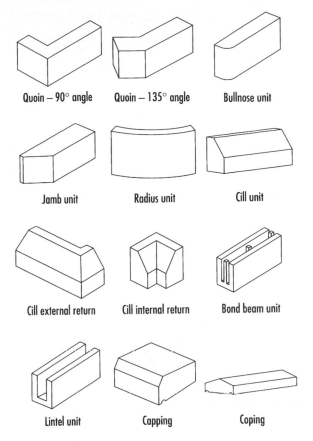

Fig. 2.6 Block specials

applied surface colour. Most blocks are uniform in colour, but there is some variability with, for example, flecked finishes. Textures range from polished, smooth and weathered (sand- or shot-blasted) to striated and split face (Fig. 2.7); the latter intended to give a random variability associated more with natural stone.

Glazed masonry units are manufactured by the application of a thermosetting material to one or more faces of lightweight concrete blocks which are then heat-treated to cure the finish. The glazed blocks are available in an extensive range of durable bright colours and are suitable for interior or exterior use. Where required, profiled blocks to individual designs can be glazed by this system. Most manufacturers produce a range of specials to co-ordinate with their standard fairfaced blocks although, as with special bricks, they may be manufactured from a different batch of mix, and this may give rise to slight variations. In specific cases, such as individual lintel blocks, specials are made by cutting standard blocks to ensure exact colour matching.

Fig. 2.7 Split and polished architectural masonry finishes. Photographs: Courtesy of Lignacite Ltd

Clay blocks

FIRED CLAY BLOCKS

Masonry clay honeycomb-insulating blocks can be used as a single skin for external load-bearing construction as an alternative to standard cavity construction. These fired clay honeycomb blocks combine structural strength, insulation and, when externally rendered, moisture protection. The internal surface is normally finished directly with gypsum plaster. Blocks for monolithic construction are 260 mm long × 40 mm high and either 300 or 365 mm thick, giving wall U-values of 0.36 and 0.30 W/m² K respectively when rendered and plastered. For internal walls, blocks are 400 mm long and range in widths from 100 to 125 and 150 mm. Horizontal joints require 10 mm of a lightweight mortar, but the vertical joint edges, if tongued and grooved, remain dry. The British Standard (BS EN 771–1: 2003) illustrates a selection of high density (HD) vertically perforated units and a range of low density (LD) fired-clay masonry units. The LD units may be vertically or horizontally perforated, with butt jointing, mortar pockets or a tongue and groove system (Fig. 2.8). Special blocks are available for corners, lintels, door and window openings, but individual blocks can also be cut.

Fairfaced fired-clay blocks, as illustrated in Figure 2.9, offer an alternative to traditional brickwork. They are manufactured in a selection of colours including terracotta red, buff and blue and in a range of unit sizes giving scope for architectural scaling effects. Where used as infill, rather than load-bearing, alternative bonding is possible including stack bond. Typical work sizes are 440 × 215, 390 × 240 and 390 × 190 mm with a width of 90 mm. A standard 10 mm mortar joint is appropriate, which may match or contrast to the block colour.

UNFIRED CLAY BLOCKS

Unfired blocks manufactured from clay and sometimes incorporating straw may be used for non-load-bearing partition walls. Blocks (typically 500 mm × 250 mm and 450 mm × 225 mm × 100 mm thick) may be tongued and grooved or square edged, but only the horizontal joints require fixing with a thin layer of cellulose-based adhesive or clay mortar. Blocks are easily cut to create architectural features, and are usually finished with a skim coat of clay plaster although they may be painted directly. Internal walls are sufficiently strong to support shelving and other fixtures. Unfired clay block walls are recyclable or biodegradable and have the advantage of absorbing odours and stabilising internal humidity and temperature by their natural absorption and release of moisture and heat. A 100-mm-thick wall gives a 45 dB sound reduction and 90 minutes' fire resistance. (The

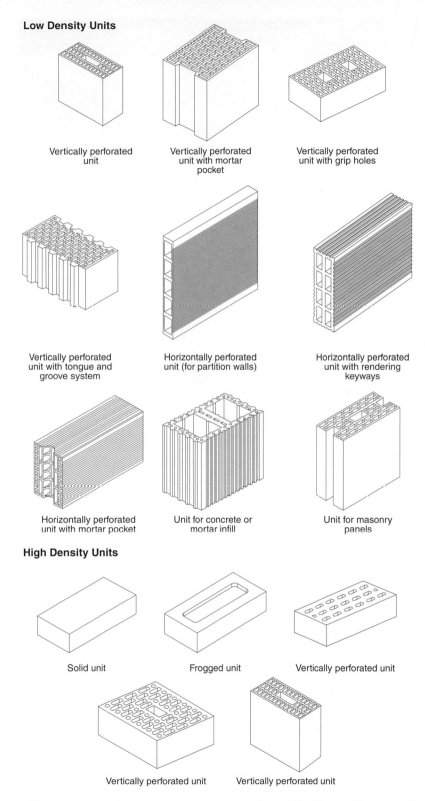

Fig. 2.8 Low density and high density units. Permission to reproduce extracts from BS EN 771–1: 2003 is granted by the British Standards Institute

Fig. 2.8 Continued. Hollow clay blocks in Greece. Photograph: Arthur Lyons

thermal conductivity of perforated unfired clay blocks is typically 0.24 W/m K.)

Blockwork

FAIRFACED CONCRETE BLOCKWORK

Within fairfaced blockwork, an appropriate choice of size is important to both co-ordination and visual scale. Whilst blocks can be cut with a masonry cutter, the addition of small pieces of block, or the widening of perpends over the 10 mm standard, is unacceptable. The insertion of a thin *jumper* course at floor or lintel height may be a useful feature in adjusting the coursing. Curved blockwork may be constructed from standard blocks, the permissible curvature being dependent upon the block size. The oversail between alternate courses should not normally exceed 4 mm in fairfaced work. If the internal radius is exposed, then the perpends can be maintained at 10 mm with uncut blocks, but if the external radius is exposed, the blocks will require cutting on a splay. For tighter curves *specials* will be required.

THIN LAYER MORTAR MASONRY SYSTEMS

Thin layer mortar blockwork may be constructed with mortar joints of only 2–3 mm, providing that the aircrete or equivalent blocks have been manufactured to fine tolerances and on-site workmanship is good. The special rapid-setting mortar sets typically within 30 minutes and the full bond strength is achieved after only two hours, allowing more courses to be laid each day. In the case of brick and block cavity construction, the inner leaf is built first, providing a weatherproof enclosure as quickly as possible. The outer skin of brickwork can subsequently be built up, using wall ties fixed to the face, either screwed or hammered into the completed blockwork. Bed joints in thin layer mortar blockwork do not co-ordinate with those of the brickwork, so conventional cavity wall ties can only be used if they are slope-tolerant.

Usually inner leaf construction commences with a line of 440 × 215 mm standard height blocks, with normal bedding mortar to compensate for variations in the foundation level, followed by the larger 440 or 620 × 430 mm high blocks, which should weigh less than 20 kg for repeated lifting by one

Fig. 2.9 Fairfaced blockwork – IDP Offices, Glasgow. Architect: IDP. Photograph: Courtesy of Ibstock Brick Ltd

operative. Heavier blocks require mechanical lifting or two-person handling. Thin-joint mortars, consisting of polymer-modified 1 : 2 cement : sand mix with water-retaining and workability admixtures, are factory pre-mixed and require only the addition of water, preferably mixed in with an electrically-powered plasterer's whisk. The mortar is applied manually with a serrated scoop or through a pumped system to achieve uniformity.

The main advantages of thin-joint systems over traditional 10-mm-joint blockwork are:

- increased productivity allowing storey-height inner leaves to be completed in one day;
- up to 10% improved thermal performance due to reduced thermal bridging by the mortar;
- improved airtightness of the construction;
- the accuracy of the wall allows internal thin-coat sprayed plaster finishes to be used;
- higher quality of construction and less wastage of mortar.

The acoustic properties of thin-joint mortar walls differ slightly from walls constructed with 10-mm-mortar joints. Resistance to low frequency noise is slightly enhanced, whilst resistance to high frequency sound is slightly reduced.

Completed thin-joint blockwork acts as a monolithic slab, which if unrestrained may crack at the weaker points, such as near openings. To avoid this, the block units should be laid dry to avoid shrinkage, and bed-joint reinforcement (1.5 mm thick) should be appropriately positioned. Larger structures require movement joints at 6 m centres.

Certain extruded multi-perforated clay and calcium silicate blocks, available in Europe, are designed for use with thin mortar bed-joints and dry interlocking vertical joints. Whilst this reduces the initial construction time, both sides of the units subsequently require plaster or cement render to minimise heat loss by air leakage.

BOND

A running half-block bond is standard, but this may be reduced to a quarter bond for aesthetic reasons. Blockwork may incorporate banding of concrete bricks, but because of differences in thermal and moisture movement it is inadvisable to mix clay bricks with concrete blocks. Horizontal and vertical stack bond and more sophisticated variations, such as basket weave bond, may be used for infill panels within framed structures (Fig. 2.10). Such panels will require reinforcement within alternate horizontal bed-joints, to compensate for the lack of normal bonding.

REINFORCEMENT

Blockwork will require bed-joint reinforcement above and below openings where it is inappropriate to divide the blockwork up into panels, with movement joints at the ends of the lintels. Bed-joint reinforcement would be inserted into two bed joints above and below such openings (Fig. 2.11). Cover to bed reinforcement should be at least 25 mm on the external faces and 13 mm on the internal faces. Combined vertical and horizontal reinforcement may be incorporated into hollow blockwork in accordance with BS 5628–2: 2000, where demanded by the calculated stresses. Typical situations would be within retaining basement walls, and large infill panels to a framed structure.

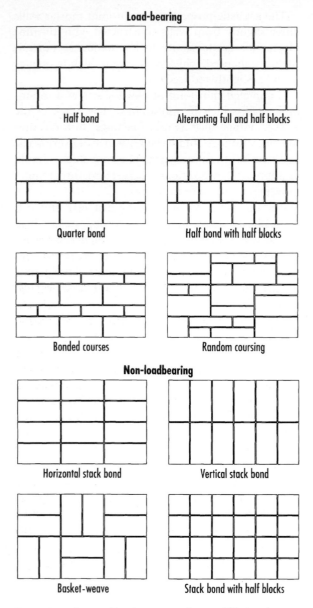

Load-bearing

Half bond

Alternating full and half blocks

Quarter bond

Half bond with half blocks

Bonded courses

Random coursing

Non-loadbearing

Horizontal stack bond

Vertical stack bond

Basket-weave

Stack bond with half blocks

Fig. 2.10 Selection of bonding patterns for visual blockwork

MOVEMENT CONTROL

Concrete blockwork is subject to greater movements than equivalent brickwork masonry. Therefore the location and form of the movement joints require greater design-detail consideration, to ensure that inevitable movements are directed to the required locations and do not cause unsightly stepped cracking or fracture of individual blocks. Blockwork walls over 6 m in length must be separated into a series of panels with movement-control joints at approximately 6 m centres. Ideally, such movement joints should be

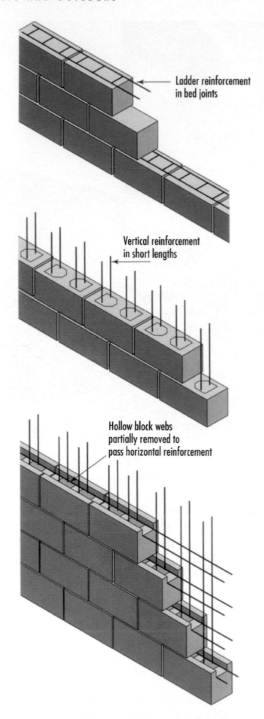

Ladder reinforcement in bed joints

Vertical reinforcement in short lengths

Hollow block webs partially removed to pass horizontal reinforcement

Fig. 2.11 Reinforced blockwork

located at intersecting walls, or other points of structural discontinuity, such as columns. Additionally, movement joints are required at changes in thickness, height or loading of walls, above and below wall openings, and adjacent to movement joints in the adjoining structure (Fig. 2.12). Wall ties should allow for differential movement between the leaves in cavity construction and should be spaced at 900 mm horizontally and 450 mm vertically, for 50–75 mm cavities.

MORTARS

The mortar must always be weaker than the blocks to allow for movement. The usual mixes for standard 10 mm joints are by volume:

cement/lime/sand	1 : 1 : 5 to 1 : 1 : 6
cement/sand + plasticiser	1 : 5 to 1 : 6
masonry cement/sand	1 : 4 to 1 : 5

Below DPC level a stronger mix is required and sulfate-resisting cement may be necessary depending upon soil conditions.

cement/sand	1 : 4
cement/lime/sand	1 : 1½ : 4½

Where high-strength blockwork is required, stronger mortars may be necessary. Mortar joints should be slightly concave, rather than flush. Bucket-handle and weathered or struck joints are suitable for external use, but recessed joints should only be used internally. Coloured mortars should be ready-mixed or carefully gauged to prevent colour variations. Contraction joints should be finished with a bond breaker of polythene tape and flexible sealant. For expansion joints, a flexible filler is required, e.g. bitumen-impregnated fibreboard with a polythene foam strip and flexible sealant. Where blockwork is to be rendered, the mortar should be raked back to a depth of 10 mm for additional key.

FINISHES

Internal finishes

Plaster should be applied normally in two coats to 13 mm. Blocks intended for plastering have a textured surface to give a good key. Dry lining may be fixed with battens or directly with adhesive dabs to the blockwork. Blockwork to be tiled should be first rendered with a cement/sand mix. Fairfaced blockwork may be left plain or painted.

External finishes

External boarding or hanging tiles should be fixed to battens, separated from the blockwork with a breather membrane. For external rendering a spatterdash coat should be applied initially on dense blockwork, followed by two coats of cement/lime/sand render. The first 10 mm coat should be the stronger mix (e.g. 1 : 1 : 6), the 5 mm second coat must be weaker (e.g. 1 : 2 : 9). Cement/sand mixes are not recommended as they are more susceptible to cracking and crazing than mixes incorporating lime. The render should terminate at damp-proof course level with a drip or similar weathering detail.

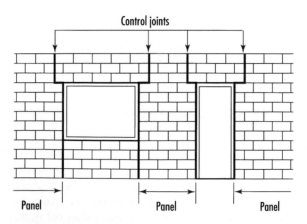

Fig. 2.12 Blockwork movement joints

FOUNDATIONS

Foundation blocks laid flat offer an alternative to trench fill or cavity masonry. Portland cement blocks should not be used for foundations where sulfate-resisting cement mortar is specified, unless they are classified as suitable for the particular sulfate conditions. Sulfate and other chemically adverse ground conditions are classified in the BRE Special Digest 1 (2005) from DS1 (Design Sulphate Class 1) to the most aggressive, DS5. Foundation blocks can be either of dense or appropriate lightweight concrete; the latter providing enhanced floor edge insulation. Interlocking foundation blocks, with a tongue and groove vertical joint, slot together with only bed-joint mortar being required. A hand-hold makes manipulating these blocks on site much easier than lifting standard rectangular blocks.

Beam and block flooring

Inverted T-beam and concrete block construction (Fig. 2.13) offers an alternative flooring system to traditional solid ground floors within domestic construction. The infill blocks may be standard 100 mm blocks with a minimum transverse crushing strength of 3.5 MPa. Insulation will be required to achieve a U-value between 0.20 and 0.25 W/m^2 K. For first and subsequent floors, the infill may be full-depth solid blocks or hollow pots and may additionally require a screed to comply with the Building Regulations.

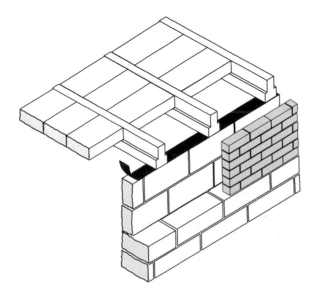

Fig. 2.13 Beam and block flooring

The following material combination achieves a U-value of 0.20 W/m^2 K.

18 mm particleboard ($\lambda = 0.13$ W/m K)
100 mm continuous insulation ($\lambda = 0.030$ W/m K)
100 mm concrete block ($\lambda = 0.46$ W/m K)
dense concrete beam inverted T beam at 515 mm centres ($\lambda = 1.65$ W/m K)
underfloor ventilated space

Landscape blockwork

BLOCK PAVING

Concrete block paving units are manufactured to a wide range of designs as illustrated in Figure 2.14. Blocks may be of standard brick form (200 × 100 mm) to thicknesses of 60, 80 or 100 mm depending upon the anticipated loading. Alternative designs include tumbled blocks, which emulate granite setts, and various interlocking forms giving designs based on polygonal and curvilinear forms. Colours range from red, brindle, buff, brown, charcoal and grey through to silver and white, with smooth, textured or simulated stone finishes. For most designs, a range of kerb blocks, drainage channels, edging and other accessory units are available. Concrete paving blocks are usually laid on a compacted sub-base with 50 mm of sharp sand. Blocks are frequently nibbed to create a narrow joint to be filled with kiln-dried sand. For the wider joints that occur between the simulated stone setts a coarser grit can be used to prevent loss by wind erosion.

Where the appearance of grass is required, but with the traffic-bearing properties of a concrete block pavement, a selection of hollow blocks is available which can be filled with soil and seeded to give the required effect. Different block depths and sub-base can be specified according to the anticipated traffic loading. Sulfate-resisting blocks are available if dictated by the soil conditions.

EARTH-RETAINING BLOCKWORK

A range of precast-cellular concrete-interlocking blocks is manufactured for the construction of dry-bed retaining walls. Soil is placed in the pockets of each successive course to allow for planting. The rear is backfilled with granular material to allow for drainage. The size of the block determines the maximum construction height, but over 20 m can be

Fig. 2.14 Selection of concrete pavers and hard landscape to the Gateshead Millennium Bridge. Photographs: Courtesy of Marshalls plc

achieved with very deep units. A face angle of 15° to 22° is typical to ensure stability, but other gradients are possible with the appropriate block systems. Limited wall curvature is possible without cutting the standard blocks. The systems are used both for earth retention and to form acoustic barriers.

References

FURTHER READING

British Cement Association. *BCA guide to materials for masonry mortar.* Camberley: BCA.

Hugues, T. Grellich, K. and Peter, C. 2004: *Detail practice: Building with large clay blocks.* Basle: Birkhauser.

Roper, P.A. 1987: *A practical guide to blockwork.* London: International Thomson.

Taylor, E.S.Oliver- 1995: Mason bricklayer: Brickwork and blockwork. Stem Systems.

STANDARDS

BS 743: 1970 Materials for damp-proof courses.

BS 1243: 1978 Specification for metal ties for cavity wall construction.

BS 5628 Code of practice for use of masonry:
Part 1: 1992 Structural use of unreinforced masonry.
Part 2: 2000 Structural use of reinforced and prestressed masonry.
Part 3: 2001 Materials and components, design and workmanship.

BS 5977 Lintels:
Part 1: 1981 Method for assessment of load.

BS 6073 Precast concrete masonry units:
Part 1: 1981 Specifications for precast concrete masonry units.
Part 2: 1981 Method for specifying precast concrete masonry units.

BS 6100 Glossary of building and civil engineering terms:
Part 5 Masonry.
Sec. 5.1: 1992 Terms common to masonry.
Sec. 5.3: 1984 Brick and blocks.

BS 6457: 1984 Specification for reconstructed stone masonry units.

BS 6717: 2001 Precast, unreinforced concrete paving blocks.

BS 7533 Part 10: 2004 Pavements constructed of clay, natural stone or concrete pavers.

BS 8000 Workmanship on building sites:
Part 3: 2001 Code of practice for masonry.

BS 8208 Guide to assessment of suitability of external cavity walls for filling with thermal insulation:
Part 1: 1985 Existing traditional cavity construction.

BS 8215: 1991 Code of practice for design and installation of damp-proof courses in masonry construction.

BS EN 413–1: 2004 Masonry cement. Composition, specifications and conformity criteria.

BS EN 771 Specification for masonry units:
Part 3: 2003 Aggregate concrete masonry units.
Part 4: 2003 Autoclaved aerated concrete masonry units.
Part 5: 2003 Manufactured stone masonry units.

BS EN 845 Specification for ancillary components for masonry:
Part 1: 2003 Ties, tension straps, hangers and brackets.
Part 2: 2003 Lintels.
Part 3: 2003 Bed joint reinforcement of steel meshwork.

BS EN 934–3: 2003 Admixtures for masonry mortar.

BS EN 998–2: 2003 Specification for mortar for masonry. Masonry mortar.

BS EN 1745: 2002 Masonry and masonry products. Methods for determining design thermal values.

BS EN 1806: 2000 Chimneys – clay/ceramic flue blocks for single wall chimneys.

BS EN 1858: 2003 Chimneys – components – concrete flue blocks.

BS EN 1996 Eurocode 6: Design of masonry structures:
Part 1.1: 2005 Rules for reinforced and unreinforced masonry.
Part 1.2: 2005 Structural fire design.

BUILDING RESEARCH ESTABLISHMENT PUBLICATIONS

BRE Special digests

BRE SD1: 2005 Concrete in aggressive ground.

BRE SD4: 2003 Masonry walls and beam and block floors: U-values and building regulations.

BRE Digests

BRE Digest 359: 1991 Repairing brick and block masonry.

BRE Digest 360: 1991 Testing bond strength of masonry.

BRE Digest 362: 1991 Building mortars.

BRE Digest 380: 1993 Damp-proof courses.

BRE Digest 401: 1995 Replacing wall ties.

BRE Digest 432: 1998 Aircrete: thin joint mortar system.

BRE Digest 460: 2001 Bricks, blocks and masonry made from aggregate concrete (Parts 1 and 2).

BRE Digest 461: 2001 Corrosion of metal components in walls.

BRE Digest 468: 2002 AAC 'aircrete' blocks and masonry.

BRE Digest 487: 2004 Structural fire engineering design. Part 4 Materials behaviour: masonry.

BRE Good building guides

BRE GBG 14: 1994 Building simple plan brick or blockwork free-standing walls.

BRE GBG 27: 1996 Building brickwork or blockwork retaining walls.

BRE GBG 44: 2000 Insulating masonry cavity walls (Parts 1 and 2).

BRE GBG 50: 2002 Insulating solid masonry walls.

BRE GBG 54: 2003 Construction site communication. Part 2 Masonry.

BRE GBG 58: 2003 Thin layer masonry mortar.

BRE GBG 62: 2004 Retro-installation of bed joint reinforcement in masonry.

BRE GBG 66: 2005 Building masonry with lime-based bedding mortars.

BRE Information papers

BRE IP 10/96 Reinforced autoclaved aerated concrete planks.

BRE IP 2/98 Mortars for blockwork: improved thermal performance.

BRE IP 14/98 Blocks with recycled aggregate: beam-and-block floors.

BRE IP 1/99 Untied cavity party walls – structural performance when using AAC blockwork.

BRE IP 7/05 Aircrete tongue and grooved block masonry.

ADVISORY ORGANISATIONS

British Concrete Masonry Association, Grove Crescent House, 18 Grove Place, Bedford MK40 3JJ (01234 353745).

Concrete Block Association. 60 Charles Street, Leicester LE1 1FB (0116 253 6161).

Mortar Industry Association, 156 Buckingham Palace Road, London SW1W 9TR (020 7730 8194).

3

LIME, CEMENT AND CONCRETE

Introduction

In the broadest sense, the term *cement* refers to materials which act as adhesives. However, in this context, its use is restricted to that of a binding agent for sand, stone and other aggregates within the manufacture of mortar and concrete. Hydraulic cements and limes set and harden by internal chemical reactions when mixed with water. Non-hydraulic materials will only harden slowly by absorption of carbon dioxide from the air.

Lime was used throughout the world by the ancient civilisations as a binding agent for brick and stone. The concept was brought to Britain in the first century AD by the Romans, who used the material to produce lime mortar. Outside Britain, the Romans frequently mixed lime with volcanic ashes, such as pozzolana from Pozzuoli in Italy, to convert a non-hydraulic lime into a hydraulic cement suitable for use in constructing aqueducts, baths and other buildings. However, in Britain, lime was usually mixed with artificial pozzolanas, for example crushed burnt clay products such as pottery, brick and tile. In the eighteenth century, a so-called *Roman cement* was manufactured by the burning of *cement stone* (argillaceous or clayey limestone), collected from the coast around Sheppey and Essex.

In 1824 Joseph Aspdin was granted his famous patent for the manufacture of *Portland cement*, from limestone and clay. Limestone powder and clay were mixed into a water slurry which was then evaporated by heat in *slip pans*. The dry mixture was broken into small lumps, calcined in a kiln to drive off the carbon dioxide, burnt to clinker and finally ground into a fine powder for use. The name Portland was used to enhance the prestige of the new concrete material by relating it to Portland stone which, to some degree, it resembled. Early manufacture of Portland cement was by intermittent processes within bottle, and later chamber, kilns. The introduction in 1877 of the rotating furnace offered a continuous burning process with consequent reductions in fuel and labour costs. The early rotating kilns formed the basis for development of the various production systems that now exist. In 1989, the peak production year, 18 million tonnes of cement were manufactured within the UK. About half of this was required by the ready-mixed concrete industry; the remainder was divided roughly equally between concrete-product factories and bagged cement for general use.

Lime

MANUFACTURE OF LIME

Lime is manufactured by calcining natural calcium carbonate, typically hard-rock carboniferous limestone. The mineral is quarried, crushed, ground, washed and screened to the required size range. The limestone is burnt at approximately 950°C in either horizontal rotary kilns or vertical shaft kilns which drive off the carbon dioxide to produce quicklime or lump lime (calcium oxide). Quicklimes include calcium limes (CL) and dolomitic limes (DL) depending upon the composition of the starting mineral as defined by BS EN 459–1: 2001.

$$\overset{950°C}{CaCO_3 \longrightarrow CaO + CO_2}$$

calcium carbonate   lime   carbon dioxide

Slaking of lime

Slaking – that is the addition of water to quicklime – is a highly exothermic reaction. The controlled addition of water to quicklime produces hydrated lime (S) (mainly calcium hydroxide) as a dry powder.

$$CaO \ + \ H_2O \ \longrightarrow \ Ca(OH)_2$$
$$\text{lime} \quad \text{water} \qquad \text{calcium hydroxide}$$

It is suitable for use within mortars or in the manufacture of certain aerated concrete blocks. Generally, the addition of lime to cement mortar, render or plaster increases its water-retention properties, thus retaining workability, particularly when the material is applied to absorbent substrates such as porous brick. Lime also increases the cohesion of mortar mixes allowing it to spread more easily. Lime-based mortars remain sufficiently flexible to allow movement but additionally, due to the presence of uncarbonated lime, any minor cracks are subsequently healed by the action of rainwater. Hydrated lime absorbs moisture and carbon dioxide from the air, and should therefore be stored in a cool, draught-free building and used whilst still fresh.

Lime putty

Lime putty is produced by slaking quicklime with an excess of water for a period of several weeks until a creamy texture is produced. Alternatively, it can be made by stirring hydrated lime into water, followed by conditioning for at least 24 hours. However, the traditional direct slaking of quicklime produces finer particle sizes in the slurry; the best lime putty is produced by maturing it for at least six months. Lime putty may be blended with Portland cement in mortars where its water-retention properties are greater than that afforded by hydrated lime. Additionally, lime putty, often mixed with sand to form *coarse stuff*, is used directly as a pure lime mortar particularly in restoration and conservation work. It sets, not by reaction with sand and water, but only by carbonation and is therefore described as *non-hydraulic.* Lime wash, as a traditional surface coating, is made by the addition of sufficient water to lime putty to produce a thin creamy consistency.

Carbonation

Lime hardens by the absorption of carbon dioxide from the air, which gradually reconverts the calcium oxide back to calcium carbonate.

$$CaO \ + \ CO_2 \ \xrightarrow{\text{slow}} \ CaCO_3$$
$$\text{lime} \quad \text{carbon dioxide} \quad \text{calcium carbonate}$$

The carbonation process is slow, being controlled by the diffusion of carbon dioxide into the bulk of the material. When sand or stone dust aggregate is added to the lime putty to form a mortar or render, the increased porosity allows greater access of carbon dioxide and a speedier carbonation process. The maximum size of aggregate mixed into lime mortars should not exceed half the mortar-joint width. Typical lime mortar mixes are within the range 1 : 2½ and 1 : 3, lime putty : aggregate ratio. Because of the slow carbonation process, masonry lifts are limited, and the mortar must be allowed some setting time to prevent its expulsion from the joints.

HYDRAULIC LIMES

Hydraulic limes are manufactured from chalk or limestone containing various proportions of clay impurities. The materials produced have some of the properties of Portland cement, and partially harden through hydration processes, rather than solely through carbonation, as happens with non-hydraulic pure calcium oxide lime. Hydraulic limes rich in the clay impurities are more hydraulic and set more rapidly than those with only a low silica and alumina content. Hydraulic limes are categorised as *feebly*, *moderately* or *eminently hydraulic* depending upon their clay content, which is in the ranges 0–8%, 8–18% and 18–25% respectively. *Eminently hydraulic* lime mortar is used for masonry in exposed situations, *moderately hydraulic* lime mortar for most normal masonry applications and *feebly hydraulic* lime mortar is appropriate for conservation work and solid wall construction. Grey semi-hydraulic lime is still produced within the UK in small quantities from chalk containing a proportion of clay. It is used with very soft bricks and for conservation work. Natural hydraulic limes (NHL) are produced by burning chalk and limestone, but hydraulic limes (HL) are produced by blending the constituents in appropriate proportions.

Hydraulic lime, usually imported from France, is mainly used for the restoration of historic buildings, where the use of modern materials would be inappropriate. It is gauged with sand only, giving a mix which develops an initial set within a few hours, but which hardens over an extended period of time. The workable render or mortar mixes adhere well and, because the material is flexible, the risks of cracking and poor adhesion are reduced. The dried mortar is off-white in colour and contains very little alkali, which in Portland cement mortars can cause staining, particularly on limestone. Hydraulic lime may be used for interior lime washes, and also for fixing glass bricks where a flexible binding agent with minimum shrinkage is required. Unlike hydrated lime, hydraulic lime is little affected by exposure to air during storage.

HEMPCRETE

Hemp is grown, particularly in France, for its fibre, which is used in the manufacture of certain grades of paper. The remaining 75% of the hemp stalks, known as *hemp hurd* is a lightweight absorbent material which has the appearance of fine wood chips. When mixed with hydraulic lime it produces a cement mixture which sets within a few hours and gradually 'petrifies' to a lightweight solid due to the high silica content of hemp hurd. The set material, sometimes referred to as *hempcrete*, which has good thermal insulation properties and a texture similar to cork, has been used for the construction of floors, walls using plywood formwork, and also blocks for blockwork. The material is also used as a solid infill for timber-frame construction. In this case the combination of the moisture-absorbing properties of the hemp with the nature of lime, affords some protection to the timber-framing which it encloses.

EXTERNAL LIME RENDERING

External lime rendering is usually applied in a two- or three-coat system, to give an overall thickness of up to 30 mm. In exposed situations, hydraulic lime is used and the thicker initial coat may be reinforced with horse-hair. The final coat can be trowelled to receive a painted finish; alternatively, pebble dash or rough cast may be applied.

Cement

MANUFACTURE OF PORTLAND CEMENT

Portland cement is manufactured from calcium carbonate in the form of crushed limestone or chalk and an argillaceous material such as clay, marl or shale. Minor constituents such as iron oxide or sand may be added depending upon the composition of the raw materials and the exact product required. In principle, the process involves the decarbonisation of calcium carbonate (chalk or limestone) by expulsion of the carbon dioxide, and sintering, at the point of incipient fusion, the resulting calcium oxide (lime) with the clay and iron oxide. Depending upon the raw materials used and their water content at extraction, four key variations in the manufacturing process have been developed. These are the *wet*, *semi-wet*, *semi-dry* and the *dry* processes.

Wet process

The wet process (Fig. 3.1), which was the precursor to the other developments, is still used in some areas for processing chalk and marl clay. Clay is mixed with water to form a slurry, when any excess sand is removed by settlement. An equivalent slurry is prepared from the chalk which is then blended with the clay slurry, screened to remove any coarse material, and stored in large slurry tanks. After final blending, the

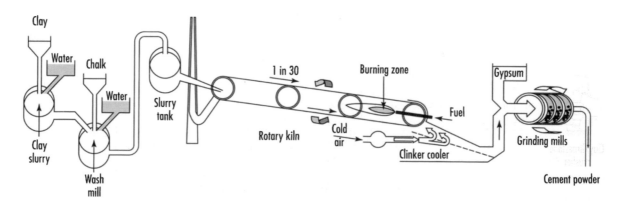

Fig. 3.1 Manufacture of Portland cement – the wet process

slurry is fed into the top of large slowly rotating kilns. The kilns, which are refractory brick-lined steel cylinders up to 200 m long, are fired to approximately 1450°C, usually with pulverised coal. The slurry is dried, calcined and finally sintered to hard grey/black lumps of cement clinker.

A major development in energy conservation has been the elimination or reduction in the slurry water content required in the manufacturing process, as this consumed large quantities of heat energy during its evaporation.

Semi-wet process

In the semi-wet process, chalk is broken down in water and blended into a marl clay slurry. The 40% water content within the slurry is reduced to 19% in a filter press; the resulting *filter-cake* is nodularised by extrusion onto a travelling preheater grate or reduced in a crusher/dryer to pellets. Heating to between 900°C and 1100°C in tower cyclones precalcines the chalk; the mix is then transferred to a short kiln at 1450°C for the clinkering process.

Semi-dry process

In the semi-dry process, dry shale and limestone powders are blended. About 12% water is added to nodularise the blend, which is then precalcined and clinkered as in the semi-wet process.

Dry process

In the dry process (Fig. 3.2) limestone, shale and sand (typically 80%, 17% and 3% respectively), are milled to fine powders, then blended to produce the *dry meal*, which is stored in silos. The meal is passed through a series of cyclones, initially using recovered kiln gases to preheat it to 750°C, then with added fuel to precalcine at 900°C, prior to passage into a fast-rotating 60 m kiln for clinkering at 1450°C. In all processes an intimately mixed feedstock to the kiln is essential for maintaining quality control of the product. Most plants operate primarily with powdered coal, but additionally other fuels including petroleum coke, waste tyre chips, smokeless fuel plant residues, or reclaimed spoil heap coal are used when available. Oil, natural gas and landfill gas have also been used when economically viable. The grey/black clinker manufactured by all processes is cooled with full heat recovery and ground up with 5% added gypsum (calcium sulfate) retarder to prevent excessively rapid *flash* setting of the cement.

The older cement grinding mills are *open-circuit* allowing one pass of the clinker, which produces a wide range of particle size. This product is typically used for concrete production. The newer cement mills are closed circuit, with air separators to extract fine materials and with recycling of the oversize particles for regrinding. This product is frequently used in the ready-mixed market, as it can be controlled to produce cement with higher later strength. To reduce grinding costs, manufacturers accept load shedding and use off-peak electrical supplies where possible. The Portland cement is stored in silos prior to transportation in bulk, by road or rail, or in palletised packs. The standard bag is 25 kg for reasons of health and safety.

With the dry processing and additional increases in energy efficiency, a tonne of pulverised coal can now produce in excess of six tonnes of cement clinker compared to only three tonnes with the traditional

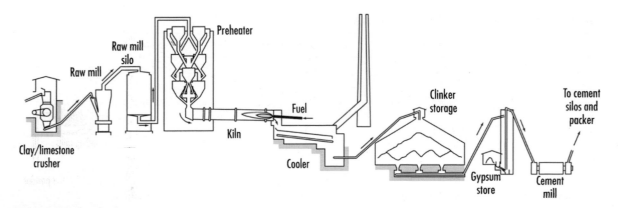

Fig. 3.2 Manufacture of Portland cement – the dry process

wet process. Because the cement industry is so large, the combined output of carbon dioxide to the atmosphere from fuel and the necessary decarbonation of the limestone or chalk, represents about 2% of the carbon dioxide emissions in Britain. Emissions of oxides of sulfur from the fuel are low as these gases are trapped into the cement clinker; however, the escape of oxides of nitrogen and dust, largely trapped by electrostatic precipitators, can only be controlled by constantly improving process technologies. On the basis of the final production of concrete, cement manufacture releases considerably less carbon dioxide per tonne than does primary steel manufacture; however, the relative masses for equivalent construction and the recycling potential of each should also be considered.

COMPOSITION OF PORTLAND CEMENT

The starting materials for Portland cement are chalk or limestone and clay, which consist mainly of lime, silica, alumina and iron oxide. Table 3.1 illustrates a typical composition.

Table 3.1 Typical composition of starting materials for Portland cement manufacture

Component	Percentage
Lime	68
Silica	22
Alumina	5
Iron oxide	3
Other oxides	2

Minor constituents, including magnesium oxide, sulfur trioxide, sodium and potassium oxides amount to approximately 2%. (The presence of the alkali oxides in small proportions can be the cause of the *alkali-silica* reaction, which leads to cracking of concrete when certain silica-containing aggregates are used.) During the clinkering process, these compounds react together to produce the four key components of Portland cement (Table 3.2).

The relative proportions of these major components significantly affect the ultimate properties of the cements and are therefore adjusted in the manufacturing process to produce the required product range. Typical compositions of Portland cements are shown in Table 3.3.

A small reduction in the lime content within the initial mix will greatly reduce the proportion of the tricalcium silicate and produce an equivalent large increase in the dicalcium silicate component of the product. The cement produced will harden more slowly, with a slower evolution of heat. As tricalcium aluminate is vulnerable to attack by soluble sulfates it is the proportion of this component that is reduced in the manufacture of sulfate-resisting cement.

Under the British Standard BS EN 197–1: 2000, except in the case of sulfate-resisting cement, up to 5% of minor additional constituents may be added to cement. These fillers must be materials that do not increase the water requirements of the cement, reduce the durability of the mortar or concrete produced, or cause increased corrosion to any steel reinforcement. In the UK, typical fillers include limestone powder and dry meal or partially calcined material from the cement manufacturing processes.

Table 3.2 Major constituents of Portland cement and their specific properties

Compound	Chemical formula	Cement notation	Properties
Tricalcium silicate	$3CaO.SiO_2$	C_3S	Rapid hardening giving early strength and fast evolution of heat
Dicalcium silicate	$2CaO.SiO_2$	C_2S	Slow hardening giving slow development of strength and slow evolution of heat
Tricalcium aluminate	$3CaO.Al_2O_3$	C_3A	Quick setting which is retarded by gypsum Rapid hardening and fast evolution of heat but lower final strength; vulnerable to sulfate attack
Tetracalcium aluminoferrite	$4CaO.Al_2O_3.Fe_2O_3$	C_4AF	Slow hardening; causes grey colour in cement

Table 3.3 Typical compositions of Portland cements

Cement Type	Class	Composition				Fineness (m²/kg)
		%C$_3$S	%C$_2$S	%C$_3$A	%C$_4$AF	
Portland cement	42.5	55	20	10	8	340
	52.5	55	20	10	8	440
White Portland cement	52.5	65	20	5	2	400
Sulfate-resisting Portland cement	42.5	60	15	2	15	380

SETTING AND HARDENING OF PORTLAND CEMENTS

Portland cement is hydraulic; when mixed with water it forms a paste, which sets and hardens as a result of various chemical reactions between the cementitious compounds and water. Setting and hardening are not dependent upon drying out; indeed, Portland cement will harden under water. Only a small proportion of the added water is actually required for the chemical hydration of the cementitious constituents to hydrated calcium silicates. The additional water is needed to ensure the *workability* of the mix when aggregates are added, so that concrete, for example, can be successfully placed within formwork containing steel reinforcement. Water in excess of that required for hydration will ultimately evaporate leaving capillary pores in the concrete and mortar products. Typically, an increase in void space of 1% reduces crushing strength by 6%. It is therefore necessary to control carefully the water content of the mix by reference to the water/cement ratio. A minimum water/cement ratio of 0.23 is required to hydrate all the cement, although as the cement powder is hydrated it expands, and thus a ratio of 0.36 represents the point at which cement gel fills all the water space. However, a water/cement ratio of 0.42 more realistically represents the minimum water content to achieve full hydration without the necessity for further water to be absorbed during the curing process.

The setting and hardening processes should be distinguished. Setting is the stiffening of the cement paste, which commences immediately the cement is mixed with water. Because the major cementitious constituents set at different rates it is convenient to refer to *initial set* and *final set*. Typically, initial set, or the formation of a plastic gel, occurs after one hour and final set, or the formation of a rigid gel, within 10 hours. The setting process is controlled by the quantity of gypsum added to the cement in the final stages of production. Hardening is the gradual gain in strength of the set cement paste. It is a process which continues, albeit at a decreasing rate, over periods of days, months and years. The rate of hardening is governed partially by the particle-size distribution of the cement powder. Finely ground cement hydrates more rapidly, and therefore begins to set and harden more quickly. Furthermore, the relative proportions of tricalcium silicate and dicalcium silicate have a significant effect upon the rate of hardening as indicated in Table 3.2.

During hydration, any sodium and potassium salts within the Portland cement are released into the pore water of the concrete, giving rise to a highly alkaline matrix. This effectively inhibits corrosion of any reinforcing steel embedded within the concrete, but if active silica is present in any of the aggregates it may react to form an alkali-silica gel which absorbs water, swells and causes cracking of the concrete. This alkali-silica reaction can however be effectively prevented by limiting the total alkali content in the cement to less than 3 kg/m³. (Cement manufacturers normally specify alkali content in terms of equivalent percentage of sodium oxide.)

TYPES OF CEMENT

Cements are classified primarily on the main constituents such as Portland cement or blastfurnace cement. (In addition there may be minor constituents up to 5% and also additives up to 1% by weight.)

The standard BS EN 197–1: 2000 lists five main types of cement:

- CEM I   Portland cement
- CEM II   Portland-composite cement
- CEM III   Blastfurnace cement
- CEM IV   Pozzolanic cement
- CEM V   Composite cement

Within these five main types of cement a wide range of permitted additional constituents, including silica fume, natural or industrial pozzolanas, calcareous or siliceous fly ash and burnt shale, may be incorporated.

The full range of products is listed in Table 3.4. High alumina cement (also known as *calcium aluminate* cement) has a totally different formulation from the range of Portland cements based on calcium silicates.

Table 3.4   Cements to European Standard EN 197–1: 2000 and sulfate resistance to BRE Special Digest 1: 2005

Cement	Type	Notation	Portland cement clinker content (%)	Additional main constituent (%)	Sulfate resistance group
Portland cement	I	CEM I	95–100	0	A
Portland slag cement	II	CEM II/A-S	80–94	6–20	A
		CEM II/B-S	65–79	21–35	A
Portland silica fume cement	II	CEM II/A-D	90–94	6–10	A
Portland pozzolana cement	II	CEM II/A-P	80–94	6–20	
		CEM II/B- P	65–79	21–35	
		CEM II/A-Q	80–94	6–20	A
		CEM II/B- Q	65–79	21–35	
Portland fly ash cement	II	CEM II/A-V	80–94	6–20	A
		CEM II/B- V	65–79	21–35	A for 21–24%
					D for ≥ 25%
		CEM II/A-W	80–94	6–20	
		CEM II/B-W	65–79	21–35	
Portland burnt shale cement	II	CEM II/A-T	80–94	6–20	
		CEM II/B-T	65–79	21–35	
Portland limestone cement	II	CEM II/A-L	80–94	6–20	B for class 32.5
					C for class ≥ 42.5
		CEM II/B-L	65–79	21–35	
		CEM II/A-LL	80–94	6–20	B for class 32.5
					C for class ≥ 42.5
		CEM II/B-LL	65–79	21–35	
Portland composite cement	II	CEM II/A-M	80–94	6–20	
		CEM II/B-M	65–79	21–35	
Blastfurnace cement	III	CEM III/A	35–64	36–65	A, or D for strictly controlled levels of tricalcium aluminate
		CEM III/B	20–34	66–80	A, or F for strictly controlled levels of
		CEM III/C	5–19	81–95	tricalcium aluminate
Pozzolanic cement	IV	CEM IV/A	65–89	11–35	
		CEM IV/B	45–64	36–55	E
		VLH IV/B	45–64	36–55	E (BS EN 14216)
Composite	V	CEM V/A	40–64	36–60	
		CEM V/B	20–39	61–80	
Sulfate-resisting Portland cement		SRPC			G (BS 4027: 1996)

The code letters used in the European Standard are:

D silica fume	F filler	K Portland cement clinker
L/LL limestone	M mixed	P natural pozzolana
Q industrial pozzolana	S granulated blastfurnace slag	T burnt shale
V siliceous fly ash	W calcareous fly ash.	

(Limestone LL has a total organic carbon content limit of 0.2%, limestone L has a total organic carbon content limit of 0.5%.)

Grouping with respect to sulfate resistance A (low) to G (high) resistance (BRE Special Digest 1: 2005). VLH refers to very low heat cements (BS EN 14216: 2004).

STRENGTH CLASSES OF CEMENT

The standard strength classes of cement are based on the 28-day compressive strength of mortar prisms, made and tested to the requirements of BS EN 196–1: 2005. The test uses specimens which are 40 × 40 × 160 mm, cast from a mix of 3 parts of CEN (European Committee for Standardisation) standard sand, 1 part of cement and 0.5 part of water. The sample is vibrated and cured for the appropriate time, then broken into halves and compression tested across the 40 mm face. Three specimens are used to determine a mean value from the six pieces.

Each cement strength class (32.5, 42.5 and 52.5) has sub-classes associated with the high early (R) and the ordinary (N) development of early strength (Table 3.5). The strength classes and sub-classes give production standards for cements, but do not specify how a particular mix of cement, aggregate and admixtures will perform as a concrete; this needs to be determined by separate testing.

The most commonly used cement within the UK (formerly ordinary Portland cement or OPC) is currently designated to the standard BS EN 197–1: 2000 as:

CEM I	42.5	N	CEM I 42.5N
type of cement	strength class	ordinary early strength development	

High early strength Portland cement is designated:

CEM I	42.5	R	CEM I 42.5R
type of cement	strength class	high early strength development	

Low early strength low heat blastfurnace cement with a granulated blastfurnace slag content between 81 and 95% and a strength class of 32.5 is designated:

CEM III/C	32.5	L – LH	CEM III/C 32.5L–LH
type of cement	strength class	low early strength development & low heat	

Portland limestone cement with between 6 and 20% limestone of 0.5% total organic content, a strength class of 32.5 and normal early strength is designated:

CEM II/A	L	32.5	N	CEM II/A–L 32.5N
type of cement /proportion of cement clinker	sub-type, lime-stone	strength class	ordinary early strength development	

Portland cements

Portland cements – strength classes 32.5, 42.5 and 52.5
The Portland cement classes 32.5, 42.5, and 52.5 correspond numerically to their lower characteristic strengths at 28 days. The 32.5 and 42.5 classes have upper characteristic strengths which are 20 MPa greater than the lower characteristic strengths, as designated by the class number. The class 52.5 has no upper strength limit. Statistically the tested strengths must fall with no more than 5% of the tests below the lower limit or 10% of the tests above the upper limit.

Table 3.5 Strength classes of cements to European Standard BS EN 197–1: 2000

Strength class	Compressive strength (MPa)			
	Early strength		Standard strength	
	2 day minimum	7 day minimum	28 day minimum	28 day maximum
32.5N		16.0	32.5	52.5
32.5R	10		32.5	52.5
42.5N	10		42.5	62.5
42.5R	20		42.5	62.5
52.5N	20		52.5	
52.5R	30		52.5	

The code letters in the standards are: N ordinary early strength development, R high early strength development.

Thus class 42.5 Portland cement has a strength within the range 42.5 MPa to 62.5 MPa, with a maximum of 5% of test results being below 42.5 MPa and a maximum of 10% of the test results being above 62.5 MPa.

Each class also has lower characteristic strength values at 2 days, except for class 32.5, which has a lower characteristic strength at 7 days. Where high early strength is required, for example to allow the early removal of formwork in the manufacture of precast concrete units, class 52.5 or class 42.5R is used. These Portland cements are more finely ground than class 42.5 to enable a faster hydration of the cement in the early stages. Class 32.5 cements, for general-purpose and DIY use, frequently contain up to 1% additives to improve workability and frost resistance, together with up to 5% minor additional constituents such as pulverised-fuel ash, granulated blastfurnace slag or limestone filler. Portland cement of strength class 42.5 accounts for approximately 90% of the total cement production within the UK.

White Portland cement

White Portland cement is manufactured from materials virtually free of iron oxide and other impurities, which impart the grey colour to Portland cement. Generally, china clay and limestone are used and the kiln is fired with natural gas or oil rather than pulverised coal. Iron-free mills are used for the grinding process to prevent colour contamination. Because of the specialist manufacturing processes, it is approximately twice the price of the equivalent grey product. To further enhance the whiteness, up to 5% of white titanium oxide pigment may be added. The standard product is to strength class 52.5N. Typical applications include renderings, cast stone, precast and *in-situ* structural concrete and pointing.

Sulfate-resisting Portland cement

Sulfate-resisting Portland cement (BS 4027: 1996) is suitable for concrete and mortar in contact with soils and groundwater containing soluble sulfates up to the maximum levels (measured as sulfur trioxide) of 2% in soil or 0.5% in groundwater. In normal Portland cements the hydrated tricalcium aluminate component is vulnerable to attack by soluble sulfates, but in sulfate-resisting Portland cement this component is restricted to a maximum of 3.5%. For maximum durability a high-quality, dense, non-permeable concrete is required. Many sulfate-resisting cements are also defined as low alkali (LA) to BS 4027: 1996, containing less than 0.6% alkali (measured as sodium oxide). Thus durable concrete, without the risk of subsequent alkali-silica reaction, can be manufactured with alkali-reactive aggregates, using up to 500 kg/m^3 of cement, providing no other alkalis are present.

Very low heat special cements

Low heat Portland cement (BS 1370: 1979) is appropriate for use in mass concrete, where the rapid internal evolution of heat could cause cracking. It contains a higher proportion of dicalcium silicate, which hardens and evolves heat more slowly. The range of very low heat special cements, listed in Table 3.6, includes products based on blastfurnace, pozzolanic and composite cements (BS EN 14216: 2004). Very low heat special cements are appropriate for use only in massive

Table 3.6 Very low heat special cements to BS EN 14216: 2004

Cement	Type	Notation	Portland cement clinker content (%)	Additional main constituent (%)
Blastfurnace cement	III	VLH III/B	20–34	66–80
		VLH III/C	5–19	81–95
Pozzolanic cement	IV	VLH IV/A	65–89	11–35
		VLH IV/B	45–64	36–55
Composite cement	V	VLH V/A	40–64	18–30 blastfurnace slag
				18–30 pozzolana & fly ash
		VLH/VB	20–38	31–50 blastfurnace slag
				31–50 pozzolana & fly ash

constructions such as dams, but not for bridges or buildings.

Blended Portland cements

Blended Portland cements include not only masonry cement, with its specific end use, but also the wide range of additional materials classified within the European Standard EN 197–1: 2000.

Masonry cements

Portland cement mortar is unnecessarily strong and concentrates any differential movement within brickwork or blockwork into a few large cracks, which are unsightly and may increase the risk of rain penetration. Masonry cement produces a weaker mortar, which accommodates some differential movement and ensures a distribution of hairline cracks within joints, thus preserving the integrity of the bricks and blocks. Masonry cements contain water-retaining mineral fillers, usually ground limestone, and air-entraining agents to give a higher workability than unblended Portland cement. They should not normally be blended with further admixtures but mixed with building sand in ratios between 1 : 4 and 1 : 6½ depending upon the degree of exposure of the brick or blockwork. The air entrained during mixing increases the durability and frost resistance of the hardened mortar. Masonry cement is also appropriate for use in renderings but not for floor screeds or concreting. It is therefore generally used as an alternative to Portland cement with hydrated lime or plasticiser. Inorganic pigments, except those containing carbon black, may be incorporated for visual effect. The strength classes for masonry cement are listed in Table 3.7.

Table 3.7 Strength classes of masonry cements to European Standard BS EN 413–1: 2004

Strength classes	Compressive strength (MPa)	
	7 day early strength	28 day standard strength
MC 5	–	5–15
MC 12.5	≥7	12.5–32.5
MC 12.5X	≥7	12.5–32.5
MC 22.5 X	≥10	22.5–42.5

Masonry cement is designated by MC, the X refers to cements which do not incorporate air-entraining agent.

Portland slag and blastfurnace cements

Granulated blastfurnace slag, formerly termed *ground granulated blastfurnace slag* (GGBS), is a cementitious material, which in combination with Portland cement and appropriate aggregates, makes a durable concrete. The material is a by-product of the iron-making process within the steel industry. Iron ore, limestone and coke are fed continuously into blastfurnaces, where at 1500°C they melt into two layers. The molten iron sinks, leaving the blastfurnace slag floating on the surface, from where it is tapped off at intervals. The molten blastfurnace slag is rapidly cooled by water quenching in a granulator or pelletiser to produce a glassy product. After drying, the blastfurnace slag granules or pellets are ground to the fine off-white powder – granulated blastfurnace slag. The composition of the material is broadly similar to that of Portland cement as illustrated in Table 3.8.

Granulated blastfurnace slag may be intimately ground with Portland cement clinker in the cement mill, although usually it is mixed with Portland cement on site. The British Standard (BS 6699: 1992) gives the specification for ground granulated blastfurnace slag. The standards BS 146: 2002 and BS EN 197–4: 2004 refer to blastfurnace cements with mixes from 36% to 80% and 95% of granulated blastfurnace slag respectively (Table 3.9).

Table 3.8 Typical compositions of granulated blastfurnace slag and Portland cement

	Granulated blastfurnace slag (%)	Portland cement (%)
Lime	41	68
Silica	35	22
Alumina	11	5
Iron Oxide	1	3
Other	12	2

Table 3.9 Composition of low early strength blastfurnace cements to British Standard BS EN 197: 2004

Composition	Type	CEM III/A (%)	CEM III/B (%)	CEM III/C (%)
Portland cement clinker		35–64	20–34	5–19
Blastfurnace slag		36–65	66–80	81–95
Minor constituents		0–5	0–5	0–5

Concrete manufactured from a blend of Portland and granulated blastfurnace slag cements has a lower permeability than Portland cement alone; this enhances resistance to attack from sulfates, weak acids and to the ingress of chlorides which can cause rapid corrosion of steel reinforcement, for example in marine environments and near roads subjected to de-icing salts. Sulfate attack is also reduced by the decrease in tricalcium aluminate content. The more gradual hydration of granulated blastfurnace slag cement evolves less heat and more slowly than Portland cement alone; thus a 70% granulated blastfurnace slag mix can be used for mass concrete, where otherwise a significant temperature rise could cause cracking. The slower evolution of heat is associated with a more gradual development of strength over the first 28-day period.

However, the ultimate strength of the mature concrete is comparable to that of the equivalent Portland cement. The initial set with granulated blastfurnace slag blends is slower than for Portland cement alone, and the fresh concrete mixes are more plastic, giving better flow for placing and full compaction. The risk of alkali-silica reaction caused by reactive silica aggregates can be reduced by the use of granulated blastfurnace slag to reduce the active alkali content of the concrete mix to below the critical 3.0 kg/m^3 level. The classes for low early strength blastfurnace cements are listed in Table 3.10.

Portland fly ash and pozzolanic cements
Pozzolanic materials are natural or manufactured materials containing silica, which react with the calcium hydroxide produced in the hydration of Portland cement to produce further cementitious products. Within the UK, natural volcanic pozzolanas are little used, but fly ash, formerly termed *pulverised-fuel ash* (PFA), the waste product from coal-fired electricity-generating stations, is used either factory mixed with Portland cement or blended in on site.

Portland fly ash cement, cures and evolves heat more slowly than Portland cement; it is therefore appropriate for use in mass concrete to reduce the risk of thermal cracking. Additions of up to 25% fly ash in Portland cement are often used; the concrete produced is darker than with Portland cement alone. Concrete made with blends of 25–40% by weight of fly ash in Portland cement has good sulfate-resisting properties. However, in the presence of groundwater with high magnesium concentrations, sulfate-resisting Portland cement should be used. Fly ash concretes also have enhanced resistance to chloride ingress, which is frequently the cause of corrosion to steel reinforcement.

The fly ash produced in the UK by burning pulverised bituminous coal is siliceous, containing predominantly reactive silica and alumina. In addition to siliceous fly ash, the European Standard EN 197–1: 2000 does allow for the use of calcareous fly ash, which additionally contains active lime, giving some self-setting properties. The range of fly ash suitable for concrete is defined in the standard BS EN 450–1: 2005. Natural pozzolanas of volcanic origin and industrial pozzolanas from other industrial processes in Europe are used with Portland cement and are categorised under EN 197–1: 2000 as pozzolanic cements.

Portland limestone cement
The addition of up to 5% limestone filler to Portland cement has little effect on its properties. The addition of up to 25% limestone gives a performance similar to that of Portland cement with a proportionally lower cementitious content; thus, if equivalent durability to Portland cement is required, then cement contents must be increased. The two categories of limestone for Portland limestone cement are defined by their total organic carbon (TOC) content; LL refers to a maximum of 0.20% and L to a maximum of 0.50% by mass.

Silica fume
Silica fume or microsilica, a by-product from the manufacture of silicon and ferro-silicon, consists of ultra-fine spheres of silica. The material, because of its high surface area, when blended as a minor addition to Portland cement, increases the rate of hydration, giving the concrete a high early strength and also a reduced permeability. This in turn produces greater resistance to chemical attack and abrasion. Silica fume may be added up to 5% as a filler, or in Portland silica fume cement to between 6 and 10%.

Table 3.10 Strength classes of low early strength blastfurnace cements to British Standard BS EN 197–4: 2004

Strength classes	Compressive strength (MPa)		
	2 day early strength	7 day early strength	28 day standard strength
32.5 L	–	≥12	32.5–52.5
42.5 L	–	≥16	42.5–62.5
52.5 L	≥10	–	≥52.5

Burnt shale

Burnt shale is produced by heating oil shale to 800°C in a kiln. It is similar in nature to blastfurnace slag, containing mainly calcium silicate and calcium aluminate, but also silica, lime and calcium sulfate. It is weakly cementitious. The European Standard EN 197–1: 2000 allows for the use of burnt shale as a filler to 5%, or between 6 and 35% in Portland burnt shale cement.

Fillers

Fillers up to 5% by weight of the cement content may be added to cements to the standard EN 197–1: 2000. They should be materials which do not increase the water requirements of the cement. Fillers may be any of the permitted alternative main constituents (e.g. granulated blastfurnace slag, pozzolanas, fly ash, burnt shale, silica fume or limestone), or other inorganic materials, providing that they are not already present as one of the main constituents. The most common fillers are limestone and either raw meal or partially calcined material from the cement-making process.

CEMENT ADMIXTURES

Admixtures may be defined as materials that are added in small quantities to mortars or concretes during mixing, in order to modify one or more of their physical or visual properties.

Plasticisers

Plasticisers, or water-reducing admixtures, are added to increase the workability of a mix, thus enabling easier placing and compaction. Where increased workability is not required, water-reducers may be used to lower the water/cement ratio, giving typically a 15% increase in strength and better durability. The plasticisers, which are usually lignosulfonates or hydroxylated polymers, act by dispersing the cement grains. Some air-entrainment may occur with the lignosulfonates, causing a 6% reduction in crushing strength for every 1% of air entrained.

Superplasticisers

Superplasticisers, such as sulfonated naphthalene or sulfonated melamine formaldehyde, when added to a normal 50 mm slump concrete produce a flowing, self-levelling or self-compacting concrete (SCC) which can be placed, even within congested reinforcement, without vibration. Alternatively, significantly reduced water contents can be used to produce early- and ultimately higher-strength concretes. As the effect of superplasticisers lasts for less than an hour, the admixture is usually added to ready-mixed concrete on site prior to discharge and placing. Standard concrete additives, fillers and steel or polypropylene fibres may be incorporated into self-compacting concrete which can be pumped or delivered by skip or chute. Good-quality off-the-form surface finishes can be achieved, especially with timber formwork. Self-levelling mixes for screeds between 3 and 20 mm thick can be adjusted to take light foot traffic after 3 to 24 hours. Renovation mixes, usually incorporating fibre-mat reinforcement, can be used over a range of existing floor surfaces to thicknesses usually in the range 4–30 mm.

Accelerators

Accelerators increase the rate of reaction between cement and water, thus increasing the rate of set and development of strength. This can be advantageous in precasting, where early removal of the formwork is required, and in cold weather when the heat generated speeds up the hardening processes and reduces the risk of frost damage. Only chloride-free accelerators, such as calcium formate, should be used in concrete, mortar or grout where metal will be embedded, because calcium chloride accelerators can cause extensive metallic corrosion. Accelerators producing a rapid set are not normally used within structural concrete.

Retarders

Retarders, typically phosphates or hydroxycarboxylic acids, decrease the rate of set, thus extending the time between initial mixing and final compaction, but they do not adversely affect 28-day strength. Retarders may be applied to formwork, to retard the surface concrete where an exposed aggregate finish is required by washing after the formwork is struck. Retarders are also frequently used in ready-mixed mortars to extend their workable life up to 36 hours. The mortars are usually delivered on site in date marked containers of 0.3 m³ capacity.

Air-entraining admixtures

Air-entraining admixtures, typically wood resins or synthetic surfactants, stabilise the tiny air bubbles which become incorporated into concrete or mortar as it is mixed. The bubbles, which are between 0.05 and 0.5 mm in diameter, do not escape during transportation or vibration. They improve the workability of the mix, reduce the risk of segregation and greatly enhance frost resistance. However, the incorporation of void space within concrete decreases its crushing strength by 6% for every 1% of air entrained, thus for a typical 3% addition of entrained air, a reduction of 18% in crushing strength is produced. This is partially offset by the increase in plasticity, which generally produces a higher-quality surface and allows a lower water content to be used. The increased cohesion of air-entrained concrete may trap air against moulded vertical formwork reducing the quality of the surface.

Water-resisting admixtures

The water penetration through concrete can be reduced by incorporation of hydrophobic materials, such as stearates and oleates, which coat the surface of pores and by surface-tension effects discourage the penetration of damp. The use of water-reducing admixtures also reduces water penetration by reducing the water/cement ratio, thus decreasing pore size within the concrete. For mortars and renders a styrene-butadiene latex emulsion admixture can be used to reduce permeability.

Foaming agents

Foamed concrete or mortar contains up to 80% by volume of void space, with densities as low as 400 kg/m^3 and 7-day strengths between 0.5 and 20 MPa. It is typically produced by blending cement, sand or fly ash and water into a preformed foam or by mechanically foaming the appropriate mix using a foaming surfactant. Foamed concrete is free flowing, can be pumped, and requires no compaction. It is therefore used for trench reinstatement, filling cellars or to provide thermal insulation under floors or in flat roofs.

Pumping agents

Not all concrete mixes are suitable for pumping. Mixes low in cement or with some lightweight aggregates tend to segregate, and require thickening with a pumping agent. Conversely, high-cement content mixes require plasticising to make them pumpable. A range of pumping agents is therefore produced to suit the requirements of various concrete mixes.

Lightweight aggregate concrete is often pumped into place for floor slabs.

Pigments

A wide range of coloured pigments is available for incorporation into concrete and mortars (BS EN 12878: 2005). Titanium oxide can be added to enhance the whiteness of white cement. Carbon black is used with grey Portland cement, although the black loses intensity with weathering. The most common colours are the browns, reds and yellows produced with synthetic iron, chromium and manganese oxides, also with complexes of cobalt, aluminium, nickel and antimony. Additionally, ultramarine and phthalocyanine extend the range of the blues and greens. The depth and shade of colour depends upon the dose rate (between 1 and 10%), and on the colour of the sand and any other aggregates. To produce pastel shades, pigments can be added to white Portland cement.

CALCIUM ALUMINATE CEMENT

Calcium aluminate cement, also known as high alumina cement (HAC), is manufactured from limestone and bauxite (aluminium oxide). The ores in roughly equal proportions are charged together into a vertical furnace which is heated to approximately 1600°C (Fig. 3.3). The mixture melts and is continuously run off into trays, where it cools to produce the clinker, which is then milled, producing calcium aluminate BS EN 14647: 2005 cement to EN 14647: 2005. The dark grey cement composition differs from that of Portland cement as it is based on calcium aluminates rather than calcium silicates. Although calcium aluminate cement can be produced over a wide range of compositions, the standard product has a 40% alumina content.

Calcium aluminate cement should not be used for structural purposes; however, it is useful where rapid strength gain is required allowing the fast removal of formwork within 6 to 24 hours. The fast evolution of heat allows concreting to take place in low temperatures. The material also has good heat-resistant properties, so may be used to produce refractory concrete. When mixed with Portland cement it produces a rapid-setting concrete, suitable for non-structural repairs and sealing leaks. Good quality calcium aluminate cement is generally resistant to chemical attack by dilute acids, chlorides and oils, but not alkalis.

Some structural failures associated with calcium aluminate cement have been caused by *conversion* of

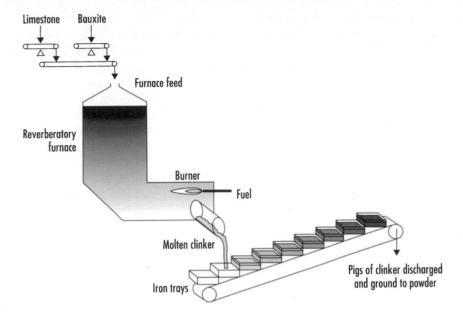

Fig. 3.3 Manufacture of calcium aluminate (high alumina) cement

the concrete, in which changes in the crystal structure, accelerated by high temperatures and humidity, have caused serious loss of strength, increased porosity and subsequent chemical attack. Depending upon the degree of conversion, calcium aluminate cement becomes friable and a deeper brown in colour; the exact degree of conversion can only be determined by chemical analysis of a core sample. It is now recognised that such failures can be prevented by using a minimum cement content of 400 kg/m³, limiting the water/cement ratio to a maximum of 0.4, and by ensuring controlled curing during the 6- to 24-hour initial hardening stage. The concrete should be covered or sprayed to prevent excessive water loss, particularly where substantial increases in temperature may occur.

Additionally, in order to prevent alkaline hydrolysis of the concrete, aggregates containing soluble alkalis should not be used; hard limestone is generally considered to be the best aggregate. Coloured calcium aluminate cement concrete has the advantage that it is free from calcium hydroxide, which causes efflorescence in Portland cements. The BRE Special Digest SD3 (2002) gives methods for assessing existing calcium aluminate cement concrete [high alumina cement concrete (HACC)] constructions and suggests appropriate remedial actions. In some cases where the depth conversion of HACC structural members is significant, with time there is an increasing risk of reinforcement corrosion.

Concrete

Concrete is a mixture of cement, aggregates and water, with any other admixtures which may be added to modify the placing and curing processes or the ultimate physical properties. Initially when mixed, concrete is a *plastic* material, which takes the shape of the mould or formwork. When hardened it may be a dense load-bearing material or a lightweight thermally insulating material, depending largely on the aggregates used. It may be reinforced or prestressed by the incorporation of steel.

Most concrete is crushed and recycled at the end of its useful life, frequently as hard core for new construction work. However, a growth in the use of recycled aggregates for new concrete can be anticipated, as this will have a significant environmental gain in reducing the demand on new aggregate extraction.

AGGREGATES FOR CONCRETE

Aggregates form a major component of concretes, typically approximately 80% by weight in cured mass concrete. Aggregate properties including crushing strength, size, grading and shape have significant effects on the physical properties of the concrete mixes and hardened concrete. Additionally, the appearance of visual concrete can be influenced by aggregate colour and surface treatments. The standard BS EN 12620: 2002 specifies the appropriate properties including materials, size, grading and shape.

Aggregates for concrete are normally classified as lightweight, dense or high-density. Standard dense aggregates are classified by size as fine (i.e. sand) or coarse (i.e. gravel). Additionally, steel or polypropylene fibres or gas bubbles may be incorporated into the mix for specialist purposes.

Dense aggregates

Source and shape

Dense aggregates are quarried from pits and from the seabed. In the south-east of England, most land-based sources are gravels, typically flint, whereas further north and west, both gravels and a variety of crushed quarried rocks are available. Marine aggregates are smooth and rounded, and require washing to remove deleterious matter such as salts, silt and organic debris. The shape of aggregates can significantly affect the properties of the mix and cured concrete. Generally rounded aggregates require a lower water content to achieve a given mix workability, compared to the equivalent mix using angular aggregates. However, cement paste ultimately bonds more strongly to angular aggregates with rough surfaces than to the smoother gravels, so a higher crushing strength can be achieved with crushed rocks as aggregate. Excessive proportions of long and flaky coarse aggregate should be avoided as they can reduce the durability of concrete. Recycled aggregates resulting from processed inorganic materials previously used in construction are increasingly being used in concrete construction.

Aggregate size

For most purposes the maximum size of aggregate should be as large as possible consistent with ease of placement within formwork and around any steel reinforcement. Typically, 20 mm aggregate is used for most construction work, although 40 mm aggregate is appropriate for mass concrete, and a maximum of 10 mm for thin sections. The use of the largest possible aggregate reduces the quantity of sand and therefore cement required in the mix, thus controlling shrinkage and minimising cost. Large aggregates have a low surface area/volume ratio, and therefore produce mixes with greater workability for a given water/ cement ratio, or allow water/cement ratios to be reduced for the same workability, thus producing a higher crushing-strength concrete.

Grading

To obtain consistent quality in concrete production, it is necessary to ensure that both coarse and fine aggregates are well graded. A typical *continuously graded* coarse aggregate will contain a good distribution of sizes, such that the voids between the largest stones are filled by successively smaller particles down to the size of the sand. Similarly, a well-graded sand will have a range of particle sizes, but with a limit on the proportion of fine clay or silt, because too high a content of *fines* (of size less than 0.063 mm) would increase the water and cement requirement for the mix. Usually a maximum of 3% fines is considered non-harmful. This overall grading of aggregates ensures that all void spaces are filled with the minimum proportion of fine material and expensive cement powder. In certain circumstances, coarse aggregate may be graded as *single-sized* or *gap graded*. The former is used for controlled blending in *designed mixes* whilst the latter is used particularly for exposed aggregate finishes on visual concrete. Sands are classified into three categories according to the proportion passing through a 0.500 mm sieve: coarse C (5–45%), medium M (30–70%) and fine F (55–100%). Only the coarse and medium categories of sands should be used for heavy-duty concrete floor finishes.

Sampling and sieve analysis

To determine the grading of a sample of coarse or fine aggregate, a representative sample has to be subjected to a sieve analysis. Normally at least ten samples would be taken from various parts of the stock pile, and these would be reduced down to a representative sample using a *riffle box*, which successively divides the sample by two until the required test volume is obtained (Fig. 3.4).

Aggregate gradings are determined by passing the representative sample through a set of standard sieves (BS EN 12620: 2002). Aggregate size is specified by the lower (d) and upper (D) sieve sizes. For coarse aggregates the sieve sizes are 63, 31.5, 16, 8, 4, 2 and 1 mm and for fine aggregates 4, 2, 1, 0.250 and 0.063 mm. Coarse aggregates are usually defined as having a minimum size (d) of 2 mm, while fine aggregates often have a maximum size (D) of 4 mm. The sieve analysis is determined by assessing the cumulative percentage passing through each sieve size. This is plotted against the sieve size and compared to the limits as illustrated for a typical coarse aggregate (Fig. 3.5).

Aggregates for concreting are normally *batched* from stockpiles of 20 mm coarse aggregate and concreting

Fig. 3.4 Riffle box

sand in the required proportions to ensure consistency, although *all-in aggregate*, which contains both fine and coarse aggregates, is also available as a less well controlled cheaper alternative, where a lower grade of concrete is acceptable. Where exceptionally high control on the mix is required, single-size aggregates may be batched to the customer's specification. The batching of aggregates should normally be done by weight, as free surface moisture, particularly in sand, can cause *bulking*, which is an increase in volume by up to 40% (Fig. 3.6). Accurate batching must take into account the water content in the aggregates in the calculations of both the required weight of aggregates and the quantity of water to be added to the mix.

Impurities within aggregates
Where a high-quality exposed concrete finish is required, the aggregate should be free of iron pyrites, which cause spalling and rust staining of the surface. Alkali-silica reaction (ASR) can occur when active silica, present in certain aggregates, reacts with the alkalis within Portland cement causing cracking.

High-density aggregates
Where radiation shielding is required, high-density aggregates such as barytes (barium sulfate), magnetite (iron ore), lead or steel shot are used. Hardened concrete densities between 3000 and 5000 kg/m^3 (double that for normal concrete) can be achieved.

Lightweight aggregates

Natural stone aggregate concretes typically have densities within the range 2200 to 2500 kg/m^3, but where densities below 2000 kg/m^3 are required, then an appropriate lightweight concrete must be used.

Lightweight concretes in construction exhibit the following properties in comparison with dense concrete:

- they have enhanced thermal insulation but reduced compressive strength;
- they have increased high-frequency sound absorption but reduced sound insulation;

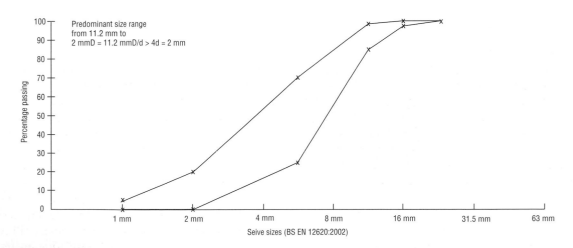

Fig. 3.5 Grading of coarse aggregates

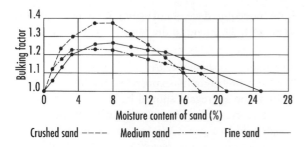

Fig. 3.6 Bulking of sands in relation to moisture content

- they have enhanced fire resistance over most dense aggregate concretes (e.g. granite spalls);
- they are easier to cut, chase, nail, plaster and render than dense concrete;
- the reduced self weight of the structure offers economies of construction;
- the lower formwork pressures enable the casting of higher lifts.

The three general categories of lightweight concrete are: lightweight aggregate concrete; aerated concrete; and no-fines concrete (Fig. 3.7).

Many of the lightweight aggregate materials are produced from by-products of other industrial processes or directly from naturally occurring minerals. The key

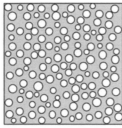

Aerated concrete – voids in a cement matrix

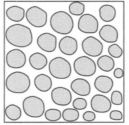

Lightweight aggregate concrete – voids in the aggregate pellets

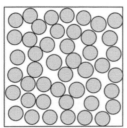

Polystyrene-bead aggregate cement – very light aggregate

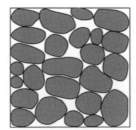

No-fines concrete – dense aggregate with voids between

Fig. 3.7 Lightweight concretes

exception is expanded polystyrene, which has the highest insulation properties, but is expensive due to its manufacture from petrochemical products.

Pulverised-fuel ash
Pulverised-fuel ash, or fly ash, is the residue from coal-fired electricity-generating stations. The fine fly ash powder is moistened, pelleted and sintered to produce a uniform lightweight PFA aggregate, which can be used in load-bearing applications.

Foamed blastfurnace slag
Blastfurnace slag is a by-product from the steel industry. Molten slag is subjected to jets of water, steam and compressed air to produce a pumice-like material. The foamed slag is crushed and graded to produce aggregate, which can be used in load-bearing applications. Where rounded pelletised expanded slag is required the material is further processed within a rotating drum.

Expanded clay and shale
Certain naturally occurring clay materials are pelletised, then heated in a furnace. This causes the evolution of gases, which expands and aerates the interior, leaving a hardened surface crust. These lightweight aggregates may be used for load-bearing applications.

Expanded perlite
Perlite is a naturally occurring glassy volcanic rock which, when heated almost to its melting point, evolves steam to produce a cellular material of low density. Concrete made with expanded perlite has good thermal insulation properties but low compressive strength and high drying shrinkage.

Exfoliated vermiculite
Vermiculite is a naturally occurring mineral, composed of thin layers like mica. When heated rapidly the layers separate, expanding the material by up to 30 times, producing a very lightweight aggregate. Exfoliated vermiculite concrete has excellent thermal insulation properties but low compressive strength and very high drying shrinkage.

Expanded polystyrene
Expanded polystyrene beads offer the highest level thermal insulation, but with little compressive strength. Polystyrene bead aggregate cement (PBAC) is frequently used as the core insulating material within precast concrete units.

Aerated concrete

Aerated concrete (aircrete) is manufactured using foaming agents or aluminium powder as previously outlined in the section on foaming agents. Densities in the range 400 to 1600 kg/m^2 give compressive strengths ranging from 0.5 to 20 MPa. Drying shrinkages for the lowest-density materials are high (0.3%), but thermal conductivity can be as low as 0.1 W/m K, offering excellent thermal insulation properties. Factory-autoclaved aerated concrete blocks have greatly reduced drying shrinkages and enhanced compressive strength over site-cured concrete. Aerated concrete is generally frost-resistant but should be rendered externally to prevent excessive water absorption. The material is easily worked on site as it can be cut and nailed.

No-fines concrete

No fines concrete is manufactured from single sized aggregate (usually between 10 and 20 mm) and cement paste. Either dense or lightweight aggregates may be used, but care has to be taken in placing the mix to ensure that the aggregate remains coated with the cement paste. The material should not be vibrated. Drying shrinkage is low, as essentially the aggregate is stacked up within the formwork, leaving void spaces; these increase the thermal-insulation properties of the material in comparison with the equivalent dense concrete. The rough surface of the cured concrete forms an excellent key for rendering or plastering which is necessary to prevent rain, air or sound penetration. Dense aggregate no-fines concrete may be used for load-bearing applications.

Fibres

Either steel or polypropylene fibres may be incorporated into concrete, as an alternative to secondary reinforcement, particularly in heavily trafficked floor slabs. The fibres reduce the shrinkage and potential cracking that may occur during the initial setting and give good abrasion and spalling resistance to the cured concrete. The low-modulus polypropylene fibres, which do not pose a corrosion risk after carbonation of the concrete, enhance the energy-absorbing characteristics of the concrete, giving better impact resistance. Steel fibres increase flexural strength as well as impact resistance but are more expensive. Alternatively, stainless steel fibres may be used where rust spots on the surface would be unacceptable. Typically, polypropylene fibres are added at the rate of 0.2% by weight (0.5% by volume) and steel at the rate of 3–4% by weight. Both polypropylene and steel fibre concretes can be pumped. (Glass-fibre reinforced cement is described in Chapter 11.)

Ultra high performance concrete

Ultra high performance concrete (UHPC) has six to eight times the compressive strength of traditional concrete. It is produced from a mixture of Portland cement, crushed quartz, sand, silica fume, superplasticiser, fibres and water with no aggregates larger than a few millimetres. Wollastonite (calcium silicate) filler may also be included in the mix. The fibres most frequently used are either high strength steel for maximum strength or polyvinyl acetate (PVA) of approximately 12 mm in length for lower load applications. The concrete can be cast into traditional moulds by gravity or pumped or even injection cast under pressure. When cast into traditional moulds, the material is self-levelling, so only slight external vibration of the formwork may be required to ensure complete filling. The material is designed for use without steel reinforcement bars.

Structural components in ultra high performance concrete may, after setting, be subjected to steam treatment for 48 hours at 90°C. This enhances durability and mechanical properties, eliminates shrinkage and reduces creep. The material does not spall under fire test conditions.

The enhanced compressive and flexural strengths of ductile fibre-reinforced ultra high performance concrete enable lighter and thinner sections to be used for structural components such as shell roofs and bridges, creating an enhanced sleek aesthetic. A high quality durable surface is produced from appropriate moulds (e.g. steel) coated with proprietary release agent.

TRANSLUCENT CONCRETE

By embedding parallel fibre-optic threads into fine concrete, the material is made translucent without any appreciable loss of compressive strength. Translucent concrete can be manufactured as blocks or panels providing that the fibres run transversely from face to face. If one face is illuminated, any shadow cast onto the bright side is clearly visible on the other face, whilst the colour of the transmitted light is unchanged. The material has many potential applications including walls, floor surfaces and illuminated pavements.

INSULATING CONCRETE FORMWORK

Large hollow interlocking polystyrene system blocks fit together to create permanent insulating formwork, which is then filled with *in-situ* concrete to produce a monolithic concrete structure. A range of units is available giving a central core of 140 to 300 mm concrete and total insulation thicknesses between 100 and 300 mm according to the structural and thermal requirements. The two faces of the insulation are connected by a matrix of polystyrene links, which become embedded into the concrete. The units, typically 250 mm high, are tongued and grooved to ensure correct location, and horizontal steel reinforcement may be incorporated if required for additional structural strength. Special blocks are available for lintels, wall ends, curved walls and fire walls. A pumpable grade of concrete (high slump) will fill the void space by gravity flow without the need for mechanical vibration. Some temporary support for the formwork is required during construction to ensure accurate alignment. Internal and external finishes may be applied directly to the polystyrene which is keyed for plaster or lightweight render. Alternatively, masonry, timber or other claddings may be used externally and dry linings (e.g. plasterboard) may be attached to the inner leaf with appropriate adhesives.

POLYMER CONCRETE

The incorporation of pre-polymers into concrete mixes, the pre-polymers then polymerising as the concrete sets and hardens, can reduce the penetration of water and carbon dioxide into cured concrete. Typical polymers include styrene-butadiene rubber. Epoxy resin and acrylic-latex modified mortars are used for repairing damaged and spalled concrete because of their enhanced adhesive properties. Similarly, polymer-modified mortars are used for the cosmetic filling of blowholes and blemishes in visual concrete.

WATER FOR CONCRETE

The general rule is that if water is of a quality suitable for drinking, then it is satisfactory for making concrete, (BS EN 1008: 2002).

CONCRETE MIXES

Concrete mixes are designed to produce concrete with the specified properties at the most economical price. The most important properties are usually strength and durability, although thermal and acoustic insulation, the effect of fire, and appearance in visual concrete may also be critical.

In determining the composition of a concrete mix, consideration is given to the *workability* or ease of placement and compaction of the fluid mix, and to the properties required in the hardened concrete. The key factor which affects both these properties is the free-water content of the mix after any water is absorbed into the aggregates. This quantity is defined by the water/cement ratio.

Water/cement ratio

$$\text{water/cement ratio} = \frac{\text{weight of free water}}{\text{weight of cement}}$$

The free water in a mix is the quantity remaining after the aggregates have absorbed water to the *saturated surface-dry* condition. The free water is used to hydrate the cement and to make the mix workable. With low water/cement ratios below 0.4, some of the cement is not fully hydrated. At a water/cement ratio of 0.4, the hydrated cement just fills the space previously occupied by the water, giving a dense concrete. As the water/cement ratio is increased above 0.4, the mix becomes increasingly workable but the resulting cured concrete is more porous owing to the evaporation of the excess water leaving void spaces. Figure 3.8 shows the typical relationship between water/cement ratio and concrete crushing strength.

Workability

Workability describes the ability of the concrete mix to be placed within the formwork, around any reinforcement, and to be successfully compacted by hand or mechanical means to remove trapped air pockets. Mixes should be cohesive, so that they do not segregate during transportation or placing. Workability is not only affected by the water/cement ratio but also the aggregate content, size, grading and shape, and the addition of admixtures. It is measured on site with the slump test (Fig. 3.9). Table 3.11 shows the relationship between water/cement ratio and workability for crushed and uncrushed aggregates at different cement contents.

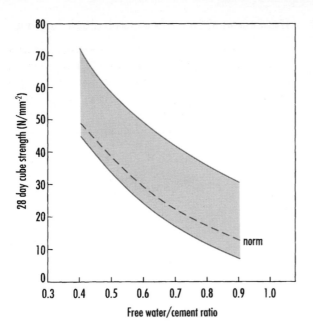

Fig. 3.8 Typical variation of crushing strengths about the published norm for the range of water/cement ratios

Free water
The workability of concrete is highly dependent upon the free water within the mix. An increase in free-water content causes a significant increase in workability, which would result in a greater slump measured in a slump test.

Aggregate shape
Rounded aggregates make a mix more workable than if crushed angular aggregates are used with the same water/cement ratio. However, because the bonding between cured cement and crushed aggregate is stronger than that to rounded aggregates, when other parameters are comparable, crushed aggregates produce a stronger concrete.

Aggregate size
The size of aggregate also affects the workability of the mix. The maximum practical size of coarse aggregate, compatible with placement around reinforcement and within the concrete section size should be used to minimise the water content necessary for adequate workability. With fine aggregates, excessive quantities of the

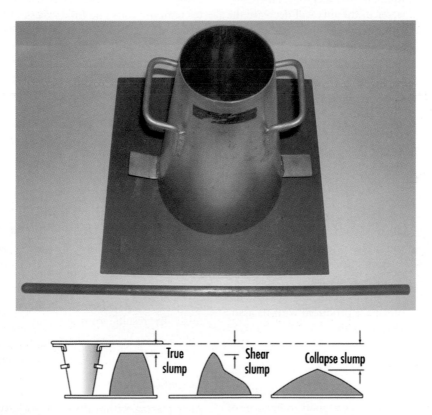

Fig. 3.9 Slump test (after Everett, A. 1994: *Mitchell's Materials*. 5th edition. Longman Scientific and Technical)

Table 3.11 Typical relationship between water/cement ratio, workability and Portland cement 42.5 content for uncrushed and crushed aggregates

water/cement ratio	Workability			
	Type of aggregate (20 mm maximum)	Low slump 10–30 mm cement content (kg/m^3)	Medium slump 25–75 mm cement content (kg/m^3)	High slump 65–135 mm cement content (kg/m^3)
0.7	uncrushed	230	260	285
	crushed	270	300	330
0.6	uncrushed	265	300	330
	crushed	315	350	380
0.5	uncrushed	320	360	400
	crushed	380	420	460
0.4	uncrushed	400	450	500
	crushed	475	525	575

fine material (passing through a 0.063 mm test sieve) would increase considerably the water requirement of a particular mix to maintain workability. This is because the smaller particles have a larger surface area/volume ratio and therefore require more water to wet their surfaces. As additional water in the mix will decrease the cured concrete strength, for good-quality dense concrete well-graded coarser sands are preferable.

Aggregate/cement ratio

For a particular water/cement ratio, decreasing the aggregate/cement ratio, which therefore increases proportionally both the cement and water content, increases workability. However, as cement is the most expensive component in concrete, cement-rich mixes are more costly than the lean mixes.

Air entraining

Workability may be increased by air entraining, although 1% voids in the cured concrete produces a decrease in compressive strength of approximately 6%. Thus in air entraining there is a balance between the increased workability and resultant improved compaction versus the void space produced with its associated reduced crushing strength.

Slump test

The slump test is used for determining the workability of a mix on site. It is gives a good indication of consistency from one batch to the next, but it is not

effective for very dry or very wet mixes. The slump test is carried out as shown in Figure 3.9. The base plate is placed on level ground and the cone filled with the concrete mix in three equal layers, each layer being tamped down 25 times with the 16 mm diameter tamping rod. The final excess of the third layer is struck off and the cone lifted off from the plate to allow the concrete to slump. The drop in level (mm) is the recorded slump, which may be a *true slump*, a *shear slump* or a *collapse slump*. In the case of a shear slump the material is retested. In the case of a collapse slump the mixture is too wet for most purposes.

Typical slump values would be zero to 25 mm for very dry mixes, frequently used in road making; 10–40 mm (low workability) for use in foundations with light reinforcement; 50–90 mm (medium workability) for normal reinforced concrete placed with vibration and over 100 mm for high-workability concrete. Typically slump values between 10 mm and 175 mm may be measured, although accuracy and repeatability are reduced at both extremes of the workability range. The slump test is not appropriate for aerated, no-fines or gap-graded concretes. The European Standard EN 206–1: 2000 classifies consistency classes of concrete mixes by results from the standard tests of slump (Table 3.12), Vebe consistency (a form of mechanised slump test), compaction and flow.

Compaction

After placing within the formwork, concrete requires compaction to remove air voids trapped in the mix

Table 3.12 Slump test classes to European Standard EN 206–1: 2000

Slump class	Slump in mm
S1	10–40
S2	50–90
S3	100–150
S4	160–210
S5	≥220

before it begins to stiffen. Air voids weaken the concrete, increase its permeability, and therefore reduce durability. In reinforced concrete, lack of compaction reduces the bond to the steel, and on exposed visual concrete, blemishes such as blowholes and honeycombing on the surface are aesthetically unacceptable and difficult to make look good successfully. Vibration, to assist compaction, may be manual by rodding or tamping for small works, but normally poker vibrators and beam vibrators are used for mass and slab concrete respectively. Vibrators which clamp on to the formwork are sometimes used when the reinforcement is too congested to allow access for poker vibrators.

The degree of compaction achieved by a standard quantity of work may be measured by the compacting factor test. In this test a fresh concrete sample is allowed to fall from one hopper into another. The weight of concrete contained in the lower hopper, when struck off flush, compared with a fully compacted sample gives the compacting factor. The compacting factor for a medium-workability concrete is usually about 0.9.

Concrete cube and cylinder tests

To maintain quality control of concrete, representative test samples should be taken, cured under controlled conditions and tested for compressive strength after the appropriate 3-, 7- or 28-day period. Steel cylinder and cube moulds (Fig. 3.10) are filled in layers with either hand or mechanical vibration. For hand tamping, a 100 mm cube would be filled in two equal layers, each tamped 25 times with a 25 mm square-end standard compacting bar; mechanical vibration would normally be with a vibrating table or pneumatic vibrator. The mix is then trowelled off level with the mould. Cubes and cylinders are cured under controlled moisture and temperature conditions for 24 hours, then stripped and cured under water at 18–20°C until required for testing.

The European standard concrete tests use cylinders 150 mm in diameter and 300 mm high rather than cubes, as they tend to give more uniform results for nominally similar concrete specimens. For a particular concrete, the characteristic compressive strength as determined by the cylinder test is lower

Fig. 3.10 Cylinder and cube test

than that obtained from the equivalent cube test. The compressive strength classes (Table 3.10) therefore have a two-number notation (e.g. C 20/25). The first number, which is used in the European structural design codes, refers to the characteristic cylinder compressive strength, and the second number is the characteristic 150 mm cube compressive strength.

DURABILITY OF CONCRETE

While good-quality well-compacted concrete with an adequate cement content and a low water/cement ratio is generally durable, concrete may be subjected to external agencies which cause deterioration or, in certain circumstances, such as alkali-silica reaction, to internal degradation. The standard BS EN 206–1: 2000 specifies requirements for the specification, constituents, composition, production and properties of concrete.

Sulfate attack

Sulfates are frequently present in soils, but the rate of sulfate attack on concrete is dependent upon the soluble sulfate content of the groundwater. Thus the presence of sodium or magnesium sulfate in solution is more critical than that of calcium sulfate, which is relatively insoluble. Soluble sulfates react with the tricalcium aluminate (C_3A) component of the hardened cement paste, producing calcium sulfoaluminate (ettringite). This material occupies a greater volume than the original tricalcium aluminate, therefore expansion causes cracking, loss of strength and increased vulnerability to further sulfate attack. The continuing attack by sulfates depends upon the movement of sulfate-bearing groundwater and in some cases delayed ettringite formation may not be apparent for 20 years. Delayed ettringite formation is sometimes observed in pre-cast concrete which has been steam cured, or when the temperature within the *in-situ* mass concrete has risen excessively during the curing process. With magnesium sulfates, deterioration may be more serious as the calcium silicates within the cured concrete are also attacked. The use of sulfate-resisting Portland cement or combinations of Portland cement and fly ash (pulverised-fuel ash [PFA]) or granulated blastfurnace slag (GGBS) reduces the risk of sulfate attack in well-compacted concrete. In the presence of high soluble sulfate concentrations, concrete requires surface protection.

The BRE Special Digest 1: 2005 describes provision for combating sulfate deterioration, including the more rapid form of attack in which the mineral thaumasite is formed. Thaumasite sulfate attack has seriously affected concrete foundations and substructures including some bridges on the UK M5 motorway. This type of sulfate attack is most active at temperatures below 15°C.

Frost resistance

Weak permeable concrete is particularly vulnerable to the absorption of water into capillary pores and cracks. On freezing, the ice formed will expand causing frost damage. The use of air-entraining agents, which produce discontinuous pores within concrete, reduces the risk of surface frost damage. Concrete is particularly vulnerable to frost damage during the first two days of early hardening. Where new concrete is at risk, frost precautions are necessary to ensure that the mix temperature does not fall below 5°C until a strength of 2 MPa is achieved. Eurocode 2 (BS EN 1992–1–1: 2004) refers to four levels of exposure class (XF1 to XF4) with respect to freeze/thaw deterioration (Table 3.15).

Fire resistance

Up to 250°C, concrete shows no significant loss of strength, but by 450°C, depending upon the duration of heating, the strength may be reduced to half and by 600°C little strength remains. However, as concrete is a good insulator, it may take four hours within a building fire for the temperature 50 mm below the surface of the concrete to rise to 650°C (Fig. 3.11).

The effect of heat on the concrete causes colour changes to pink at 300°C, grey at 600°C and to buff by 1000°C. The aggregates used within concrete have a significant effect on fire resistance. For fire protection, limestone aggregates may perform slightly better than granites and other crushed rocks, which spall owing to differential expansion. Where the concrete cover over reinforced steel is greater than 40 mm, secondary reinforcement with expanded metal gives added protection to the structural reinforcement. Lightweight-aggregates concretes, owing to their enhanced thermal properties, perform significantly better in fires with respect to both insulation and spalling.

Concrete manufactured without organic materials is Class A1 with respect to reaction to fire. If more than 1% of organic materials are incorporated into

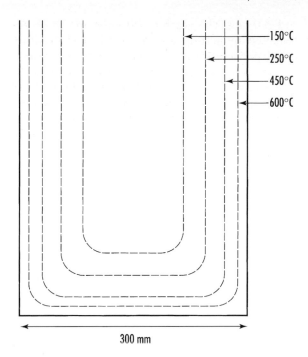

150°C
250°C
450°C
600°C

300 mm

Fig. 3.11 Temperature profile within dense concrete exposed to fire for 60 minutes

the mix, then the material will require testing to the standard (BS EN 13501–1: 2002).

Chemical attack and aggressive ground

The resistance of cured concrete to acid attack is largely dependent upon the quality of the concrete, although the addition of granulated blastfurnace slag (GGBS) or fly ash (pulverised-fuel ash [PFA]) increases the resistance to acids. Limestone-aggregate concrete is more vulnerable to acid attack than concretes with other aggregates. The resistance of cured concrete to chemical attack is defined by the design chemical class number, ranging from DC1 (low resistance) to DC 4 (high resistance). The required design chemical class (DC Class) of the concrete is calculated by combining the effects of the sulfate content of the ground, the nature of the groundwater and the anticipated working life of the construction (BRE Special Digest 1: 2005).

Determining the design chemical class required for concrete in a particular ground environment is a three-stage process. The first stage is to determine the design sulfate class (DS) of the site. This is a five-level classification based primarily on the sulfate content of the soil and/or groundwater. It takes into account the concentrations of calcium sulfate, also the more soluble

magnesium and sodium sulfates, and the presence of chlorides and nitrates if the pH is less than 5.5 (acid).

Design Sulfate Class	Limits of sulfate (mg/l)
DS1	< 500
DS2	500 – 1500
DS3	1600 – 3000
DS4	3100 – 6000
DS5	> 6000

The next stage is to determine the aggressive chemical environment for concrete (ACEC) classification. Adverse ground conditions such as acidity (low pH), often found in *brownfield* sites, and/or mobile groundwater lead to a more severe ACEC classification. Static water is more benign and leads to a less severe ACEC classification. The aggressive chemical environments for concrete classes range from AC1 (the least aggressive) to AC5 (the most aggressive), and are based on a combination of the design sulfate class, groundwater mobility and pH.

The design chemical class (DC1 to DC4) defines the qualities of the concrete required to resist chemical attack. It is determined from the ACEC class of the ground together with factors relating to the concrete, such as section size and intended working life (e.g. 100 years). As there are only four design chemical classes against five ACEC classes, for the severest grade of ACEC (i.e. AC5) there are additional protective measures (APMs) which can be specified to combat the more adverse conditions. Usually APM3 (surface protection to the concrete) is appropriate for AC5 environments, but for increasing the intended working life from 50 to 100 years under the less aggressive AC3 or AC4 conditions, any one APM may be applied.

Additional protective measures (APMs) for buried concrete:

APM1 enhance the concrete quality;
APM2 use controlled permeability formwork;
APM3 provide surface protection to the concrete;
APM4 increase the thickness of the concrete as a sacrificial layer;
APM5 reduce groundwater by drainage of the site.

Careful consideration of all these additional factors is required to ensure that a suitably durable concrete, appropriate to the job, is delivered on site for use in aggressive ground and chemical environments (BRE Special Digest 1: 2005).

Crystallisation of salts

The crystallization of salts, particularly from sea water, within the pores of porous concrete can cause sufficient internal pressure to disrupt the concrete.

Alkali-silica reaction

Alkali-silica reaction (ASR) may occur between cements containing sodium or potassium alkalis and any active silica within the aggregate. In severe cases, expansion of the gel produced by the chemical reaction causes *map* cracking of the concrete, which is characterised by a random network of very fine cracks bounded by a few larger ones. Aggregates are defined as having low, normal or high reactivity. The risk of alkali-silica reaction when using normal reactivity aggregates can be controlled by restricting the alkali content of the Portland cement to a maximum of 0.6% (low alkali cement) or the soluble alkali content of the concrete to 3 kg/m^3. Additions of limited quantities of silica fume, ground granulated blastfurnace slag (GGBS) or pulverised-fuel ash (PFA) may be used with low or normal reactivity aggregates to reduce the risk of alkali-silica reaction. Alternative methods of minimising the risk of alkali-silica reaction include the addition of lithium salts or metakaolin to the concrete mix.

Carbonation

Carbon dioxide from the atmosphere is slowly absorbed into moist concrete and reacts with the calcium hydroxide content to form calcium carbonate. The process occurs mainly at the surface and only penetrates very slowly into the bulk material. The rate of penetration is dependent on the porosity of the concrete, the temperature and humidity; it generally only becomes problematic when the concrete surrounding steel reinforcement is affected. Carbonation turns strongly alkaline hydrated cement (pH 12.5) into an almost neutral medium (pH 8.3) in which steel reinforcement will corrode rapidly if subjected to moisture.

$$Ca(OH)_2 \quad + CO_2 \quad \longrightarrow \quad CaCO_3$$
hydrated cement   carbon dioxide   calcium carbonate

Good-quality dense concrete may only show carbonation to a depth of 5–10 mm after 50 years, whereas a low-strength permeable concrete may carbonate to a depth of 25 mm within 10 years. If reinforcement is not correctly located with sufficient cover it corrodes causing expansion, spalling and rust staining. The depth of carbonation can be determined by testing a core sample for alkalinity using phenolphthalein chemical indicator, which turns pink in contact with the uncarbonated alkaline concrete. Where steel reinforcement has become exposed due to carbonation and rusting, it may be coated with a rust-inhibiting cement and the cover restored with polymer-modified mortar, which may contain fibre reinforcement. Additional protection against further attack can be achieved by the final application of an anti-carbonation coating, which acts as a barrier to carbon dioxide. Thermosetting polymers such as polyurethane and chlorinated rubber, also certain acrylic-based polymers, give some protection against carbonation.

PHYSICAL PROPERTIES OF CONCRETE

Thermal movement

The coefficient of thermal expansion of concrete varies between 7 and 14 × 10^{-6} deg C, according to the type of aggregate used, the mix proportions and curing conditions.

Moisture movement

During the curing process, concrete exhibits some irreversible shrinkage which must be accommodated within the construction joints. The extent of the shrinkage is dependent upon the restraining effect of the aggregate and is generally larger when smaller or lightweight aggregates are used. High-aggregate content mixes with low workability tend to have small drying shrinkages.

The reversible moisture movement for cured concrete is typically 2–6 × 10^{-4} deg C, depending upon the aggregate.

Creep

Creep is the long-term deformation of concrete under sustained loads (Fig. 3.12). The extent of creep is largely dependent upon the modulus of elasticity of the aggregate. Thus an aggregate with a high modulus of elasticity offers a high restraint to creep. The extent of creep may be several times that of the initial elastic deformation of the concrete under the same applied load. Where rigid cladding is applied to a concrete-frame building, compression joints at each storey must be sufficiently wide to take up any deformation due to creep, in addition to normal cyclical movements.

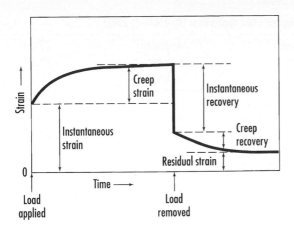

Fig. 3.12 Creep and creep recovery in concrete

CONCRETE STRENGTH CLASSES

Concrete should be specified, placed and cured according to BS EN 206–1: 2000. The preferred strength classes of concrete are shown in Table 3.13, in which the numbers refer to the test sample crushing strengths of a 150 × 300 mm cylinder and a 150 mm cube, respectively.

SPECIFICATION OF CONCRETE MIXES

There are five methods for specifying concrete described in BS 8500–1:2002. All should conform to the standards BS 8500–1: 2002 and BS EN 206–1: 2000.

The five methods are:

- designated concrete
- designed concrete
- prescribed concrete
- standardised prescribed concrete
- proprietary concrete.

If the application can be considered to be routine, then designated concrete is usually appropriate. If, however, the purchaser requires specific performance criteria and accepts the higher level of responsibility in the specification, then designed or prescribed concretes may be used. For housing and similar applications, standardised prescribed mixes should give the required performance, providing that there is sufficient control over the production and quality of materials used.

Designated concrete

Designated concretes are appropriate for most concrete construction including general-purpose work, foundations, reinforced concrete and air-entrained pavement concrete. The purchaser is responsible for correctly specifying the proposed use and the concrete mix designation. In addition the purchaser must specify whether the concrete is to be reinforced, the exposure (or soil) conditions, the nominal aggregate size if other than 20 mm, and the consistence class (slump). The producer must ensure that the mix fulfils all the performance criteria. Thus normally for foundations in design chemical class soil conditions DC3, the designated mix FND 3 would be required. This mix may be supplied with sulfate-resisting Portland cement at 340 kg/m³ and a maximum water/cement ratio of 0.5, or as Portland cement with 25% fly ash or 75% granulated blastfurnace slag. Any of these mixes will perform to the required criteria for the specified purpose. For routine work, designated mixes produced by quality assured plants offer the specifier the least risk of wrong specification. Table 3.14 illustrates typical housing applications for designated mixes.

Designed concrete

The producer is responsible for selecting a designed concrete which will meet the performance criteria

Table 3.13 Compressive strength classes for dense and lightweight concrete

Compressive strength classes for dense concrete							
C8/10	C12/15	C16/20	C20/25	C25/30	C30/37	C35/45	C40/50
C45/55	C50/60	C55/67	C60/75	C70/85	C80/95	C90/105	C100/115

Compressive strength classes for lightweight concrete							
LC8/9	LC12/13	LC16/18	LC20/22	LC25/28	LC30/33	LC35/38	LC40/44	LC45/50
LC50/55	LC55/60	LC60/66	LC70/77	LC80/88				

Within each compressive strength class the numbers indicate the 28 day crushing strength in MPa as determined by the 150 mm diameter by 300 mm cylinder and 150 mm cube test, respectively.

Table 3.14   Designated and standardized prescribed concrete for housing and other applications (BS 8500–1: 2002)

Typical application	Designated concrete	Standardized prescribed concrete	Consistence class
Foundations (Design Chemical Class1):			
Blinding and mass concrete fill	GEN 1	ST2	S3
Strip footings	GEN 1	ST2	S3
Mass concrete foundations	GEN 1	ST2	S3
Trench fill foundations	GEN 1	ST2	S4
Fully buried reinforced foundations	RC 35		S3
Foundations DC2-DC4:			
Foundations (Design Chemical Class 2)	FND 2		S3
Foundations (Design Chemical Class 3)	FND 3		S3
Foundations (Design Chemical Class 4)	FND 4		S3
Foundations (Design Chemical Class 4m)	FND 4M		S3
General applications:			
Kerb bedding and backing	GEN 0	ST1	S1
Drainage works (immediate support)	GEN 1	ST2	S1
Drainage works (other)	GEN 1	ST2	S3
Oversite below suspended slab	GEN 1	ST2	S3
Floors:			
House floors with no embedded metal for screeding	GEN 1	ST2	S2
House floors with no embedded metal - no finish	GEN 2	ST3	S2
Garage floors with no embedded metal	GEN 3	ST4	S2
Wearing surface - light foot and trolley traffic	RC 30	ST4	S2
Wearing surface - general industrial	RC 40		S2
Wearing surface - heavy industrial	RC 50		S2
Paving:			
House drives, domestic parking and external paving	PAV 1		S2
Heavy duty external paving for rubber tyre vehicles	PAV 2		S2

Note: m refers to resistance to the higher magnesium levels in the various sulfate classes.

listed by the specifier. The specifier must clearly indicate the required use, curing conditions, exposure conditions, surface finish, maximum aggregate size and any excluded materials. In addition the compressive strength class, the maximum water/cement ratio, the minimum cement content, the consistence (slump) and permitted cement types should be quoted. Within these constraints, the producer is responsible for producing a concrete which conforms to the required properties and any additional stated characteristics.

Prescribed concrete

The purchaser fully specifies all the materials by weight (kg/m^3), including admixtures and the standard strength class. The purchaser is therefore responsible for the performance characteristics of the concrete. Prescribed concretes are used particularly for specialist finishes such as exposed aggregate visual concrete.

Standardised prescribed concrete

Standardised prescribed concretes are a set of five standard mixes, which may be mixed on site, with a restricted range of materials. Standard mixes ST1 to ST5 may be made to S1, S2, S3 or S4 slump classes, giving low, medium, high or very high workability. The specification must record a maximum aggregate size and whether the concrete is to be reinforced or not.

Proprietary concrete

Proprietary concrete must conform to the standards BS 8500–2: 2002 and BS EN 206–1: 2000 and be properly identified. This category allows for a concrete supplier to produce a concrete mix with an appropriate performance but without indicating its composition.

IN-SITU CONCRETE TESTING

The compressive strength of hardened concrete may be estimated *in-situ* by mechanical or ultrasonic measurements. The Schmidt hammer or sclerometer measures the surface hardness of concrete by determining the rebound of a steel plunger fired at the surface. In the *pull-out* test, the force required to extract a previously cast-in standard steel cone gives a measure of concrete strength. Ultrasonic devices determine the velocity of ultrasound pulses through concrete. Since pulse velocity increases with concrete density, the technique can be used to determine variations within similar concretes. The test gives a broad classification of the quality of concrete, but not absolute data for concretes of different materials in unknown proportions.

Reinforced concrete

Concrete is strong in compression, with crushing strengths typically in the range 20–40 MPa, and up to 100 MPa for high-strength concretes. However, the tensile strength of concrete is usually only 10% of the compressive strength. Steel is the universally accepted reinforcing material as it is strong in tension, forms a good bond and has a similar coefficient of thermal expansion to concrete. The location of the steel within reinforced concrete is critical, as shown in Figure 3.13, to ensure that the tensile and shear forces are transferred to the steel. The longitudinal bars carry the tensile forces while the links or stirrups combat the shear forces and also locate the steel during the casting of the concrete. Links are therefore more concentrated around locations of high shear although inclined bars may also be used to resist the shear forces. Fewer or thinner steel bars may be incorporated into reinforced concrete to take a proportion of the compressive loads in order to minimise the beam dimensions.

Steel reinforcement for concrete is manufactured within the UK from recycled scrap into round, ribbed

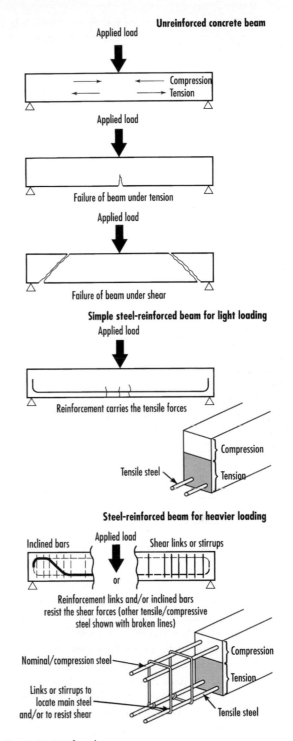

Fig. 3.13 Reinforced concrete

or ribbed and twisted bars (Fig. 3.14). Mild steel is used for the round bars, which are used mainly for the bent links. Hot rolled high-yield steel is used for the ribbed bars, and high-yield steel is cold worked to

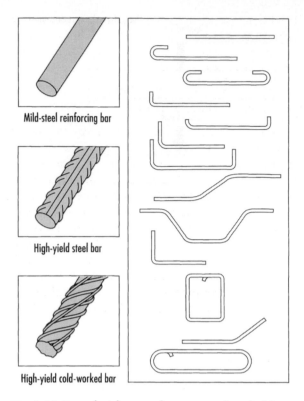

Mild-steel reinforcing bar

High-yield steel bar

High-yield cold-worked bar

Fig. 3.14 Types of reinforcement for concrete and standard forms

produce the ribbed and twisted bars. High-yield steel has a minimum yield stress of 460 Mpa, roughly double that of mild steel at 250 MPa. Welded steel mesh reinforcement is used for slabs, roads and within sprayed concrete. The British Standard BS 4449: 1997 refers plain round steel bars in grade 250 and the normal (N) and high (H) ductility classes 460A and 460B for grade 460 ribbed steel bars.

Austenitic stainless steels may be used for concrete reinforcement where failure due to corrosion is a potential risk. Grade 1.4301 (18% chromium, 10% nickel) stainless steel is used for most applications, but the higher grade 1.4436 (17% chromium, 12% nickel, 2.5% molybdenum) is used in more corrosive environments. Where long-term performance is required in highly corrosive environments, the duplex grades of stainless steel may be used. The initial cost of stainless steel reinforcement is approximately eight times that of standard steel reinforcement, but in situations where maintenance costs could be high, for example due to chloride attack from sea water or road salts, the overall life-cycle costs may be reduced by its use. Additionally, stainless steels have higher strengths than the standard carbon steels. Suitable stainless

steels for the reinforcement of concrete are specified in BS 6744: 2001.

Bond between steel and concrete

For reinforced concrete to act efficiently as a composite material the bond between the concrete and steel must be secure. This ensures that any tensile forces within the concrete are transferred to the steel reinforcement. The shape and surface condition of the steel and the quality of the concrete all affect the bond strength.

To obtain the most efficient mechanical bond with concrete, the surface of the steel should be free of flaky rust, loose scale and grease, but the thin layer of rust, typically produced by short-term storage on site, should not be removed before use. The use of hooked ends in round bars reduces the risk of the steel being pulled out under load, but high bond strength is achieved with ribbed bars, which ensure a good bond along the full length of the steel. Steel *rebars* are usually supplied either in stock lengths, or cut and bent ready for making up into cages. Sometimes the reinforcement may be supplied as prefabricated cages, which may be welded rather than fixed with iron wire as on site. Steel reinforcement although weldable, is rarely welded on site. Rebar joints can easily be made with proprietary fixings, such as steel sleeves fastened by shear bolts. Spacers are used to ensure the correct separation between reinforcement and formwork.

Good-quality dense concrete gives the strongest bond to the steel. Concrete should be well compacted around the reinforcement; thus the maximum aggregate size must not bridge the minimum reinforcement spacing.

Corrosion of steel within reinforced concrete

Steel is protected from corrosion providing that it has adequate cover of a good-quality, well-compacted and cured concrete. The strongly alkaline environment of the hydrated cement renders the steel passive. However, insufficient cover caused by the incorrect fixing of the steel reinforcement or the formwork can allow the steel to corrode. Rust expansion causes surface spalling, then exposure of the steel allows corrosion, followed by rust staining of the concrete surface. Calcium chloride accelerators should not normally be used in reinforced concrete as the residual chlorides cause accelerated corrosion of the steel reinforcement.

Additional protection from corrosion can be achieved by the use of galvanised epoxy-coated or stainless steel reinforcement. The protective alkalinity of the concrete is reduced at the surface by *carbonation*. The depth of carbonation depends upon the permeability of the concrete, the moisture content and any surface cracking. The nominal cover for concrete reinforcement therefore is calculated from the anticipated degree of exposure (Table 3.15) and the concrete strength class as in Table 3.16. The recommended cover specified relates to all reinforcement, including any wire ties and secondary reinforcement. Some reduction in carbonation rate can be achieved by protective coatings to the concrete surface. It should be noted that the choice of an adequately durable concrete for the protection of the concrete itself against attack, and for the prevention of reinforcement corrosion, may result in a higher compressive strength concrete being required than is necessary for the structural design (Table 3.17).

Where the depth of concrete cover over reinforcement is in doubt it can be measured with a *covermeter*.

If reinforcement is corroding, cathodic protection by application of a continuous direct current to the steel reinforcement may prevent further deterioration and lead to realkalisation of the carbonated concrete.

Fibre-composite reinforced concrete

In most situations steel is used for reinforcing or prestressing concrete. However, for structures in highly aggressive environments high modulus continuous fibres embedded in resin offer an alternative. The fibres, either glass, carbon or aramid are encased in a thermosetting resin and drawn through a die by pultrusion to produce the required cross-section. The extruded material is then overwound with further fibres to improve its bond with concrete. The fibre-composite rods are used as reinforcement or as prestressing tendons within standard concrete construction.

Table 3.15 Concrete exposure classes to Eurocode 2 (BS EN 1992–1–1: 2004)

Exposure classes		Typical environmental conditions
No risk of reinforcement corrosion or attack on concrete		
X0	Concrete with no reinforcement	
	Dry concrete	Dry building interiors
Corrosion induced by carbonation		
XC1	Dry or permanently wet	Interior of buildings and concrete under water
XC2	Wet and rarely dry	Foundations
XC3	Moderate humidity	Sheltered external concrete and high humidity interiors
XC4	Cyclic wet and dry	Concrete in occasional contact with water
Corrosion induced by chlorides		
XD1	Humid environment	Components exposed to airborne spray
XD2	Wet and rarely dry	Swimming pools and contact to industrial waters
XD3	Cyclic wet and dry	Exposed external concrete surfaces
Corrosion induced by sea water		
XS1	Exposure to sea air	Coastal structures
XS2	Submerged under sea water	Submerged marine structures
XS3	Tidal and sea spray zone	Parts of marine structures
Freeze/thaw deterioration		
XF1	Moderate saturation	Vertical surfaces exposed to rain and freezing
XF2	Moderate saturation with de-icing agent	Vertical surfaces exposed to rain, freezing and de-icing
XF3	High saturation	Horizontal surfaces exposed to rain and freezing
XF4	High saturation with de-icing agent	Surfaces exposed to rain, freezing and de-icing or marine spray
Chemical attack		
XA1	Slightly aggressive agencies	Soil and ground water
XA2	Moderately aggressive agencies	Soil and ground water
XA3	Highly aggressive agencies	Soil and ground water

Table 3.16   Minimum cover required to ensure durability of steel reinforcement in structural concrete for exposure classes to Eurocode 2 (BS EN 1992–1–1: 2004)

Exposure class	X0	XC1	XC2/XC3	XC4	XD1/XS1	XD2/XS2	XD3/XS3
Recommended cover (mm)	10	15	25	30	35	40	45
Minimum cover (mm)	10	10	10	15	20	25	30
Strength Class	$\geq$C30/37	$\geq$C30/37	$\geq$C35/45	$\geq$C40/50	$\geq$C40/50	$\geq$C40/50	$\geq$C45/55

Notes:

The recommended cover relates to standard production with a design working life of 50 years.

Increased cover is required for a design working life of 100 years.

The minimum cover relates to very specific conditions combining high quality control for positioning of the reinforcement and the concrete production additionally the use of 4% (minimum) air entrainment.

Table 3.17   Indicative strength classes for durability of concrete to Eurocode 2 (BS EN 1992–1–1: 2004)

Corrosion risk	XC1	XC2	XC3 and XC4	XD1 and XD2	XD3	XS1	XS2 and XS3
Indicative Strength Class	C20/25	C25/30	C30/37	C30/37	C35/45	C30/37	C35/45
Damage to concrete	X0	XF1	XF2	XF3	XA1	XA2	XA3
Indicative Strength Class	C12/15	C30/37	C25/30	C30/37	C30/37	C30/37	C35/45

Bendy concrete

Fibre-reinforced concrete of an appropriate mix may be continuously extruded into various sections to produce sheets, cylinders or tubes. The product is more flexibile and has a higher impact strength than ordinary concrete. *Bendy* concrete may be drilled, cut and nailed without damage. It is lighter than ordinary concrete and with its good fire resistance may be used as an alternative to other wall boards.

Fibre-reinforced aerated concrete

Polypropylene fibre-reinforced aerated concrete is used for making lightweight blocks, floor, wall and roofing panels, offering a combination of strength and insulation properties. The material, like standard aerated concrete, can be cut and worked with standard hand tools. Where additional strength is required, steel fibre-reinforced aerated concrete may be used for cast *in-situ* or factory-produced units. The fibre-reinforced material has a greater resilience than standard aerated concrete. Roofing membranes and battens for tiling can be directly nailed to roofing panels; whilst floor panels accept all the standard floor finishes.

Fire resistance of reinforced concrete

Concrete manufactured without organic materials is Class A1 with respect to reaction to fire. If more than

1% of organic materials are incorporated into the mix, then the material will require testing to the standard (BS EN 13501–1: 2002).

The depth of concrete cover over the steel reinforcement to ensure various periods of fire resistance is listed in Table 3.18. Where cover exceeds 40 mm, additional reinforcement will be required to prevent surface spalling of the concrete. The cover should prevent the temperature of the steel reinforcement exceeding 550°C (or 450°C for prestressing steel).

PRESTRESSED CONCRETE

Concrete has a high compressive strength but is weak in tension. Prestressing with steel wires or tendons ensures that the concrete component of the composite material always remains in compression when subjected to flexing up to the maximum working load. The tensile forces within the steel tendons act upon the concrete putting it into compression, such that only under excessive loads would the concrete go into tension and crack. Two distinct systems are employed: in pre-tensioning, the tendons are tensioned before the concrete is cured; and in post-tensioning the tendons are tensioned after the concrete is hardened (Fig. 3.15).

Table 3.18 Typical cover to concrete reinforcement for fire resistance to Eurocode 2 (BS EN 1992–1–2: 2004)

Fire resistance (minutes)		Typical cover to reinforcement (mm)	
Beams	Width (mm)	Simply supported	Continuous beams
R 30	80	25	15
R 60	120	40	25
R 90	150	55	35
R 120	200	65	45
R 180	240	80	60
R 240	280	90	75
Columns	Minimum dimensions (mm)	One face exposed	
R 30	155	25	
R 60	155	25	
R 90	155	25	
R 120	175	35	
R 180	230	55	
R 240	295	70	
Walls	Minimum dimensions (mm)	One face exposed	
REI 30	100	10	
REI 60	110	10	
REI 90	120	20	
REI 120	150	25	
REI 180	180	40	
REI 240	230	55	
Slabs	Slab thickness (mm)	One-way slabs	Two-way slabs
REI 30	60	10	10
REI 60	80	20	10–15
REI 90	100	30	15–20
REI 120	120	40	20–25
REI 180	150	55	30–40
REI 240	175	65	40–50

Fire resistance class:

R = load-bearing criterion, E = integrity criterion and I = insulation criterion in standard fire exposure.

All reinforcement cover requirements are also dependent on the dimensions and geometry of the concrete components and the degree of fire exposure (BS EN 1992–1–2: 2004).

Where low cover thicknesses are required for fire protection, a higher depth of cover may be required for corrosion protection (BS EN 1992–1–1: 2004).

Pre-tensioning

Large numbers of precast concrete units, including flooring systems, are manufactured by the pre-tensioning process. Tendons are fed through a series of beam moulds and the appropriate tension applied. The concrete is placed, vibrated and cured. The tendons are cut at the ends of the beams, putting the concrete into compression. As with precast reinforced concrete it is vital that prestressed beams are

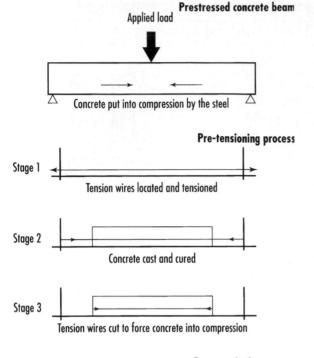

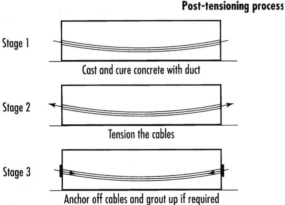

Fig. 3.15 Prestressed concrete

the anchorages, which are subject to very high localised forces. In the bonded system, after tensioning the free space within the ducts is grouted up, which then limits the reliance on the anchorage fixing; however, in the unbonded system the tendons remain free to move independently of the concrete. Tendon ducts are typically manufactured from galvanised steel strip or high-density polythene.

Post-tensioning has the advantage over pre-tensioning that the tendons can be curved to follow the most efficient prestress lines. In turn this enables long spans of minimum thickness to be constructed. During demolition or structural alteration work, unbonded post-tensioned structures should be de-tensioned, although experience has shown that if demolished under tension, structures do not fail explosively. In alteration work, remaining severed tendons may subsequently require re-tensioning and re-anchoring to recover the structural performance. However, the use of post-tensioning does not preclude subsequent structural modifications.

Visual concrete

The production of visual concrete, whether precast or *in-situ*, requires not only a high standard of quality control in manufacture, but also careful consideration to the correct specification and detailing of the material to ensure a quality finish which weathers appropriately. The exposed concrete at St John's College, Oxford (Fig. 3.16) illustrates the visual qualities of the material when designed, detailed and executed under optimum conditions.

The appearance of visual concrete is affected by four key factors:

- the composition of the concrete mix;
- the formwork used;
- any surface treatment after casting;
- the quality of workmanship.

DESIGN CONSIDERATIONS

The satisfactory production of large areas of smooth concrete is difficult due to variations in colour and the inevitability of some surface blemishes, which can be improved, but not eradicated by remedial work. Externally smooth concrete weathers unevenly due to the build-up of dirt deposits and the flow of rainwater. Therefore, if concrete is to be used externally as a visual

installed the correct way up according to the anticipated loads.

Post-tensioning

In the post-tensioning system the tendons are located in the formwork within sheaths or ducts. The concrete is placed, and when sufficiently strong, the tendons are stressed against the concrete and locked off with special anchor grips incorporated into the ends of the concrete. Usually reinforcement is incorporated into post-tensioned concrete, especially near

Fig. 3.16 High-quality visual concrete – St John's College, Oxford. Architects: MacCormac Jamieson Prichard. Photograph: Courtesy of Peter Cook

material, early design considerations must be given to the use of textured or profiled surfaces to control the flow of rainwater. Generally the range of finishes and quality control offered by precasting techniques is wider than that available for *in-situ* work, but frequently construction may involve both techniques. The use of external renderings offers an alternative range of finishes for concrete and other substrates. Figure 3.17 illustrates the range of processes available in the production of visual concrete.

PRECAST CONCRETE

Precast concrete units may be cast vertically or horizontally, although most factory operations use the latter, either face-up or face-down, as better quality control can be achieved by this method. Moulds are usually manufactured from plywood or steel. Whilst steel moulds are more durable for repeated use, plywood moulds are used for the more complex forms; they can also be more readily modified for non-standard units. Moulds are designed to be dismantled for the

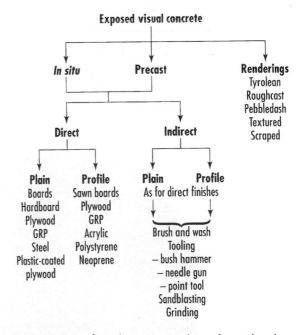

Fig. 3.17 Types of visual concrete according to formwork and surface treatment

removal of the cast unit and must be manufactured to tight tolerances to ensure quality control on the finished product. As high costs are involved in the initial production of the moulds, economies of construction can be achieved by limiting the number of variations. This can have significant effects on the overall building aesthetic. Fixing and lifting systems for transportation must be incorporated into precast units, usually in conjunction with the steel reinforcement. In addition to visual concrete panels, units faced with natural stone, brickwork or tiles extend the range of precast architectural claddings (Figs. 3.18 and 3.19).

IN-SITU CONCRETE

The quality of *in-situ* visual concrete is heavily dependent upon the formwork as any defects will be mirrored in the concrete surface. The formwork must be strong enough to withstand, without distortion, the pressure of the fresh concrete, and the joints must be tight enough to prevent leakage, which can cause honeycombing of the surface. A wide range of timber products, metals and plastics are used as formwork, depending upon the surface finish required.

The Independent Television News building, London (Fig. 3.20), built with *in-situ* concrete, is characterised by its atrium, which allows daylight to diffuse down the central core past a series of stepped terraces and cantilevered balconies in ribbed concrete. The illumination

Fig. 3.19 Reconstructed stone cladding – Experian Data Centre, Nottingham. Photograph: Courtesy of Trent Concrete Ltd

Fig. 3.18 Slate surfaced precast concrete cladding – Swansea Museum. Photograph: Courtesy of Trent Concrete Ltd

Fig. 3.20 Concrete construction – Independent Television News Headquarters, London. Architects: Foster and Partners. Photographs: Courtesy of Richard Davies

of the restrained central space is enhanced by sun and light penetration through the frosted-glass south wall. The front facade is undercut below the second floor with the exposed concrete columns creating a sense of enclosure to the main entrance.

CONCRETE FINISHES

Smooth finishes

In direct as-cast concrete, the surface texture and water absorbancy of the formwork , or any formwork lining, directly determines the final exposed fairfaced finish. A high level of quality control is therefore required to ensure a visually acceptable finish. Hard shiny non-absorbent formwork materials such as steel glass-fibre reinforced polyester (GRP) or plastic-coated plywood can give surfaces which suffer from *map crazing* due to differential shrinkage between the surface and underlying bulk material. Additionally, *blow-holes* caused by air bubbles trapped against the form face may spoil the surface if the concrete has not been sufficiently vibrated. Where the absorbency of the formwork varies, because of the mixing of new and reused formwork, or variations within the soft-wood timbers, or because of differing application of release agent to the formwork, permanent colour variations may be visible on the concrete surface. Release agents prevent bonding between the concrete and the formwork, which might cause damage to the concrete on striking the formwork. Cream emulsions and oils with surfactant are typically used as release agents for timber and steel respectively. Formwork linings with controlled porosity can improve the quality of *off-the-form* finishes, by substantially reducing the number of blow-holes. The linings allow the escape of air and excess moisture but not cement during vibration. A good-quality direct-cast concrete should exhibit only a few small blow holes and modest colour variation.

The application of paint to *off-the-form* concrete will emphasise the surface blemishes such as blow holes. These become particularly noticeable if a light colour gloss paint is used. Surface defects must therefore be made good with filler before priming and subsequent painting of the concrete.

Textured finishes

A variety of textured finishes can be achieved by the use of rough-sawn boards as formwork. The grain effect can be enhanced by abrasive blasting, and a three-dimensional effect can be achieved by using variations in board thickness. Plastic materials such as glass-fibre reinforced polyester (GRP), vacuum-formed thermoplastic sheeting, neoprene rubber and polystyrene can be used as formwork linings to give different pattern effects. Colour variations are reduced by the use of matt finishes, which retain the mould release agent during compaction of the concrete. The number of blow-holes is reduced by the use of the slightly absorbent materials such as timber and polystyrene. Concrete panels cast face-up can be textured by rolling or tamping the concrete whilst it is still plastic.

Ribbed and profiled finishes

Ribbed concrete is typically cast *in-situ* against vertical timber battens fixed to a plywood backing. In order to remove the formwork, without damage to the cured concrete, the battens must be splayed and smooth. A softer ribbed appearance is achieved by hammering off the projecting concrete to a striated riven finish. Profiled steel formwork and rope on plywood produce alternative finishes. Where deep profiles are required, expanded polystyrene and polyurethane foam can be carved out to produce highly sculptural designs.

Abraded, acid etched and polished finishes

Light abrasion with sand paper may be applied to *in-situ* or precast concrete. Acid etching is normally limited to precast concrete due to the hazards associated with using acids on site. Both techniques remove the surface laitance (cement-rich surface layer) to create a more stone-like finish with some exposure of the aggregate. Polishing with carborundum abrasives produces a hard shiny finish, imparting full colour brightness to the aggregate. It is, however, a slow and therefore expensive process.

Exposed aggregate finishes

The exposure of the coarse aggregate in concrete, by removal of the surface smooth layer formed in contact with the formwork, produces a concrete with a more durable finish and better weathering characteristics, which is frequently aesthetically more pleasing. Smooth, profiled and deeply moulded concrete can all be treated, with the visual effects being largely dependent upon the form and colour of the coarse aggregates used. While gap-graded coarse aggregates can be used both in precast and *in-situ* exposed aggregate finishes, precasting gives additional opportunities for the uniform placement of the aggregate. In face-down casting, flat stones can be laid on the lower face of the mould, which can be pretreated with retardant to slow the hardening of the surface cement. In face-up casting, individual stones can be pressed into the surface either randomly or to prescribed patterns without the use of retardants. Alternatively a special facing mix can be used on the fairfaced side of the panel, with the bulk material made up with a cheaper standard mix. The aggregate has to be exposed by washing and brushing when the concrete has cured sufficiently to be self-supporting. The use of a retarder applied to the formwork face enables the timing of this process to be less critical. The surface should be removed to a depth no more than one third of the thickness of the aggregate to eliminate the risk of it becoming detached. An alternative method of exposing the aggregate in both precast and *in-situ* concrete involves the use of abrasive blasting. Depending upon the size of grit used and the hardness of the concrete, a range of finishes including sculptural designs can be obtained.

Tooled concrete finishes

A range of textures can be obtained by tooling hardened concrete either by hand or mechanically. Generally, a high-quality surface must be tooled as blemishes can be accentuated rather than eliminated by tooling. Only deep tooling removes minor imperfections such as blow-holes and the effects of slight formwork misalignment. Hand tooling is suitable for a light finish on plain concrete and club hammering can be used on a ribbed finish. Where deep tooling is anticipated, allowance must be made for the loss of cover to the steel reinforcement. The exposed aggregate colour in tooled concrete is less intense than that produced by wash-and-brush exposure, due to the effect of the hammering on the aggregate. Standard mechanical tools are the needle-gun, the bush hammer and the point-tool (Fig. 3.21). A range of visual concrete finishes is illustrated in Figure 3.22.

Weathering of concrete finishes

The weathering of exposed visual concrete is affected by the local microclimate, the concrete finish itself and the detailing used to control the flow of rainwater over the surface. It is virtually impossible to ensure that all sides of a building are equally exposed, as inevitably there will be a prevailing wind and rain direction which determines the weathering pattern. It is therefore likely that weathering effects will differ on the various elevations of any building. Some elevations will be washed regularly, whilst others may suffer from an accumulation of dirt which is rarely washed. However, this broad effect is less likely to

Point tool

Bush hammer

Needle gun

Fig. 3.21 Tools for indirect visual concrete finishes

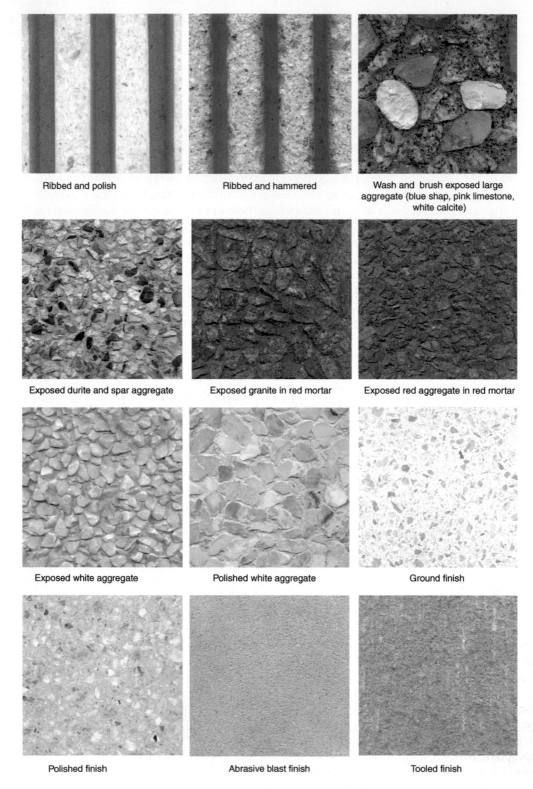

Ribbed and polish

Ribbed and hammered

Wash and brush exposed large aggregate (blue shap, pink limestone, white calcite)

Exposed durite and spar aggregate

Exposed granite in red mortar

Exposed red aggregate in red mortar

Exposed white aggregate

Polished white aggregate

Ground finish

Polished finish

Abrasive blast finish

Tooled finish

Fig. 3.22 Selection of visual concrete finishes

cause unsightly weathering than the pattern streaking on individual facades.

The choice of concrete finish can have a significant effect on the weathering characteristics. Good-quality dense uniform concrete is essential if patchy weathering is to be avoided, and generally a rougher finish is likely to perform better than a smooth as-cast finish. Profiling and the use of exposed aggregates have the advantage of dictating the flow of rainwater, rather than letting it run in a random manner, but dirt becomes embedded in the hollows. Dark aggregates and bold modelling minimise the change in appearance on weathering but, generally, exposed non-absorbent aggregates are likely to give the best weathering performance. Horizontal surfaces may be subject to organic growths and this effect is increased by greater surface permeability.

Careful detailing is necessary to ensure a dispersed and controlled flow of water over the washed areas. The water should then be collected or shed clear by bold details to prevent, pattern staining below. Water collected onto horizontal surfaces should not be allowed to run down facades below, so copings, sills and string courses all should be provided with drips to throw the water off the building face; alternatively water should be removed by gutters. Multistorey facades should be articulated with horizontal features to throw the water off, at least at each storey height. Only on seriously exposed facades where strong winds are likely to cause rain to be driven upwards, should small horizontal drip projections be avoided. Where concrete is modelled, due consideration should be given to the direction of flow and the quantity of rainwater anticipated.

EXTERNAL RENDERING

Renders are used to provide a durable and visually acceptable skin to sound but unattractive construction. Renders can reduce rain penetration and maintain the thermal insulation of walls. The finishes illustrated in Figure 3.23 are all appropriate for external use. In each case it is essential to ensure good adhesion to the background. Where a good mechanical key, such as raked-out brickwork joints is not present, an initial stipple coat of sand, cement, water and appropriate bonding agent (e.g. styrene-butadiene-rubber) is required to create a key. Bonding is also affected by the suction or absorbency of the background; where suction is very high, walls may be

lightly wetted before the rendering is applied. Metal lathing may be used over timber, steel or friable masonry to give a sound background. Two or three coats of rendering are normally applied; in either case the successive coats are weaker by a reduction in thickness or strength of the mix. Smooth renders require careful workmanship for external work, as they may craze if finished off with a steel rather than a wooden float.

Generally, permeable renders are more durable than dense impermeable renders as the latter may suffer cracking and subsequent localised water penetration. Sands for external renderings should be sharp rather than soft. The design detailing of rendering is important to ensure durability. The top edges of rendering should be protected from the ingress of water by flashings, copings or eaves details. Rendering should stop above damp-proof course level and be formed into a drip with an appropriate edging bead. Rainwater run-off from sills and opening heads should be shed away from the rendering to prevent excessive water absorption at these points, which would lead to deterioration and detachment of the rendering. The colour photograph in Figure 3.23 illustrates the striking visual effect of the rendered blockwork student halls of residence at the University of East London adjacent to the Royal Albert Dock.

Roughcast render

Roughcast, consisting of a wet mix of cement (1 part), lime (½ part), sand (3 parts) and 5–15 mm shingle or crushed stone (1½ parts) which is applied to walls by throwing from a hand scoop.

Dry-dash render

A 10 mm coat of cement (1 part), lime (1 part) and sand (5 parts) is applied to the wall and while it is still wet, calcined flint, spar or shingle is thrown onto the surface and tamped in with a wooden float.

Scraped finish

A final coat of cement (1 part), lime (2 parts) and sand (9 parts) is applied and allowed to set for a few hours, prior to scraping with a rough edge (e.g. saw blade) to remove the surface material. After it has been scraped the surface is lightly brushed over to remove loose material.

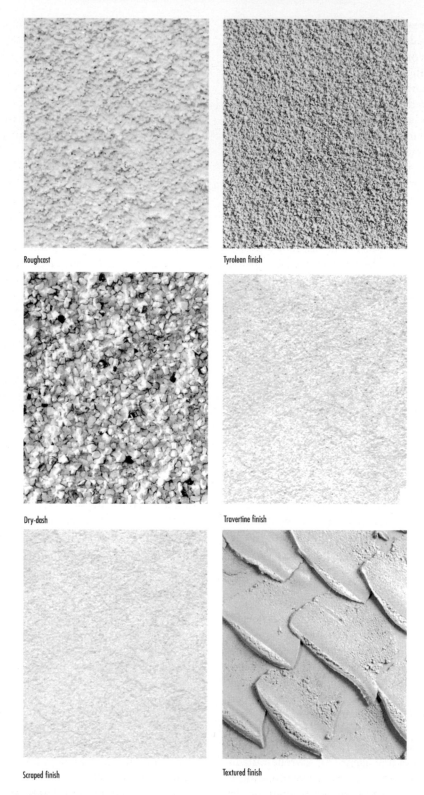

Roughcast

Tyrolean finish

Dry-dash

Travertine finish

Scraped finish

Textured finish

Fig. 3.23 Typical render finishes

Fig. 3.23 Continued. Rendered blockwork – University of East London, Docklands Campus. Architects: Edward Cullinan Architects. Photograph: Arthur Lyons

Textured finishes

A variety of finishes can be obtained by working the final rendering coat with a float, brush, comb or other tool to produce a range of standard textured patterns. Pargetting, in which more sophisticated patterns are produced, has its cultural roots in Suffolk and Essex.

Tyrolean finish

For a Tyrolean finish, cement mortar is spattered onto the wall surface from a hand-operated machine. Coloured mixes may be used.

Painted rendered finishes

Most renderings do not necessarily need painting; however, smooth renderings are frequently painted with masonry paint to reduce moisture absorption and give colour. Once painted, walls will need re-painting at regular intervals.

Concrete components

In addition to the use of concrete for the production of large *in-situ* and precast units, concrete bricks (Chapter 1) and concrete blocks (Chapter 2), the material is widely used in the manufacture of small components, particularly concrete tiles, slates and paving slabs.

CONCRETE ROOFING TILES AND SLATES

Concrete plain and interlocking slates and tiles form a group of highly competitive pitched roofing materials, with concrete interlocking tiles remaining the cheapest visually acceptable unit pitched-roof product. Plain and feature double-lap and interlocking tiles are manufactured to a range of designs, many of which emulate the traditional clay tile forms (Fig. 3.24). Concrete plain tiles may be used on pitches down to 35°, whilst the ornamental tiles are appropriate for vertical hanging and pitches down to 70°. The ranges of colours usually include both granular and through colour

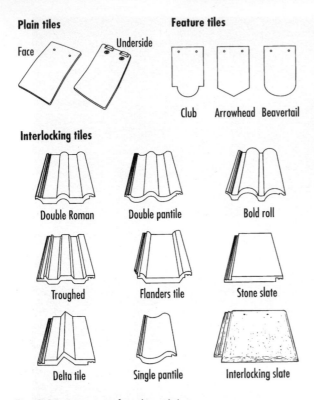

Fig. 3.24 Concrete roofing tiles and slates

finishes. Standard ranges of concrete interlocking tiles and slates can be used in certain cases down to roof pitches of 17.5°, and for some shallow-pitched roofs the concrete tiles are laid to broken bond. One interlocking product emulates the appearance of plain tiles, but can be used down to a minimum pitch of 22.5°. Colours include brown, red, rustic and grey in granular and smooth finish. A limited range of polymer-surfaced concrete interlocking tiles may be used at roof rafter pitches as low as 12.5°, providing that all tiles are clipped to prevent wind lift.

Concrete interlocking slates are manufactured with either a deep flat profile, giving a stone/slate appearance, or with a thin square or chamfered leading edge to simulate natural slate. Surfaces can be simulated riven or smooth in a range of colours including grey, blue, brown, buff and red. Matching accessories for either traditional mortar bedding or dry-fixing for ridges, hips and verges are available, together with appropriate ventilation units.

CONCRETE PAVING SLABS AND TILES

Grey concrete paving slabs are manufactured from Portland cement mixes with pigments added to produce the standard buff, pink and red colours. Standard sizes include 900 × 600 mm, 750 × 600 mm and 600 × 600 × 50 mm, but a wide range of smaller and thinner units is available for the home improvement market including 600 × 600 mm, 600 × 450 mm, 450 × 450 mm and 400 × 400 mm by 40 to 30 mm. Thicker units (65 mm and 70 mm) are manufactured to withstand light traffic. Plain pressed slabs may have slightly textured surfaces, whilst cast slabs are available with smooth, simulated riven stone, terrazzo or textured finishes. Tooled textured-finished slabs and associated products are available for use in visually sensitive locations. In addition to the standard square and rectangular units, a wide range of decorative designs including hexagonal, simulated bricks and edging units is generally available.

Tile units for roof terraces, balconies and external pedestrian areas in frost-resistant Portland cement concrete are manufactured to square and hexagonal designs in a range of standard red, brown and buff colours. They are suitable for laying on asphalt, built-up sheet roofing, inverted roofs and sand/cement screed. Typical sizes are 305 × 305 mm and 457 × 457 mm with thicknesses ranging from 25 to 50 mm.

NITROGEN OXIDE ABSORBING PAVING STONES

Titanium-coated paving stones absorb nitrogen oxides produced by road traffic and convert them by a photochemical reaction into nitrogen and oxygen, thus reducing the harmful pollution within trafficated zones. Nitrogen oxide levels can be reduced in urban areas by between 10 and 20%. Furthermore the titanium-coated slabs are easier to clean than standard concrete pavings.

References

FURTHER READING

Allen, G. 2003: *Hydraulic lime mortar for stone, brick and block masonry.* London: Donhead.

Ando, T., Arets, W., Legorreta, R. and Predock, A., 2000: *Concrete regionalism.* London: Thames & Hudson.

Beall, C. 2000: *Masonry and concrete.* Maidenhead: McGraw.

Beckett, D. 1997: *Introduction to Eurocode 2 design of structures.* London: E. & F.N. Spon.

Blackledge, G.F. and Binns, R.A. 2002: *Concrete practice.* 3rd ed. Crowthorne: British Cement Association.

British Cement Association. 1998: *National structural concrete specification for building construction.* 2nd ed. Publication no. 97.378. Crowthorne: British Cement Association.

Burkhard, F. 2002: *Concrete architecture design and construction.* Basel: Birkhäuser.

Bye, G.C. 1999: *Portland cement: Composition, production and properties.* London: Thomas Telford.

Croft, C. 2004: *Concrete architecture.* London: Laurence King Publishing.

Dawson, S. 1995: *Cast in concrete.* Leicester: Architectural Cladding Association.

Dhir, R.K. 2005: *Cement combinations for durable concrete.* London: Thomas Telford.

Eckel, E.C. 2005: *Cements, limes and plasters.* London: Donhead.

Elsener, B. 2004: *Corrosion of steel in concrete.* Weinheim: Wiley.

Gani, M.S.J. 1997: *Cement and concrete.* Oxford: Spon Press.

Gaventa, S. 2001: *Concrete design, the extraordinary nature of concrete.* London: Mitchell Beazley.

Gaventa, S. 2006: *Concrete design.* London: Mitchell Beazley.

Glass, J. 2000: *Future for precast concrete in low rise housing.* Leicester: British Precast Concrete Federation.

Goodchild, C.H. 1995: *Hybrid concrete construction.* BCA publication no. 97.337. Crowthorne: British Cement Association.

Hewlett, P.C. 2004: *Lea's chemistry of cement and concrete.* 4th ed. London: Butterworth-Heinemann.

Holland, R. 1997: *Appraisal and repair of reinforced concrete.* London: Thomas Telford.

Holmes, S. and Wingate, M. 2002: *Building with lime: a practical introduction.* London: Intermediate Technology Publications.

Institution of Structural Engineers. 1999: *Interim guidance on the design of reinforced concrete structures using fibre composite reinforcement.* London: IStructE.

Mosley, W.H., Hulse, R. and Bungey, J.H. 1996: *Reinforced concrete design to Eurocode 2.* London: Macmillan.

Neville, A.M. 1997: *Properties of concrete.* 4th ed., Harlow: Longman Higher Education.

Oehlers, D.J. 1999: *Elementary behaviour of composite steel and concrete structural members.* Oxford: Butterworth-Heinemann.

Schofield, J. 1997: *Lime in building: a practical guide.* USA: Black Dog Press.

STANDARDS

BS 146: 2002 Specification for blastfurnace cements with strength properties outside the scope of BS EN 197–1.

BS 410 Test sieves:
Part 1: 2000 Test sieves of metal wire cloth.
Part 2: 2000 test sieves of perforated metal.

BS 812 Testing aggregates:
Parts 2, 100–6, 109–114, 117–21, 123–4.

BS 915 Specification for high alumina cement:
Part 2: 1972 Metric units.

BS 1370: 1979 Specification for low heat Portland cement.

BS 1881 Testing concrete:
Parts 5, 112–13, 119, 121–2, 124–5, 127–8, 131, 201, 204, 207–9.

BS 3797: 1990 Lightweight aggregates for masonry units and structural concrete.

BS 3892 Pulverised-fuel ash:
Part 1: 1997 Specification for pulverised-fuel ash for use with Portland cement.
Part 2: 1996 Specification for pulverised-fuel ash to be used as a Type 1 addition.
Part 3: 1997 Specification for pulverised-fuel ash for use in cementitious grouts.

BS 4027: 1996 Specification for sulfate-resisting Portland cement.

BS 4248: 2004 Supersulfated cement.

BS 4449: 1997 Specification for carbon steel bars for the reinforcement of concrete.

BS 4483: 1998 Steel fabric for the reinforcement of concrete.

BS 4486: 1980 Specification for hot rolled and processed high tensile alloy steel bars for the pre-stressing of concrete.

BS 4550 Methods of testing cement:
Parts 0, 3.1, 3.4, 3.8 and 6

BS 4551 Methods of testing mortars, screeds and plasters:
Part 1: 1998 Physical testing.
Part 2: 1998 Chemical analysis and aggregate grading.

BS 4887 Mortar admixtures:
Part 1: 1986 Specification for air-entraining (plasticising) admixtures.
Part 2: 1987 Specification for set-retarding admixtures.

BS 5224: 1995 Masonry cement.

BS 5262: 1991 Code of practice for external renderings.

BS 5642 Sills and copings:

Part 1: 1978 Specification for window sills of precast concrete, cast stone, clayware, slate and natural stone.

Part 2: 1983 Specification for coping of precast concrete, cast stone, clayware, slate and natural stone.

BS 5838 Specification for dry packaged cementitious mixes:

Part 1: 1980 Prepacked concrete mixes.

BS 5896: 1980 Specification for high tensile steel wire and strand for the prestressing of concrete.

BS 5977 Lintels:

Part 1: 1981 Method for assessment of load.

BS 6073 Precast concrete masonry units:

Part 1: 1981 Specification for precast concrete masonry units.

Part 2: 1981 Method for specifying precast concrete masonry units.

BS 6089: 1981 Guide to assessment of concrete strength in existing structures.

BS 6100 Glossary of building and civil engineering terms:

Part 0: 2002 Introduction.

Part 1: 2004 General terms.

Part 10: Work with concrete and plaster.

BS 6463 Quicklime, hydrated lime and natural calcium carbonate:

Part 101: 1996 Methods of preparing samples for testing.

Part 102: 2001 Methods for chemical analysis.

Part 103: 1999 Methods for physical testing.

BS 6610: 1996 Specification for pozzolanic pulverised-fuel ash cement.

BS 6699: 1992 Specification for ground granulated blastfurnace slag for use with Portland cement.

BS 6744: 2001 Stainless steel bars for the reinforcement of and use in concrete.

BS 7542: 1992 Method of test for curing compounds for concrete.

BS 7979: 2001 Specification of limestone fines for use with Portland cement.

BS 8000 Workmanship on building sites:

Part 2: 1990 Code of practice for concrete work.

Part 9: 2003 Cementitious levelling screeds and wearing screeds.

Part 10: 1995 Code of practice for plastering and rendering.

BS 8110 Structural use of concrete:

Part 1: 1997 Code of practice for design and construction.

Part 2: 1985 Code of practice for special circumstances.

Part 3: 1985 Design charts for singly reinforced beams, doubly reinforced beams and rectangular columns.

BS 8204 Screeds, bases and *in-situ* floorings:

Part 1: 2003 Concrete bases and cement sand levelling screeds to receive floorings.

Part 2: 2003 Concrete wearing surfaces.

Part 3: 2004 Polymer modified cementitious levelling screeds and wearing surfaces.

Part 4: 1993 Cementitious terrazzo wearing surfaces.

Part 7: 2003 Pumpable self-smoothing screeds.

BS 8297: 2000 Code of practice for design and installation of non-load-bearing precast concrete cladding.

BS 8443: 2005 Specification for establishing the suitability of special purpose concrete admixtures.

BS 8500 Concrete. Complementary British Standard to BS EN 206–1:

Part 1: 2002 Method of specifying and guidance to the specifier.

Part 2: 2002 Specification for constituent materials and concrete.

Part 4 Requirements for standardised prescribed concretes.

BS EN 196 Methods of testing cement:

Part 1: 2005 Determination of strength.

Part 2: 2005 Chemical analysis of cement.

Part 3: 2005 Determination of setting time and soundness.

Part 5: 2005 Pozzolanicity test for pozzolanic cements.

Part 6: 1992 Determination of fineness.

Part 7: 1992 Methods of taking and preparing samples of cement.

BS EN 197 Cement:

Part 1: 2000 Composition specfications and conformity criteria for common cements.

Part 2: 2000 Conformity evaluation.

Part 4: 2004 Low early strength blastfurnace cements.

BS EN 206 Concrete:

Part 1: 2000 Specification, performance, production and conformity.

BS EN 413 Masonry cement:

Part 1: 2004 Composition, specifications and conformity criteria.

Part 2: 2005 Test methods.

BS EN 446: 1997 Grout for prestressing tendons – grouting procedures.

BS EN 447: 1997 Grout for prestressing tendons – specification for common grout.

BS EN 450 Fly ash for concrete:

Part 1: 2005 Definitions, specification and conformity criteria.

Part 2: 2005 Conformity evaluation.

BS EN 451 Method of testing fly ash:

Part 1: 1995 Determination of free calcium oxide content.

Part 2: 1995 Determination of fineness by wet sieving.

BS EN 459 Building lime:

Part 1: 2001 Definitions, specifications and conformity criteria.

Part 2: 2001 Test methods.

Part 3: 2001 Conformity evaluation.

BS EN 480 Admixtures for concrete, mortar and grout. Test methods.

BS EN 490: 2004 Concrete roofing tiles and fittings for roof covering and wall cladding.

BS EN 491: 2004 Concrete roofing tiles and fittings for roof covering and wall cladding.

BS EN 845 Specification for ancillary components for masonry:

Part 2: 2003 Lintels.

BS EN 934 Admixtures for concrete, mortar and grout:

Part 1: 2006 Common requirements.

Part 2: 2001 Concrete admixtures.

Part 4: 2001 Admixtures for prestressing tendons.

Part 6: 2001 Sampling, conformity control and evaluation of conformity.

BS EN 998 Specification of mortar for masonry:

Part 1: 2003 Rendering and plastering mortar.

Part 2: 2003 Masonry mortar.

BS EN 1008: 2002 Mixing water for concrete. Specification for sampling, testing and assessing suitability.

BS EN 1992 Eurocode 2: Design of concrete structures:

Part 1.1: 2004 General rules and rules for buildings.

Part 1.2: 2004 Structural fire design.

NA Part 1.2: 2004 UK National Annex to Eurocode 2. Design of concrete structures.

BS EN 1994 Eurocode 4: Design of composite steel and concrete structures:

Part 1.1: 2004 General rules and rules for buildings.

Part 1.2: 2005 Structural fire design.

BS EN 12350 Testing fresh concrete:

Part 1: 2000 Sampling.

Part 2: 2000 Slump test.

Part 3: 2000 Vebe test.

Part 4: 2000 Degree of compaction.

Part 5: 2000 Flow table test.

Part 6: 2000 Density.

Part 7: 2000 Pressure methods.

BS EN 12390: 2000 Testing hardened concrete:

Part 1: 2000 Shape, dimensions for specimens and moulds.

Part 2: 2000 Making and curing specimens for strength tests.

Part 3: 2002 Compressive strength of test specimens.

Part 4: 2000 Specification for testing machines.

Part 5: 2000 Flexural strength of test specimens.

Part 6: 2000 Tensile splitting strength of test specimens.

Part 7: 2000 Density of hardened concrete.

Part 8: 2000 Depth of penetration of water under pressure.

BS EN 12504 Testing concrete in structures:

Part 1: 2000 Cored specimens.

Part 2: 2001 Non-destructive testing.

Part 3: 2005 Determination of pull-out strength.

Part 4: 2004 Determination of ultrasonic pulse velocity.

BS EN 12620: 2002 Aggregates for concrete.

BS EN 12696: 2000 Cathodic protection of steel in concrete.

BS EN 12794: 2005 Precast concrete products – foundation piles.

BS EN 12878: 2005 Pigments for the colouring of building materials based on cement and/or lime.

BS EN 12794: 2005 Precast concrete – foundation piles.

BS EN 13055: Lightweight aggregates:

Part 1: 2002 Lightweight aggregates for concrete, mortar and grout.

BS EN 13139: 2002 Aggregates for mortar.

BS EN 13263: 2005 Silica fume for concrete.

BS EN 13369: 2004 Common rules for precast concrete products.

BS EN 13501 Fire classification of construction products and building elements:

Part 1: 2002 Classification using test data from reaction to fire tests.

Part 2: 2003 Classification using data from fire resistance tests.

BS EN 13747: 2005 Precast concrete products.

BS EN 13813: 2002 Screed materials and floor screeds. Properties and requirements.

BS EN 14216: 2004 Cement – composition, specifications and conformity criteria for very low heat special cements.

BS EN 14487: 2005 Sprayed concrete. Definitions, specifications and conformity.

BS EN 14488: 2006 Testing sprayed concrete.

BS EN 14647: 2005 Calcium aluminate cement – composition, specification and conformity criteria.

BS ISO 14656: 1999 Epoxy powder and sealing material for the coating of steel for the reinforcement of concrete.

PD 6682 Aggregates:

  Part 1: 2003 Aggregates for concrete – guidance on the use of BS EN 12620.

  Part 3: 2003 Aggregates for mortar – guidance on the use of BS EN 13139.

  Part 4: 2003 Lightweight aggregates for concrete, mortar and grout – Guidance on the use of BS EN 13055–1.

  Part 9: 2003 Guidance on the use of European test method standards.

BUILDING RESEARCH ESTABLISHMENT PUBLICATIONS

BRE Special digests

SD1: 2005 Concrete in aggressive ground.

SD3: 2002 HAC concrete in the UK: assessment, durability management, maintenance and refurbishment.

BRE Digests

BRE Digest 330: 2004 Alkali-silica reaction in concrete (Parts 1–4).

BRE Digest 357: 1991 Shrinkage of natural aggregate in concrete.

BRE Digest 361: 1991 Why do buildings crack?

BRE Digest 362: 1991 Building mortar.

BRE Digest 389: 1993 Concrete cracking and corrosion of reinforcement.

BRE Digest 405: 1995 Carbonation of concrete and its effects on durability.

BRE Digest 410: 1995 Cementitious renders for external walls.

BRE Digest 433: 1998 Recycled aggregates.

BRE Digest 434: 1998 Corrosion of reinforcement in concrete: electrochemical monitoring.

BRE Digest 444: 2000 Corrosion of steel in concrete (Parts 1, 2 and 3).

BRE Digest 451: 2000 Tension tests for concrete.

BRE Digest 455: 2001 Corrosion of steel in concrete: service life design and prediction.

BRE Digest 473: 2002 Marine aggregates in concrete.

BRE Digest 487 Part 1: 2004 Structural fire engineering design: materials behaviour – concrete.

BRE Digest 491: 2004 Corrosion of steel in concrete.

Good building guides

BRE GBG 18: 1994 Choosing external rendering.

BRE GBG 23: 1995 Assessing external rendering for replacement or repair.

BRE GBG 24: 1995 Repairing external rendering.

BRE GBG 28: 1997 Domestic floors: assessing them for replacement or repair.

BRE GBG 39: 2001 Simple foundations for low-rise housing: rule of thumb design.

BRE GBG 64 Part 2: 2005 Tiling and slating pitched roofs: plair and profiled clay and concrete tiles.

Information papers

BRE IP 1/91 Durability of non-asbestos fibre-reinforced cement.

BRE IP 2/91 Magnesian limestone aggregate in concrete.

BRE IP 11/91 Durability studies of pfa concrete structures.

BRE IP 6/92 Durability of blastfurnace slag cement concretes.

BRE IP 15/92 Assessing the risk of sulphate attack on concrete in the ground.

BRE IP 6/93 European concreting practice: a summary.

BRE IP 16/93 Effects of alkali-silica reaction on concrete foundations.

BRE IP 5/94 The use of recycled aggregates in concrete.

BRE IP 7/96 Testing anti-carbonation coatings for concrete.

BRE IP 11/97 Progress in European standardisation for the protection and repair of concrete.

BRE IP 9/98 Energy efficient concrete walls using EPS permanent formwork.

BRE IP 11/98 Assessing carbonation depth in ageing high alumina cement concrete.

BRE IP 8/00 Durability of pre-cast HAC concrete in buildings.

BRE IP 15/00 Water reducing admixtures in concrete.

BRE IP 20/00 Accelerated carbonation testing of concrete.

BRE IP 9/01 Porous aggregates in concrete: Jurassic limestones.

BRE IP 11/01 Delayed ettringite formation: *in-situ* concrete.

BRE IP 18/01 Blastfurnace slag and steel slag: their use as aggregates.

BRE IP 1/02 Minimising the risk of alkali-silica reaction: alternative methods.

BRE IP 7/02 Reinforced autoclaved aerated concrete panels.

BRE IP 15/02 Volumetric strain of concrete under uniaxial compression with reference to sustained loading and high grade concrete.

BRE IP 4/03 Deterioration of cement-based building materials: lessons learnt.

BRE IP 16/03 Proprietary renders.

BRE IP 3/04 Self-compacting concrete.

BRE IP 6/04 Porous aggregates in concrete.

BRE IP 12/04 Concrete with minimal or no primary aggregate content.

BRE IP 11/05 Innovation in concrete frame construction.

BRE IP 17/05 Concretes with high ggbs contents for use in hard/firm secant piling.

BRE IP 3/06 Reinforced concrete service life design (Parts 1, 2 and 3).

BRE Reports

BR 106: 1988 Design of normal concrete mixes.

BR 114: 1987 A review of carbonation in reinforced concrete.

BR 216: 1994 Durability of pfa concrete.

BR 243: 1993 Efficient use of aggregates and bulk construction materials Volume 1: An overview.

BR 244: 1993 Efficient use of aggregates and bulk construction materials Volume 2: Technical data and results of surveys.

BR 245: 1993 Performance of limestone-filled cements.

BR 254: 1994 Repair and maintenance of reinforced concrete.

BR 429: 2001 High alumina cement and concrete.

BR 468: 2004 Fire safety of concrete structures.

CONCRETE SOCIETY REPORTS

Technical Report 51: 1998 Guidance on the use of stainless steel reinforcement.

Technical Report 55: 2000 Design guidance for strengthening concrete structures using fibre composite materials.

Technical Report 61: 2004 Enhancing reinforced concrete durability.

Concrete Advice 07: 2003 Galvanised steel reinforcement.

Concrete Advice 14: 2003 Concrete surfaces for painting.

Concrete Advice 16: 2003 Assessing as struck *in-situ* concrete surface finishes.

CS 23: 2003 The new concrete standards – getting started.

CS 152: 2004 National structural concrete specification for building construction.

ADVISORY ORGANISATIONS

Architectural Cladding Association, 60 Charles Street, Leicester LE1 1FB (0116 253 6161).

British Precast Concrete Federation Ltd., 60 Charles Street, Leicester LE1 1FB (0116 253 6161).

Cement Admixtures Association, 38a Tilehouse Green Lane, Knowle, West Midlands B93 9EY (01564 776362).

Concrete Information Ltd., Riverside House, 4 Meadows Business Park, Station Approach, Camberley, Surrey GU17 9AB (01276 608770).

Concrete Society, Riverside House, 4 Meadows Business Park, Station Approach, Camberley, Surrey GU17 9AB (01276 607140).

Concrete Society Advisory Service, Riverside House, 4 Meadows Business Park, Station Approach, Camberley, Surrey GU17 9AB (01276 607140).

Construct Concrete Structures Group Ltd., Riverside House, 4 Meadows Business Park, Station Approach, Camberley, Surrey GU17 9AB (01276 38444).

Lime Centre, Long Barn, Morestead, Winchester, Hants. SO21 1LZ (01962 713636).

Mortar Industry Association, 156 Buckingham Palace Road, London SW1W 9TR (020 7730 8194).

Prestressed Concrete Association, 60 Charles Street, Leicester LE1 1FB (0116 251 4568).

Quarry Products Association, Riverside House, 4 Meadows Business Park, Station Approach, Camberley, Surrey GU17 9AB (01276 33144).

Sprayed Concrete Association, Association House, 99 West Street, Farnham, Surrey GU9 7EN (01252 739153).

Structural Precast Association, 60 Charles Street, Leicester LE1 1FB (0116 253 6161).

4

TIMBER AND TIMBER PRODUCTS

—

Introduction

Timber, arguably the original building material, retains its prime importance within the construction industry because of its versatility, diversity and aesthetic properties. About 20% of the earth's land mass is covered by forests, divided roughly two-thirds as hardwoods in temperate and tropical climates and one third as softwoods within temperate and colder regions. Approximately a third of the annual world-wide timber harvest is used in construction, and the rest is consumed for paper production, as a fuel, or wasted during the logging process.

Environmental issues, raised by the need to meet the current and future demands for timber, can only be resolved by sustainable forest developments. In temperate climate forests, clear cutting, in which an area is totally stripped, followed by replanting, is the most economical, but the shelterwood method, involving a staged harvest over several years, ensures that replacement young trees become established as the mature ones are felled. The managed forests of North America and Scandinavia are beginning to increase in area due to additional planting for future use. The deforestation of certain tropical regions has allowed wind and rain to erode the thin topsoil, leaving inhospitable or desert conditions; furthermore, the overall reduction in world rain forest areas is contributing significantly to the *greenhouse effect* by reducing the rate of extraction of carbon dioxide from the atmosphere.

Compared to the other major construction materials, timber as a renewable resource is environmentally acceptable. As illustrated in Figure 4.1 brick, steel, plastics and particularly aluminium all use more energy in their production, thus contributing considerably to carbon dioxide emissions. Trees require little energy for their conversion into usable timber, and young replacement trees are particularly efficient at absorbing carbon dioxide and releasing oxygen into the atmosphere. Temperate and tropical hardwoods, suitably managed, can be brought to maturity within a human lifespan; softwoods within half that period. Timber products manufactured from reconstituted and waste wood add to the efficient use of forestry. Increasingly, emphasis is being placed upon timber certification schemes, which track the material from source to user, to ensure the accuracy of environmental claims being made. For UK-produced timber, the Forest Stewardship Council ensures that timber with their label has been harvested from properly managed sustainable sources.

Timber

The Study Centre at Darwin College, Cambridge (Fig. 4.2), which occupies a narrow site overlooking the River Cam, is designed to accommodate both books and computers. It is a load-bearing masonry and timber building which features the extensive use of English oak, including massive paired columns to the first-floor reading room which is partly cantilevered over the river. The columns in green oak have characteristic shakes and splits giving an impression of great age, and these contrast with the refined oak and oak veneer of the floors, windows frames and furniture. Joints in the green oak are held by stainless steel fixings, which can be tightened as the timber dries and shrinks. The use of oak throughout gives unity to the building, which sits comfortably within its highly sensitive location.

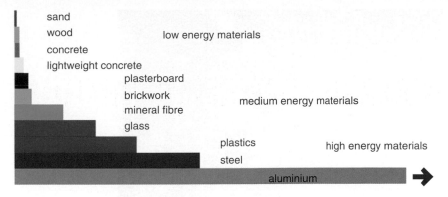

Relative embodied CO$_2$ of various materials by weight
Includes CO$_2$ generated through extraction, manufacture and construction processes

Fig. 4.1 Embodied energy in building materials. Diagram: Courtesy of Make Architects

METABOLISM OF THE TREE

The tree, a complex living organism, can be considered in three main sections: the branches with their leaves, the trunk (or bole) and the roots (Fig. 4.3). The roots anchor the tree to the ground and absorb water with dissolved minerals from the soil. The leaves absorb carbon dioxide from the air and in the presence of sunlight, together with chlorophyll as a catalyst, combine carbon dioxide with water to produce sugars. The oxygen, a by-product of the process, diffuses out of the leaves. The sugars in aqueous solution are transported down the branches and trunk to be subsequently converted, where required for growth, into the cellulose of the tree. The trunk gives structural strength to the tree, and acts as a store for minerals and food, such as starch, and also as a two-way transport medium.

The tree is protected from extremes of temperature and mechanical damage by the bark, inside which is the bast layer, which transports downwards the sugars synthesised in the leaves. Radial rays then move the food into the sapwood cells for storage. Inside the bast is the thin and delicate cambium, which is the growing layer for the bark and sapwood. Growth only takes place when the cambium layer is active, which in temperate climates is during the spring and summer seasons.

A transverse section through the bole shows the growth rings. These are sometimes referred to as annual rings, but unusual growth patterns can lead to multiple rings within one year, and in tropical climates, where seasonal changes are less pronounced, growth rings may be indistinct and not annual. The growth rings are apparent because the *early wood*

produced at the start of the growing season tends to be made from larger cells of thinner walls, and is thus softer and more porous than the *late wood* produced towards the end of the growing season. Each year as the tree matures with the production of an additional growth ring, the cells of an inner ring are strengthened by a process of *secondary thickening*. This is followed by lignification in which the cell dies. These cells are no longer able to act as food stores, but now give increased structural strength to the tree. The physical changes are often associated with a darkening of the timber due to the incorporation into the cell walls of so-called *extractives*, such as resins in softwoods or tannins in oak. These are natural wood preservatives which make heartwood more durable than sapwood.

CONSTITUENTS OF TIMBER

The main constituents of timber are cellulose, hemicellulose and lignin, which are natural polymers. Cellulose, the main constituent of the cell walls, is a polymer made from glucose, a direct product from the photosynthesis within the leaves of the tree. Glucose molecules join together to form cellulose chains containing typically 10 000 sugar units (Fig. 4.4). Alternate cellulose chains, running in opposite directions to each other, form a predominantly well-ordered crystalline material. It is this crystalline chain structure which gives cellulose its fibrous properties, and accounts for approximately 45% of the dry weight of the wood.

Hemicelluloses, which account for approximately 25% of the weight of wood, have more complex

Fig. 4.2 Green Oak construction – Darwin College Study Centre, Cambridge. Architects: Jeremy Dixon • Edward Jones. Photographs: Courtesy of Dennis Gilbert

partially crystalline structures, being composed of a variety of other sugars. The molecular chains are shorter than those in cellulose, producing a more gelatinous material. Lignin (approximately 25% by weight of the timber) is an insoluble non-crystalline polymeric material. Its main constituents are derivatives of benzene combined to form a complex branched-chain structure.

The three major components are combined to form *microfibrils*, which are in turn the building

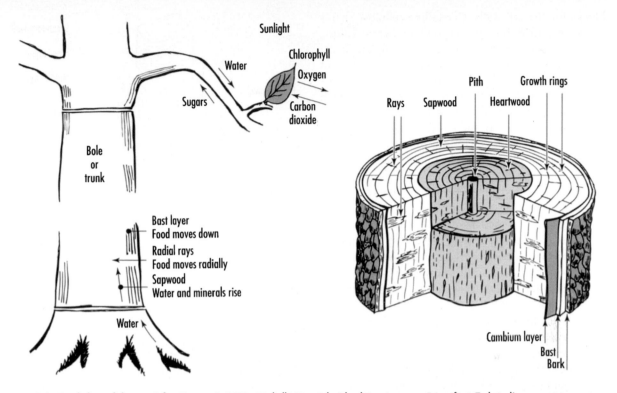

Fig. 4.3 Metabolism of the tree (after Everett, A. 1994: *Mitchell's Materials.* 5th edition. Longman Scientific & Technical)

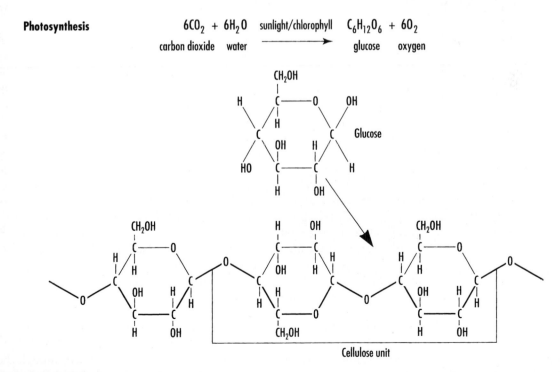

Photosynthesis

$$6CO_2 + 6H_2O \xrightarrow{\text{sunlight/chlorophyll}} C_6H_{12}O_6 + 6O_2$$

carbon dioxide   water   glucose   oxygen

Fig. 4.4 Structure of cellulose

blocks for the cell walls. Crystalline cellulose chains are surrounded by semi-crystalline hemicellulose, then a layer of non-crystalline cellulose and are finally cemented together with lignin (Fig. 4.5). Millions of these microfibrils are built up in layers to form the individual cell walls. It is this composite structure which gives timber its physical strength, with the cellulose contributing mainly to the tensile properties and the hemicellulose and lignin to the compressive strength and elasticity.

In addition to the three major constituents and significant quantities of water, timbers contain many minor constituents; some, such as resins, gums and tannins, are associated with the conversion of sapwood to heartwood. Starch present in sapwood is attractive to fungi, and inorganic materials such as silica make working certain tropical hardwoods, such as teak, difficult. The various colours present in different timbers arise from these minor constituents, as the various celluloses and lignin are virtually colourless. Some colours are fixed to the polymeric chains, but others are light-sensitive natural dyes, which fade on prolonged exposure to sunlight unless the timber is coated with an ultraviolet-absorbing finish.

HARDWOODS AND SOFTWOODS

Commercial timbers are defined as hardwoods or softwoods according to their botanical classification rather than their physical strength. Hardwoods (*angiosperms*) are from broad-leafed trees, which in temperate climates are deciduous, losing their leaves in autumn, although in tropical climates, where there is little seasonal variation, old leaves are constantly being replaced by new. Softwoods (*gymnosperms*) are from conifers, characteristically with needle-shaped leaves, and growing predominantly in the northern temperate zone. Mostly they are evergreen, with the notable exception of the European Larch (*Larix decidua*) and they include the Californian redwood (*Sequoia sempervirens*), the world's largest tree with a height of over 100 metres.

Although the terms *hardwood* and *softwood* arose from the physical strength of the timbers, paradoxically balsa (*Ochroma lagopus*), used for model-making, is botanically a hardwood, whilst yew (*Taxus baccata*), a strong and durable material, is defined botanically as a softwood. Under microscopic investigation, softwoods show only one type of cell, which varies in size between the rapid growth of spring and early summer (early

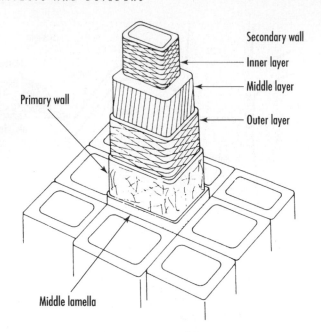

Fig. 4.5 Cell structure of timber (after Desch, H.E. 1981: *Timber: its structure properties and utilisation*, 6th edition. Macmillan Education – Crown Copyright)

wood) and the slow growth of the late summer and autumn (late wood). These cells, or *tracheids*, perform the food and water conducting functions and give strength to the tree. Hardwoods, however, have a more complex cell structure with large cells or vessels for the conducting functions and smaller cells or wood fibres which provide the mechanical support. According to the size and distribution of the vessels, hardwoods are divided into two distinct groups. Diffuse-porous hardwoods, which include beech (*Fagus sylvatica*), birch (*Betula pendula*) and most tropical hardwoods, have vessels of a similar diameter distributed approximately evenly throughout the timber. Ring-porous hardwoods, however, including oak (*Quercus robur*), ash (*Fraxinus excelsior*) and elm (*Ulmus procera*), have large vessels concentrated in the earlywood, with only small vessels in the latewood (Fig. 4.6). The Jerwood Library of Trinity Hall, Cambridge (Fig. 4.7) illustrates the visual quality of limed oak as an architectural feature within the context of a sensitive built environment.

TIMBER SPECIES

Any specific timber can be defined through the correct use of its classification into family, genus and species. Thus oak, and beech are members of the

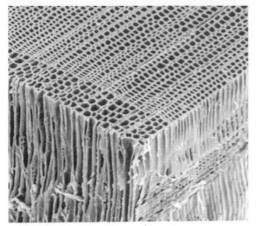

Softwood – Scots Pine (*Pinus sylvestris*)

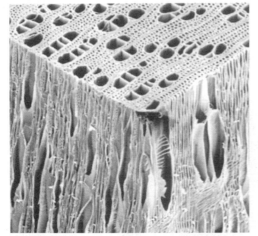

Diffuse-porous hardwood – Birch (*Betula pendula*)

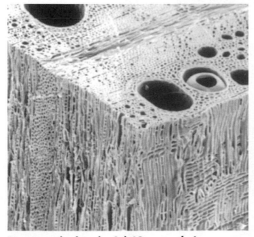

Ring-porous hardwood – Oak (*Quercus robur*)

Fig. 4.6 Cell structures of hardwoods and softwoods

Fig. 4.7 Limed Oak – Jerwood Library, Trinity Hall, Cambridge. Architects: Freeland Rees Roberts. Photograph: Arthur Lyons

Fagaceae family; beech is one genus (*Fagus*) and oak (*Quercus*) another. The oak genus is subdivided into several species, including the most common, the pedunculate oak (*Quercus robur*) and the similar but less common sessile oak (*Quercus petraea*). Such exact timber nomenclature is, however, considerably confused by the use of lax terminology within the building industry; for example, both Malaysian meranti and Philippine lauan are frequently referred to as Philippine mahogany, and yet they are from a quite different family and genus to the true mahogany (*Swietenia*) from the West Indies, or Central America. This imprecision can cause the erroneous specification or supply of timber, with serious consequences. Where there is the risk of confusion, users should specify the correct family, genus and species.

Softwood accounts for approximately 80% of the timber used in the UK construction industry. Pine (European redwood) and spruce (European whitewood) are imported from Northern and Central Europe, whilst western hemlock, spruce, pine, and fir are imported in quantity from North America. Forest management in these areas ensures that supplies will continue to be available. Smaller quantities of western red cedar, as a durable lightweight cladding material, are imported from North America, together with American redwood from California, pitch pine from Central America and parana pine from Brazil. Increasingly, New Zealand, South Africa and Chile are becoming significant exporters of renewable timber. The UK production of pine and spruce provides only about 10% of the national requirements while Ireland plans to be self-sufficient early in the next century.

Over 100 different hardwoods are used in the UK, although together beech, oak, meranti, lauan, elm, American mahogany and ramin account for over half of the requirements. Approximately half of the hardwoods used in the UK come from temperate forests in North America and Europe including Britain, but the remainder, including the durable timbers such as iroko, mahogany, sapele and teak, are imported from the tropical rain forests. The Great Oak Hall at Westonbirt Arboretum, Gloucestershire (Fig. 4.8) illustrates the use of 'medieval' construction systems within a modern building by using 'green' oak fixed with dowels and wedges.

Since 1965, 6.5% of the Amazon forest has been lost, but much of this deforestation has been for agricultural purposes, with more than three quarters of the timber felled used as a local fuel rather than exported as timber. With the growing understanding of the environmental effects of widespread deforestation, some producer governments are now applying stricter controls to prevent clear felling, and to encourage sustainable harvesting through controlled logging. Other imported naturally durable hardwoods, available in long lengths, include ekki, greenheart and opepe, whilst UK-produced sweet chestnut is durable and an appropriate structural timber.

Fig. 4.8 Traditional oak construction – Great Oak Hall, Westonbirt Arboretum, Gloucestershire. Architects: Roderick James Architects. Photograph: Arthur Lyons

CONVERSION

Conversion is the process of cutting boles or logs into sections prior to seasoning. Subsequent further cutting into usable sizes is called *manufacture*. Finishing operations, involving planing and sanding, produce a visually smooth surface but reduce the absorption of penetrating wood stains. Timber for solid sections is sawn, whereas thin layers for plywood are peeled and veneers are usually sliced across the face of the log to maximise the visual effect of colour and figure, which is the pattern effect seen on the longitudinal surface of cut wood.

Types of cut

The two main types of cut – *plain sawn* and *quarter sawn* – refer to the angle between the timber face and the growth rings. This is best observed from the end of the timber, as in Figure 4.9. If the cut is such that the growth rings meet the surface at less than 45° then the timber is plain sawn. Timber with this type of cut tends to have a more decorative appearance but a greater tendency to distort by *cupping*. Timber cut with the growth rings meeting the surface at not less than 45° is quarter sawn. Such timber is harder wearing, weather-resistant and less likely to flake. If a log is cut *through and through*, which is most economical, then a mixture of plain and quarter-sawn timber is produced. Quarter

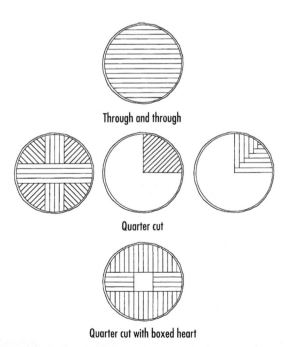

Fig. 4.9 Conversion of timber

Through and through

Quarter cut

Quarter cut with boxed heart

sawing is more expensive as the log requires resetting for each cut and more waste is produced; however, the larger sections will be more dimensionally stable. The centre of the tree, the pith, is frequently soft and may be weakened by splits or shakes. In this case the centre is removed as a *boxed heart*.

Sizes

BS EN 1313–1: 1997 defines the standard sizes of sawn softwood timbers at a 20% moisture content (Table 4.1). Widths over 225 mm and lengths over 5 m are scarce and expensive, but finger jointing (BS EN 385: 2001), which can be as strong as the continuous timber, does allow longer lengths to be specified. Regularising, which ensures uniformity of width of a sawn timber, reduces the nominal section by 3 mm (5 mm over 150 mm) and planing on all faces or 'processed all round' (PAR) reduces, for example, a 47 × 100 mm section to 44 × 97 mm (Table 4.2). Hardwood sizes are more variable due to the diversity of hardwood species, but preferred sizes to BS EN 1313–2: 1999 are specified in Table 4.1. Hardwoods are usually imported in random widths and lengths; certain structural hardwoods such as Iroko (*Chlorophora excelsa*) are available in long lengths (6–8 m) and large sections. Tolerances for acceptable deviations from target sizes for softwood are given in BS EN 1313–1: 1997 and BS EN 336: 2003 (Table 4.3). The latter defines two tolerance levels for sawn surface dimensions (Tolerance Class 1, T1 and Tolerance Class 2, T2) with T2 specifying the smaller tolerance limits, also appropriate to planed timber. Customary lengths for structural softwood timber and hardwood are given in Table 4.3.

MOISTURE CONTENT AND SEASONING

As a tree is a living organism, the weight of water within it is frequently greater than the dry weight of wood itself. The water content of a tree is equal in winter and in summer, but one advantage of winter felling is that there is a reduced level of insect and fungal activity. After felling, the wood will loose the water held within the cell cavities without shrinkage, until the *fibre saturation point* is reached when the cells are empty. Subsequently, water will be removed from the cell walls, and it is during this process that the timber becomes harder and shrinkage occurs. As cellulose is a hygroscopic material, the timber will eventually equilibrate at a moisture content dependent upon the atmospheric conditions. Subsequent reversible

Table 4.1   Standard sizes of softwoods and hardwoods

Standard sizes of sawn softwood (20% moisture content) to BS EN 1313–1: 1997.

Thickness (mm)	Width (mm)											
	75	100	115	125	138	150	175	200	225	250	275	300
16	V	V		V		V						
19	V	V		V		V						
22	V	V		V		V	V	V	V			
25	V	V		V		V	V	V	V	V	V	V
32		V	V	V	V	V	V	V	V	V	V	V
38	x	✓	x	✓	x	✓	x	x	x	x	x	x
47	x	x		x		x	x	x	x	x		x
50	x	✓		✓		✓	✓	✓	✓	x		x
63		✓		✓		✓	✓	x	x			
75		x		x		✓	✓	✓	✓	x	x	x
100		x				x		✓	x	x	x	x
150						x		x				x
250										x		
300												x

Sizes marked with a tick indicate preferred EU sizes.
Sizes marked with a cross are the complementary UK preferred sizes.
Sizes marked with a V are the additional UK customary sizes.

Customary lengths of structural timber to BS EN1313–1: 1997

			Length m			
1.80	2.10	3.00	4.20	5.10	6.00	7.20
	2.40	3.30	4.50	5.40	6.30	
	2.70	3.60	4.80	5.70	6.60	
		3.90			6.90	

Lengths over 5.70 m may not be readily available without finger jointing.

Standard sizes of sawn hardwood (20% moisture content) to BS EN1313–2: 1999

Preferred thicknesses										
EU	20	27	32	40	50	60	65	70	80	100 mm

Complementary thicknesses
UK   19   26   38   52   63   75 mm

Preferred widths
EU   10 mm intervals for widths between 50 mm and 90 mm,
20 mm intervals for widths of 100 mm or more.

Preferred lengths
EU   100 mm intervals for lengths between 2.0 m and 6.0 m,
50 mm intervals for lengths less than 1.0 m.

Table 4.2 Maximum permitted reduction from target sawn sizes of softwoods and hardwoods by planing two opposed faces

Maximum reductions from sawn softwood sizes by planing two opposed faces (BS EN 1313–1: 1997)

Typical application	Reduction from basic size (mm)			
	15–35	36–100	101–150	over 150
Constructional timber	3	3	5	5
Matching and interlocking boards (not flooring)	4	4	6	6
Wood trim	5	7	7	9
Joinery and cabinet work	7	9	11	13

Maximum reductions from sawn hardwood sizes by planing two opposed faces (BS EN 1313–2: 1999)

Typical application	Reduction from basic size (mm)				
	15–25	26–50	51–100	101–150	151–300
Flooring, matchings, interlocked boarding and planed all round	5	6	7	7	7
Trim	6	7	8	9	10
Joinery and cabinet work	7	9	10	12	14

Table 4.3 Permitted deviations on structural timber sizes to BS EN 336: 2003

Maximum deviations from target sizes	Tolerance Class T1	Tolerance Class T2
Thicknesses and widths $\leq$ 100 mm	− 1 to + 3 mm	− 1 to + 1 mm
Thicknesses and widths > 100 mm	− 2 to + 4 mm	− 1.5 to + 1.5 mm

changes in dimension are called *movement*. The controlled loss of moisture from green timber to the appropriate moisture content for use is called *seasoning*.

$$\text{Moisture content} = \frac{\text{weight of wet specimen} - \text{dry weight of specimen}}{\text{dry weight of specimen}} \times 100\,\%$$

The primary aim of seasoning is to stabilise the timber to a moisture content that is compatible with the equilibrium conditions under which it is to be used, so that subsequent movement will be negligible. At the same time, the reduction in water content to below 20% will arrest any incipient fungal decay, which can only commence above this critical level. Drying occurs with evaporation of water from the surface, followed by movement of moisture from the centre of the timber outwards due to the creation of a vapour-pressure gradient. The art of successful seasoning is to control the moisture loss to an appropriate rate. If the moisture loss is too rapid then the outer layers shrink while the centre is still wet and the surface sets in a distended state (case hardening) or opens up in a series of cracks or checks. In extreme cases as the centre subsequently dries out and shrinks it may honeycomb.

Air seasoning

Timber, protected both from the ground and from rain, is stacked in layers separated by strips of wood called *stickers* which, depending on their thickness, control the passage of air. The air, warmed by the sun and circulated by the wind, removes moisture from the surface of the timbers. The timber ends are protected by waterproof coatings (bituminous paint) to prevent rapid moisture loss, which would cause splitting. Within the UK a moisture content of between 17 and 23% may be achieved within a few months for softwoods, or over a period of years for hardwoods.

Kiln drying

Kiln drying or seasoning is effected by heating within a closed chamber, which can be programmed to a precise schedule of temperature and humidity. Thus, drying to any desired moisture content can be achieved without significant degradation of the timber, although some early examples of kiln-dried timber showed serious damage through the use of inappropriate drying schedules. For economic reasons, timber is frequently air seasoned to fibre saturation point, followed by kiln drying to the required moisture content. This roughly halves the necessary kiln time and fuel costs. A typical softwood load would be dried from fibre saturation point within a few days and hardwood within two to three weeks.

Seasoned timber, if exposed to rain on site, will re-absorb moisture. Good site management is therefore necessary to protect timber both from physical damage and wetting prior to its use. The heating up of new buildings by central heating systems can cause rapid changes in the moisture content of joinery timber and lead to shrinkage, cracking and splitting.

MOISTURE MOVEMENT

Wood is an anisotropic material, with differing moisture movements along the three principal axes: tangential, radial and longitudinal (Fig. 4.10). The highest moisture movement is tangential to the grain, next being radial, with the least along the grain. Typical figures are given in Table 4.4. The larger the ratio between tangential and radial movement, the greater the distortion. Moisture movements are conventionally

Table 4.4 Moisture movement of some hardwoods and softwoods normally available in the UK

Hardwoods

Small moisture movement (less than 3.0%)	Medium moisture movement (3.0–4.5%)	Large moisture movement (over 4.5%)
Afzelia	Ash	Beech, European
Agba	Cherry	Birch
Iroko	Elm, European	Keruing
Jelutong	Maple	Ramin
Lauan	Oak, American	
Mahogany, African	Oak, European	
Mahogany, American	Sapele	
Meranti	Utile	
Merbau	Walnut, European	
Obeche		
Teak		

Softwoods

Small moisture movement (less than 3.0%)	Medium moisture movement (3.0–4.5%)	Large moisture movement (over 4.5%)
Douglas fir	European redwood	
Sitka spruce	European whitewood	
Western hemlock	Parana pine	
Western red cedar		

Moisture movement is assessed on the sum of the radial and tangential movements for a change in environmental conditions from 60 to 90% relative humidity.

quoted for a change in relative humidity from 90% to 60% at 25°C. The BRE classifies woods into three categories according to the sum of radial and tangential movement effected by this standard change in relative humidity. Small movement is defined as less than 3%, medium between 3% and 4.5%, and large over 4.5%. Large movement timbers are not recommended for use as cladding.

TIMBER DEFECTS

Timber, as a natural product, is rarely free from blemishes or defects, although in some instances, such as knotty pine, waney-edge fencing timber or burr veneers, the presence of the imperfections enhances the visual quality of the material. Timber imperfections can be divided into the three main categories: natural, conversion and seasoning defects, according to whether

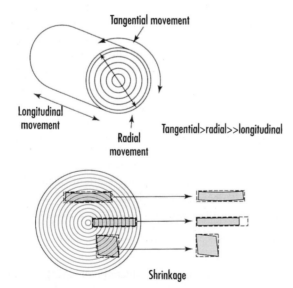

Fig. 4.10 Moisture movement and initial drying shrinkage

they were present in the living tree, or arose during subsequent processing. Additionally, timber may be subject to deterioration by weathering, fungal and insect attack, and fire. These latter effects are discussed later in the chapter.

Natural defects

Knots

Knots are formed where branches of the tree join the trunk (Fig. 4.11). Where the wood fibres of the branch are continuous with the trunk, then a live knot is produced. If, however, the branch is dead, or bark becomes incorporated into the trunk, a dead knot is produced. This is liable to be loose, lead to incipient decay and cause structural weakness.

Knots are described as face, edge, splay, margin or arris, dependent upon how they appear on the faces of converted timber. Additionally, knots may appear as clusters, and range in size from insignificant to many millimetres across. Frequently they are hard to work, and in softwoods contain quantities of resin, which will continue to seep out unless the wood is sealed before painting.

Natural inclusions

Many minor defects occur to varying degrees in different varieties of timber. Bark pockets occur where pieces of bark have been enclosed within the timber as a result of earlier damage to the cambium or growth layer. Pitch pockets and resin streaks, containing fluid resin, are frequently seen along the grain of softwoods; their extent in usable timber is limited by BS EN 942: 1996.

Compression and tension wood

Trees leaning owing to sloping ground, or subject to strong prevailing winds, produce reaction wood to counteract these forces. In softwoods, compression wood is produced which is darker in colour due to an increased lignin content. In hardwoods, tension wood is produced, which is lighter in colour owing to the presence of an extra cellulose layer in the cell walls. Both types of reaction wood have an abnormally high longitudinal shrinkage, causing distortion on seasoning; furthermore, tension wood tends to produce a rough surface when it is machined.

Abnormal growth rings

The width of the growth rings is an indicator of the growth rate and timber strength, with the optimum ranged around five rings per centimetre for softwoods and three rings per centimetre for hardwoods depending on the species. Excessively fast or slow growth rates give rise to weaker timber owing to a reduction in the proportion of the stronger late wood or its production with thinner-walled fibres.

Conversion defects

Sloping grain

For maximum strength, timber should be approximately straight grained, as with increasing slope of the grain (Fig. 4.12) there is a proportionate reduction in bending strength, ranging typically from 4% at 1 in 25 to 19% at 1 in 10. The British Standard (BS 5756: 1997) limits the slope of grain in visual strength graded structural tropical hardwood (HS) to 1 in 11. BS 4978: 1996 limits the slope of grain in visual strength graded softwoods to 1 in 6 for the general

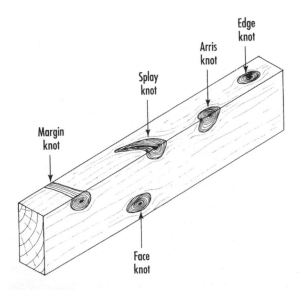

Fig. 4.11 Knots (after Porter, B. and Rose, R. 1996: *Carpentry and joinery: Bench and site skills.* Arnold)

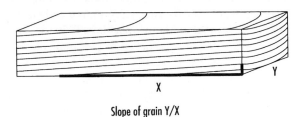

Slope of grain Y/X

Fig. 4.12 Sloping grain

structural grade (GS) and 1 in 10 for the special structural grade (SS). The slope of grain in timber for internal and external windows, doors and doorframes is limited to 1:10.

Wane

Wane is the loss of the square edge of the cut timber due to the incorporation of the bark or the curved surface of the trunk. A degree of wane is acceptable in structural and floor timbers (BS 4978: 1996 and BS 1297: 1987), and is a special feature in waney-edge fencing.

Seasoning defects

Some of the commonest defects in timber are associated with the effects of seasoning. During the seasoning process, the contraction of the timber is different in the three major directions; furthermore, as described in the 'Moisture movement' section above, the outside of the timber tends to dry out more rapidly than the interior. These combined effects cause distortion of the timber including warping and the risk of rupture of the timber to produce surface checks and splits (Fig. 4.13).

Shakes

Major splits within timber are termed *shakes*, and these may result from the release of internal stresses within the living tree on felling and seasoning; however, some fissures may be present within the growing timber. Commonly, shakes are radial from the exterior of the trunk, but star shakes which originate at the centre or pith may be associated with incipient decay. Ring shakes follow round a particular growth ring and are frequently caused by the freezing of the sap in severe winters.

SPECIFICATION OF TIMBER

The building industry uses timber for a wide range of purposes from rough-sawn structural members to claddings, trim and highly machined joinery. The specification of timber for each use may involve defining the particular hardwood or softwood, where particular visual properties are required. However, for the majority of general purposes, where strength and durability are the key factors, timber is specified either by a strength class, or a combination of timber species and strength grade.

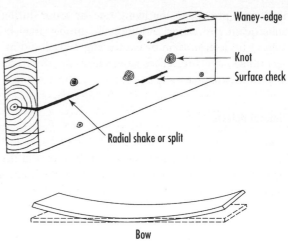

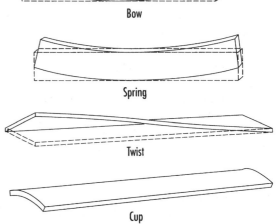

Fig. 4.13 Warping, splits and checks

In addition to strength class or grade, the specification of structural timber should include: lengths and cross-section sizes; surface finish or tolerance class; moisture content; and any preservative or special treatments (BRE Digest 416: 1996).

Strength grading

Strength grading is the measurement or estimation of the strength of individual timbers, which allows each piece to be used to its maximum efficiency. It may be done visually, a slow and skilled process, or within a grading machine which tests flexural rigidity. The standards for the visual grading of hardwood and softwood are BS 5756: 1997 and BS 4978: 1996 respectively (BRE Digest 492: 2005).

Visual strength grading

Each piece of timber is inspected for distortions, growth ring size and slope of grain, then checked against the allowed standards for the number and severity of the natural defects such as knots, waney-edge, and fissures. The timber is then assigned to a grade and stamped accordingly. Softwood timber is assessed as special structural grade (SS), general structural grade (GS) or reject. Hardwoods are graded to THA or THB (heavy structural temperate hardwood), TH1 or TH2 (general structural temperate hardwood) or reject as appropriate. (The higher grades THA and TH1 in each category have fewer natural defects such as knots or sloping grain.) The one grade for tropical hardwood is HS (structural tropical hardwood).

Machine strength grading

Each piece of timber is quickly inspected for any distortions which may cause it to be rejected manually, or any serious defects within 500 mm of either end, at which points machine testing is ineffective. It is then tested, usually by one of two systems of contact or bending-type grading machines. In both techniques, the timber is moved through a series of rollers and either the machine measures the load required to produce a fixed deflection, or it measures the deflection produced by a standard load. Either technique is measuring stiffness, which is then related to timber strength and therefore a grading standard.

The three grading machines used within the UK are shown in Figure 4.14. In the constant deflection machine, the timber is then moved through a series of rollers which press it firmly against a curved metal plate. The force required to bend the timber to this standard deflection is determined by a series of transducers and from this data the timber strength is computed. However, a second pass through this machine is required to eliminate the effects of bow. The constant load system applies a defined lateral load, depending on the sample thickness, and the resulting deflection, with automatic adjustment for bow, indicates the timber grade. A more sophisticated system measures the forces necessary to bend the timber into an S-shape with two fixed deflections, thus neutralising the effects of any natural bow in the material.

As the timber leaves the machine it is stamped with the European/British Standard number and strength class, together with information on its species group, wet/dry graded state and the certification body. For machine-graded timber, the timber is graded directly to a strength class and marked accordingly. Timber for trussed rafters may be colour coded according to its strength class. The machine classes and associated dye colour codes are given in Table 4.5.

Table 4.5 Strength classes and the associated machine-graded colour codings

Strength class	Colour code
C27	red
C24	purple
C22	blue
C16	green

Recent non-contact techniques for strength grading include X-ray and stress wave systems. X-ray machines assess the density of the timber, which is then related to strength. Stress wave techniques measure either the speed through the timber, or the natural frequency of a stress wave produced by a small impact, and relate this to strength. Both techniques offer the potential for faster throughput than conventional contact strength grading systems although they are sensitive to timber moisture content. Some grading machines combine physical bending techniques with the use of X-ray or microwave systems for the detection of natural defects such as knots or sloping grain respectively.

Strength classes

Strength classes to BS EN 338: 2003, see Table 4.6, are defined as C14 to C50 and D30 to D70, where the prefix C refers to softwoods (coniferous) and D to hardwoods (deciduous). The number refers to the characteristic bending strength in newtons per square millimetre. The full specification of the strength classes gives characteristic values for density and a wide range of strength and stiffness properties all based on sample test values. The data do not take into account any safety factors to be included in the design process. For trussed rafters, the grades TR20 and TR26 defined in BRE Digest 445: 2000 are applicable.

Table 4.7 shows softwood species and strength grade combinations in relation to strength classes for visual and machine-graded softwood timber.

Service class

The service class defines the conditions in which the timber will be used and thus the anticipated moisture content. There are three categories defined within Eurocode 5: Design of timber structures; Part 1–1;

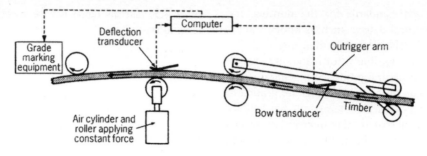

The MPC Computermatic grading machine

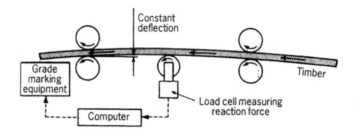

The MPC Cook-Bolinder grading machine

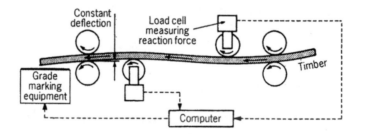

The Raute Timgrader grading machine

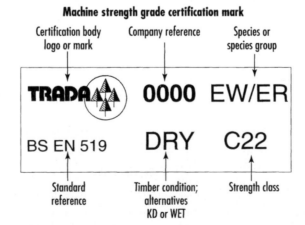

Fig. 4.14 Strength-grading machines and timber certification mark. (Grading machine diagram reproduced from Digest 476 by permission of BRE)

Table 4.6 Relationship between strength classes and physical properties

Strength classes to BS EN 338: 2003 – Characteristic values

	C14	C16	C18	C20	C22	C24	C27	C30	C35	C40	C45	C50	D30	D35	D40	D50	D60	D70
Strength properties MPa Bending	14	16	18	20	22	24	27	30	35	40	45	50	30	35	40	50	60	70
Stiffness properties MPa Mean modulus of elasticity Parallel to the grain	7	8	9	9.5	10	11	11.5	12	13	14	15	16	10	10	11	14	17	20
Density kg/m³ Average density	350	370	380	390	410	420	450	460	480	500	520	550	640	670	700	780	840	1080

Strength classes to BS 5268–2: 2002 – Grade stresses for permissible stress design code

	C14	C16	C18	C22	C24	C27	C30	C35	C40	D30	D35	D40	D50	D60	D70
Strength properties MPa Bending parallel to the grain	4.1	5.3	5.8	6.8	7.5	10	11	12	13	9	11	12.5	16	18	23
Stiffness properties MPa Minimum modulus of elasticity parallel to the grain	4.6	5.8	6.0	6.5	7.2	8.2	8.2	9.0	10.0	6.0	6.5	7.5	12.6	15.6	18.0

C refers to coniferous softwoods and D refers to deciduous hardwoods.

(BS EN 1995–1–1: 2004). Timber to be used in Service Classes 1 and 2 must be adequately protected from the weather when on site.

Service classes of wood in use:

Service Class 1
Timber with a moisture content corresponding to an ambient temperature of 20°C and a relative humidity of the surrounding air only exceeding 65% for a few weeks each year.
Average moisture content not exceeding 12% (e.g. internal walls, internal floors except ground floor, warm roofs).

Service Class 2
Timber with a moisture content corresponding to an ambient temperature of 20°C and a relative humidity of the surrounding air only exceeding 85% for a few weeks each year.
Average moisture content not exceeding 20% (e.g. ground floors, inner leaf of cavity walls, single leaf walls with external cladding).

Service Class 3
Timber exposed to conditions leading to higher moisture contents than in Service Class 2.

Average moisture content 20% and above (e.g. exposed parts of buildings and marine structures).

Limit state design

Eurocode 5 represents a significant change for designers in timber and timber products as it is based on *limit state design*, rather than permissible stress. This brings timber into line with steel and concrete for which this approach has already been taken. There are generally two limit states to be considered: first, the *ultimate limit state* beyond which parts of the structure may fail or collapse; and secondly the *serviceability limit state* beyond which excessive deformation, deflection or vibration would render the structure unfit for its purpose. The ultimate limit states are determined from the *characteristic values of the loads* or actions and the material properties, to which partial safety factors are applied. Generally, the characteristic values of the material properties in BS EN 338: 2003 are higher than in the BS 5268–2: 2002 grade stresses, as they are derived from laboratory tests without reductions for long-term loading or safety factors which become the responsibility of the designer. The characteristic values are used in limit state design (Eurocode 5) and the grade stresses apply to the BS 5268–2: 2002 permissible stress design code.

Table 4.7 Softwood species/grade combinations which satisfy the requirements of BS EN 338: 1995 and BS 5268 Part 2: 2002 strength classes as listed in Eurocode 5 Part 1:1

Species	Origin	Grading Standard	European Standard BS EN 338: 1995 Strength Classes						
			C14	C16	C18	C22	C24	C27	C30
British grown softwoods									
Douglas fir	UK	BS EN 519	✓	✓	✓	✓	✓		
		BS 4978	GS		SS				
Pine	UK	BS EN 519	✓	✓	✓	✓	✓	✓	
		BS 4978	GS			SS			
Spruce	UK	BS EN 519	✓	✓	✓	✓	✓		
		BS 4978			SS				
Larch	UK	BS EN 519	✓	✓	✓	✓	✓	✓	
		BS 4978		GS			SS		
Imported softwoods									
Redwood	Europe	BS EN 519	✓	✓	✓	✓	✓	✓	✓
		BS 4978		GS			SS		
Whitewood	Europe	BS EN 519	✓	✓	✓	✓	✓	✓	✓
		BS 4978		GS			SS		
Sitka spruce	Canada	BS 4978	GS		SS				
		US/Can	1,2		Sel				
Hem-fir	Canada	BS EN 519	✓	✓	✓	✓	✓	✓	✓
Spruce pine-fir	and US	BS 4978		GS			SS		
Douglas fir-larch		US/Can		1,2			Sel		
Southern pine	US	BS EN 519	✓	✓	✓	✓	✓	✓	✓
		BS 4978			GS		SS		
		US/Can		3		1,2			Sel
Western white woods	US	BS 4978	GS		SS				
		US/Can	1,2		Sel				
Pitch pine	Caribbean	BS 4978			GS			SS	
Parana pine		BS 4978		GS			SS		
Radiata pine	New Zealand Chile	BS EN 519	✓	✓	✓	✓	✓	✓	✓
Zimbabwian pine	Zimbabwi	BS EN 519	✓	✓	✓	✓	✓	✓	✓
S. African pine	S. Africa	BS EN 519	✓	✓	✓	✓	✓	✓	✓
Western red cedar		BS 4978	GS		SS				

Grading: BS EN 519 refers to machine grading, BS 4978 refers to visual grading,
✓ indicates available machine graded to the class indicated.
GS and SS are General Structural and Special Structural visual grades respectively.
US/Can refers to the US and Canadian visual grading standards for lumber.
The Canadian and US grades 1, 2, 3, and Sel refer to No.1, No.2, No.3 and Select respectively.

GRIDSHELL CONSTRUCTION

The Weald and Downland Open-Air Museum in Sussex (Fig. 4.15) illustrates a new approach to timber construction combining modern computing technologies with traditional craftsmanship. The large barn-like construction (50 m long × 12 m wide × 10 m high) is formed from a double skin grid of 35 × 50 mm × 40 m long green oak laths at 1 m centres, which generate an undulating envelope of curved walls blending into three domes. Continuous curvature of the walls and roof are necessary for structural integrity.

Freshly sawn green oak was used as it is supple and easily formed. Initially it was bolted into a flat grid with stainless steel bolts. The supporting scaffolding was

Fig. 4.15 Gridshell construction – Weald and Downland Open-Air Museum. Architects: Edward Cullinan Architects. Photograph: Courtesy of Edward Cullinan Architects

then gradually removed, allowing the construction to settle into its design form, which was finally fixed around the perimeter. Once the correct form is established, the geometry is locked to ensure stability against wind and snow loading. The construction requires no interior supports, which would have inhibited the free use of the internal space. The structure was glazed with polycarbonate clerestory panels and clad in western red cedar vertical boarding. Gridshell construction has been used previously in Germany and Japan, but the Weald and Downland Museum by Edward Cullinan Architects is the first of its type within the UK.

TIMBER PILES AND FOUNDATIONS

Timber piled foundations have been used for many centuries and have a good record of durability. The city of Venice is largely built on timber piles and their use as an alternative construction system is current within North America for the foundations of bridges and other significant structures. The use of timber pile foundations, in appropriate ground conditions, offers an economical alternative to concrete, with the environmental advantage of creating carbon dumps to reduce global warming.

Historically, a range of softwoods and hardwoods has been used for timber piles, but in the UK Douglas fir, up to 500 mm square and 12–15 m long, is a standard material (BS 8004: 1986). Other suitable timbers are treated Scots pine or larch, oak in non salt water soils, elm, beech and sycamore. Untreated timber below the water table is virtually immune to decay, but it is at risk from biological degradation above this level. It is therefore appropriate to treat timber with preservatives if it is to be used above the water table. Preservative-treated timber piles, cut off below ground level and capped with concrete, should have a service life of 100 years.

TIMBER POLE CONSTRUCTION

Forest thinnings, which are too small to be converted into rectangular standard sections for construction, have the potential to be used directly for certain low-technology forms of building. Currently much of this material is used for paper and particleboard production or burnt as firewood.

The advantage of this material is that is a renewable resource with a relatively short production cycle and rapid carbon dioxide sequestration. Timber poles are naturally tapered, but the effects on mechanical properties of defects such as knots and sloping grain, which are significant in converted timber, are virtually eliminated. Also, as little machining is required energy and labour costs are low.

Forest thinnings up to 200 mm in diameter are generally available, but typical small-scale construction usually requires poles in the 50 mm to 150 mm range, with lengths of between 3 m and 15 m. Figure 4.16 illustrates an experimental building at Hooke Park using Norway spruce timber pole construction.

SOFTWOOD CLADDING

Western red cedar has long been the preferred timber for external timber cladding because of its durability and warm colour. However, recently the popularity of larch and Douglas fir as softwood cladding has increased as greater emphasis is placed on the use of renewable resources from sustainable forests. Large quantities of these materials, which are classified as moderately durable to decay, will become available as

Fig. 4.16 Timber pole construction – Hooke Park, Dorset. Lodge (Edward Cullinan Architects and Buro Happold), and workshop interior (Frei Otto, Ahrends Burton & Koralek and Buro Happold). Photographs: Courtesy of Allan Glennie

plantation-grown stocks reach maturity. Both larch and Douglas fir are both more resistant to impact damage than western red cedar, and for cladding purposes should not need additional preservative treatment. The timbers have been used successfully for school and health centre buildings both as vertical and horizontal cladding. As the timbers are acidic, all fixings must be in corrosion-resistant materials; also some resin bleed can be expected which will penetrate any applied surfaces finishes.

HARDWOOD FLOORING

Hardwood flooring has a proven track record for durability and aesthetic impact. Both solid timber and plywood laminates with a 4 mm hardwood-wearing layer are commercially available. The standard timbers are the European oak, beech, birch, ash, chestnut, walnut and maple, but additionally some imported hardwoods with darker grain colours are available and interesting effects are produced with bamboo. Frequently the timbers are offered with minimal knots and uniform graining or as *rustic* with knots and a larger variation of colour. Laminates are usually prefinished but solid timber may be sealed with oil or lacquer after installation on site.

JOINERY TIMBER

The term *joinery* applies to the assembly of worked timber and timber panel products, using timber which has been planed to a smooth finish. By contrast, carpentry refers to the assembly of the structural carcase of a building usually with rough sawn timbers. Joinery work, including the production of windows, doors, staircases, fitted furniture, panelling and mouldings, requires timber that is dimensionally stable, appropriately durable with acceptable gluing properties, and which can be machined well to a good finish. Joinery grade timber is categorised into five quality classes (Table 4.8) according to number and size of natural defects, particularly knots. These classes are sub-divided into two surface categories – visible and concealed – according to whether the timber is to be visible in use. (BRE Digest 407: 1995 and BS EN 942: 1996 list the softwoods and hardwoods suitable for joinery.) Softwood flooring, cladding and profiled boards should not be specified as joinery.

DETERIORATION OF TIMBER

The major agencies causing the deterioration of timber in construction are weathering, fungi, insects and fire. The natural durability of timber is defined into five categories in relation to the resistance of the heartwood to wood-decaying fungi (BS EN 350–1: 1994).

Natural durability of timber:

Class 1	Very durable
Class 2	Durable
Class 3	Moderately durable
Class 4	Slightly durable
Class 5	Not durable (perishable)

Weathering

On prolonged exposure to sunlight, wind and rain, external timbers gradually lose their natural colours and turn grey. Sunlight and oxygen break down some of the cellulose and lignin into water-soluble materials which are then leached out of the surface leaving it grey and denatured. Moisture movements, associated with repeated wetting and drying cycles, raise the surface grain, open up surface checks and cracks and increase the risk of subsequent fungal decay. Providing the weathering is superficial, the original appearance of the timber can be recovered by removing the denatured surface.

Table 4.8 Classes of timber for joinery use and maximum knot sizes (BS EN 942: 1996)

Class	Visible faces					Concealed faces
	J2	J10	J30	J40	J50	
Maximum knot size	2 mm	10 mm	30 mm	40 mm	50 mm	These knots are all permitted
Maximum percentage of width of finished piece		30%	30%	40%	50%	

Note:
The standard also refers to shakes, resin pockets, bark, discoloured sapwood, pith and Ambrosia beetle damage.

Fungal attack

Fungi are simple plants, which unlike green plants, cannot synthesise chlorophyll, and therefore must obtain their nutrients by metabolising organic material, breaking it down into soluble forms for absorption into their own system. For growth they need oxygen and a supply of food and water, a minimum moisture content of 20% being necessary for growth in timber. The optimum temperature for growth is different for the various species of fungi, but usually within the range 20–30°C. Little growth takes place below 5°C and fungi will be killed by prolonged heating to 40°C. Some timbers, particularly the heartwoods of certain hardwoods are resistant to attack because their minor constituents or *extractives* are poisonous to fungi.

All fungi have a similar life-cycle (Fig. 4.17), commencing with the microscopic spores which are always present in quantity in the air. Under favourable conditions, spores within the surface cracks of timber will germinate and produce fine filaments or hyphae, which feed on the cellulose of the timber. The hyphae branch and grow through the timber cells feeding on both the walls and their contents. With increasing colonisation of the timber, the fine hyphae combine to produce a white matrix or mycelium, which is then visible to the eye. After a period of growth, the mycelium at the surface produces fruiting bodies which generate many thousands of spores to continue the life-cycle. The spores, which are less than 10 microns in size, are readily distributed by air movement.

Moulds and stains

Moulds and stains are fungi that metabolise only the starch and sugar food reserves stored within the timber cells; therefore sapwoods are generally more vulnerable than heartwoods, since during the conversion of sapwood to heartwood the stored food is removed. Generally, there is little loss of strength associated by such an attack, although one variety, *blue-stain* aesthetically degrades large quantities of timber and its presence may indicate conditions for incipient wood-rotting fungal attack. This is best prevented by kiln drying to quickly reduce the surface moisture content, unless infection has already occurred within the forest. Generally softwoods are more susceptible to attack than hardwoods. However, the light-coloured hardwoods ramin, obeche and jelutong are sometimes affected.

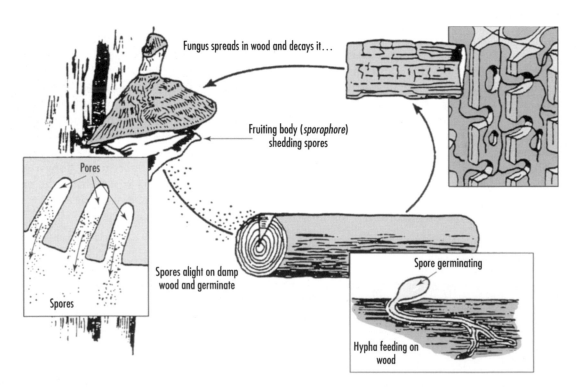

Fig. 4.17 Fungus life-cycle

Wet and dry rots

The name *dry rot*, attributed to one variety of fungus, is a misnomer, as all fungal growths require damp conditions before they become active. Destructive fungi can be categorised as soft, brown, or white rots.

Soft rots, which belong to a group of micro-fungi, are restricted to very wet conditions such as timbers buried in the ground and are therefore not experienced within normal construction. They are usually found only within the timber surface, which becomes softened when wet and powdery when dry.

The brown rots preferentially consume the cellulose within timber, leaving more of the lignin, tannin and other coloured extractives; thus the timber becomes progressively darker. In contrast, the white rots consume all the constituents of the cells, so the timber becomes lighter in colour as the attack proceeds.

A major cause of deterioration of timber within buildings is *Serpula lacrymans*, the so-called *dry rot*. Under damp conditions, above 20% moisture content, the mycelium forms cotton-wool-like masses over the surface of the timber, which becomes wet and slimy. The mycelium strands, up to 20 mm in diameter, can grow through brickwork and past inert materials to infect otherwise dry timber. Under drier conditions the mycelium forms a grey-white layer over the timber, with patches of bright yellow and occasionally lilac. The fruiting bodies, or fructifications, are plate-like forms which disperse the rust-red spores. In some circumstances the fruit bodies may be the first signs of attack by dry rot. After an attack by dry rot the timber breaks up both along and across the grain into cube-shaped pieces, becoming dry and friable, hence the name *dry rot*.

Wet rot or cellar fungus (*Coniophora puteana*) is the most common cause of timber decay within buildings in the UK. It requires a higher moisture content than dry rot (40–50%) and is therefore frequently associated with water ingress due to leaks or condensation. The decayed timber is darkened and tends to crack mainly along the grain. The thin individual strands or hyphae are brown or black, and the fruit bodies, rarely seen, are olive green in colour. Frequently the decay is internal without significantly affecting the exposed faces of the timber.

Phellinus contiguus (*Poria contigua*) causes decay to external softwood joinery, particularly window frames, causing the timber to decompose into fibrous lengths. Another variety, *Phellinus megaloporus*, is known to attack oak timbers ultimately leaving a white mass.

Figure 4.18 illustrates the relative vulnerability of sapwood compared to the naturally more resistant darker heartwood, which has been partially protected from rot by secondary thickening and the inclusion of extractives.

Insect attack

Insect attack on timber within the UK is limited to a small number of species, and tends to be less serious than fungal attack. This is the reverse of the situation in hotter climates where termites and other insects can cause catastrophic damage, although the recent unintentional importation and subsequent establishment of subterranean termites in North Devon shows that this species may pose a future threat to UK buildings. The main damage by insects within the UK comes from beetles, which during their larval stage bore through the timber, mainly within the sapwood, causing loss of mechanical strength. For other species, such as the pinhole borers (*Platypus cylindricus*), the adult beetle bores into the timber to introduce a fungus on which the larvae live. The elm bark beetle (*Scolytus scolytus*) was responsible for the spread of *Dutch elm disease* in the 1970s. The larvae tunnelled

Fig. 4.18 Deterioration of sapwood timber illustrating the relative durability of heartwood over sapwood

under the bark, within the bast and cambium layers, preventing growth and spreading the destructive fungus, which eventually killed large numbers of the trees across the UK.

The typical life-cycle (Fig. 4.19) commences with eggs laid by the adult beetle in cracks or crevices of timber. The eggs develop into the larvae which tunnel through the timber leaving behind their powdery waste or frass. Dependent on the species, the tunnelling process can continue for up to several years before the development of a pupa close to the surface of the timber, prior to the emergence of the fully developed adult beetle, which eats it way out leaving the characteristic flight hole. The insects which attack well-seasoned timber within the UK are the common furniture beetle, death-watch beetle, the house longhorn beetle and the powder-post beetle. Wood-boring weevils only attack timber that has been previously affected by fungal decay (Fig. 4.20).

Common furniture beetle
The common furniture beetle (*Anobium punctatum*) attacks mainly the sapwood of both hardwoods and softwoods. It can be responsible for structural damage in the cases of severe attack, and is thought to be present in up to 20% of all buildings within the UK. The brown beetle is 3–5 mm long and leaves flight holes of approximately 2 mm in diameter. Both waterborne and organic-solvent insecticides offer effective treatments.

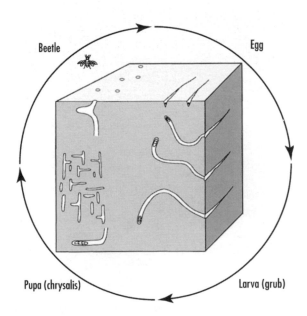

Fig. 4.19 Life-cycle of wood-boring beetles

Death-watch beetle

The death-watch beetle (*Xestobium rufovillosum*) characteristically attacks old hardwoods, particularly oak, and is therefore responsible for considerable damage to historic buildings. Attack is normally on the sapwood, but heartwood softened by moisture and fungal decay will attract infestation; adjacent softwood may also be affected. The brown beetle is approximately 8 mm long and leaves a flight hole of 3 mm diameter. Remedial measures should include the eradication of damp and the application of organic-solvent insecticides.

House longhorn beetle

The house longhorn beetle (*Hylotrupes bajulus*) is a serious pest in some parts of southern England, particularly in Surrey, and it is referenced in the Building Regulations 2000 – Approved Document to support regulation 7: 1999 (amended 2000). House longhorn beetle can infest and cause serious structural damage to the sapwood of seasoned softwood roof timbers. With an average life-cycle of six years and a larva that is up to 35 mm long, this beetle can cause serious damage before evidence of the infestation is observed. The affected timbers bulge where tunnelling occurs just below the surface, and the eventual flight holes of the black beetle are oval and up to 10 mm across. Where sufficient serviceable timber remains, remedial treatment with organic solvent or paste formulations is appropriate.

Powder-post beetle

The powder-post beetle (*Lyctus brunneus*) attacks the sapwood of certain hardwoods, particularly oak and ash. The sapwood of large-pored tropical hardwoods, such as ramin and obeche may also be affected. Only timbers with sufficient starch content within the sapwood are vulnerable to attack as the larvae feed on starch rather than the cell walls. The eggs are laid by the adult female beetle into the vessels, which are the characteristically large cells within hardwoods. Timbers with low starch content or fine vessels are immune, and the extended soaking of vulnerable timbers in water can reduce the risk of attack, but owing to the long timescale involved, this is not commercially viable. The 4 mm reddish brown beetle leaves a flight hole of approximately 1.5 mm diameter. Timbers are attacked only until all the sapwood is consumed, so in older buildings damage is usually extinct. In new

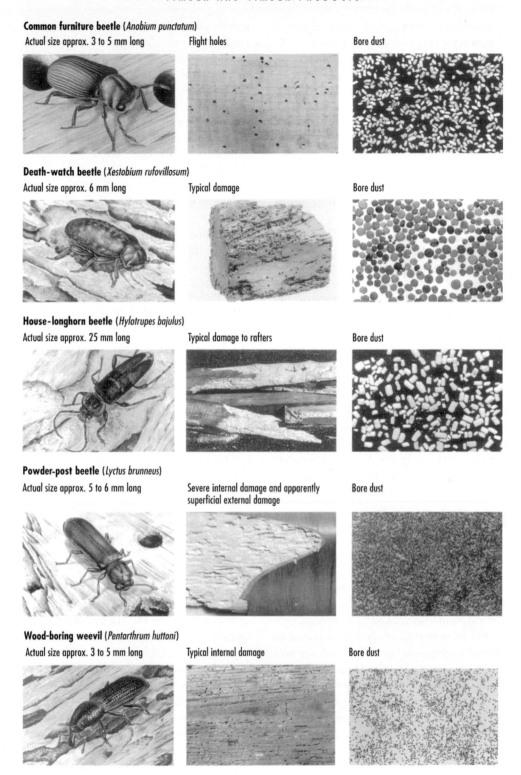

Common furniture beetle (*Anobium punctatum*)

Actual size approx. 3 to 5 mm long

Flight holes

Bore dust

Death-watch beetle (*Xestobium rufovillosum*)

Actual size approx. 6 mm long

Typical damage

Bore dust

House-longhorn beetle (*Hylotrupes bajulus*)

Actual size approx. 25 mm long

Typical damage to rafters

Bore dust

Powder-post beetle (*Lyctus brunneus*)

Actual size approx. 5 to 6 mm long

Severe internal damage and apparently superficial external damage

Bore dust

Wood-boring weevil (*Pentarthrum huttoni*)

Actual size approx. 3 to 5 mm long

Typical internal damage

Bore dust

Fig. 4.20 Wood-boring beetles common within the UK

buildings, coatings of paint or varnish make treatment impractical so replacement is the usual option.

Ambrosia beetle

A large number of ambrosia beetle species attack freshly felled hardwood and softwood logs, in both temperate and tropical regions. A high moisture content of over 35% is necessary for this beetle attack which is therefore eliminated on seasoning. The circular pinholes range from 0.5 to 3 mm depending on the particular beetle species and the tunnelling is across the grain of the timber.

Wood-boring weevils

Wood-boring weevils attack only timber previously softened by fungal decay. The most common weevil (*Pentarthrum huttoni*) produces damage similar in appearance to the common furniture beetle, but removal of decayed timber will eliminate the secondary infestation.

Termites

Termites are social insects, similar in size to ants (4–5 mm), which live in colonies containing millions of individuals. Most species of termites are beneficial to nature in breaking down organic matter, but a few varieties can cause catastrophic damage to buildings. The subterranean termite family *Rhinotermitidae* is the major cause of building damage, with the genus *Reticulitermes* being a significant threat in the UK. The species *R. santonensis* is already widely distributed in Europe and the species *R. lucifugus* is now established at Saunton, North Devon. The primary source of food for termites is the cellulose in wood, particularly from structural softwood timbers near to ground level or in partially decayed timber. Timber may be only slightly affected or it may be heavily excavated leaving only the surface and any protective coatings. In areas at risk of termites, termiticide protection may be appropriate, but where attack has already occurred, specialist advice from the Forestry Commission is essential. Colony elimination by physical, chemical or biological techniques is a slow process taking months or even years to complete.

Preservation of timber

Wood preservatives contain pesticides in the form of insecticides and fungicides. Their use is therefore strictly controlled to limit unnecessary or accidental environmental damage. Preservative treatments should involve only materials in current approval by the Control of Pesticides Regulations (1986 as amended 1997) and they should be used in accordance with the COSHH (Control of Substances Hazardous to Health) regulations, the manufacturers' instructions and by operatives wearing appropriate protective clothing. Timber treatments may be divided into the application of preservatives to new timber, and remedial treatments used to eradicate or reduce an existing problem.

Preservative treatments for new timber

A wide range of effective timber preservatives is commercially available for use under controlled industrial conditions. However, within the current climate of increasing health and environmental awareness, the drive towards more ecologically friendly products is leading to considerable changes within industrial timber preservation processes. Already creosote is only available for industrial use and may not be used within buildings or on outside furniture.

Two industrial processes involve the use of vacuum and pressure impregnation, a third process involves dip-diffusion. Chemicals are either water-borne, solvent based or micro-emulsions.

The double-vacuum process, using organic solvent-borne preservatives, is suitable for low- and medium-risk timber such as external joinery. The timber, at less than 28% moisture content, is loaded into a low-pressure vessel which is evacuated to extract the air from within the timber. The vessel is flooded with preservative and a low positive pressure applied for between several minutes and one hour depending upon the permeability of the timber. The vessel is then drained and evacuated to remove excess preservative from the timber surface. Formulations consist of either fungicides or insecticides, or both, dissolved in volatile organic solvents. The solvents penetrate well into the timber but have a strong odour and are highly flammable. Pentachlorophenol (PCP) and tri-butyl tin oxide (TBTO) are the standard fungicides with lindane (gamma hexachlorocyclohexane) used as the insecticide. A water repellent may also be incorporated into the preservative formulation. Organic-solvent preservatives will eventually be limited in use to timbers where it is critical that the dimensions are not affected by the preservative treatment. To reduce the environmental effects of volatile organic compound (VOC) emissions, organic solvents are being replaced by micro-emulsion formulations, with significantly reduced organic-solvent content.

The pressure/vacuum process is similar to the double-vacuum process, but uses water-borne preservatives and the application of high pressure within a pressure vessel to ensure deep penetration. The standard water-borne wood preservative is a copper/chromium/arsenate (CCA) mixture. The copper salt acts as the fungicide, the arsenate component as the insecticide and the sodium dichromate fixes the active ingredients within the timber, preventing their loss through leaching. Timber treated with proprietary products such as *Tanalith* is coloured slightly green but can be directly painted. However, under new legislation CCA preservative cannot be used for timber to be used in residential or domestic structures where repeated skin contact is possible. Copper/chromium/boron (CCB), which is good for hardwoods, has been used as an alternative to CCA, but is also likely to be restricted in its use in future. Fluor/chrome/ arsenate/ phenol (FCAP) is effective against attack by termites and house longhorn beetle and sodium pe tachlorophenate prevents attack by *Serpula lacrymans*. Creosote (BS 144: 1997) may also be applied to timber by this high pressure process. Timbers to be built into high risk situations, such as industrial roofs, frames and floors, and timbers embedded in masonry, sole plates, sarking boards, tile battens, etc. should be treated by this process. Long-term field tests (BRE IP 14/01) have confirmed that CCA and creosote both remain very effective preservative treatments for softwoods in contact with groundwater.

The use of pentachlorophenol (PCP) and tri-butyl tin oxide (TBTO) is normally restricted to controlled industrial use where environmental hazards are minimised. With the trend away from using environmentally hazardous wood preservatives, organic biocides are gradually replacing chromium and arsenic in the CCA formulations.

Dip-diffusion treatments involve the immersion or spraying of the freshly sawn green timber using boron derivatives (disodium octaborate tetrahydrate). Two or three immersions are used to ensure complete coverage of all faces of the timber, and larger sections require a second treatment by spraying or immersion. After treatment the timber should then be stored for an appropriate period to allow the diffusion of the preservative into the timber to the required depth of penetration.

Remedial treatment for timber

Remedial treatments to existing buildings should be limited to those strictly necessary to deal with the fungal or insect attack. The use of combined fungicides and insecticides is not advised when the attack is by one agent only. Within the UK, much timber decay is caused by building failures. As fungal decay can only occur in damp conditions, the first remedial measure must be to restore dry conditions. This should remove the necessity for frequent chemical applications.

The orthodox approach to the eradication of fungal and wood-boring beetle attack involves the removal of severely decayed or affected timber, followed by appropriate preservative treatment to the remaining timber. For fungal attack, 300 mm of apparently sound timber should be removed beyond the last visible sign of decay, and the adjoining timbers treated with fungicide. For wood-boring beetle, unless the infestation is widespread, preservative treatment should be applied only up to 300 mm beyond the visible holes. Organic-solvent fungicides and insecticides applied by brush or spray offer some protection from further attack, but applications of pastes which deliver higher quantities of the active ingredients are usually more effective. Insecticidal smoke treatments need to be repeated annually as they are only effective against emerging beetles, but they may be useful in situations where brushing or spraying is impracticable.

The environmental approach to the eradication of fungal decay relies heavily on the removal of the causes of damp. On the basis that fungal attack will cease when timber is at less than 20% moisture content, increased ventilation and the rectification of building defects should prevent further attack. Only seriously affected timbers need to be replaced, and affected masonry sterilised; however, continual monitoring is required as dormant fungal decay will become active if the timber moisture content rises again above 20%. Rothounds (specially trained sniffer dogs), fibre optics and chemical detection systems offer non-destructive methods for locating active dry rot before it becomes visible.

Pesticides used professionally for remedial treatment include permethrin and cypermethrin as insecticides, with boron esters, copper naphthenate and acypetacs zinc as fungicides. In remedial work, copper-based products should not be used near aluminium and, during permethrin treatment, sarking felt, electrical wiring and roof insulation should be protected. These pesticides are currently considered acceptable for treatment in areas inhabited by bats, which are a protected species under the Wildlife and Countryside Act 1981.

Guidance on timber treatments

Under the European Standards EN 351–1: 1996 and BS EN 335–1: 1992, timber preservative treatments against wood-destroying organisms are categorised by performance standards and not to the individual chemical preservative treatments. The standards define wood preservatives according to their effectiveness in a range of environmental conditions.

Hazard classes of wood locations against biological attack (BS EN 335: 1992):

Hazard Class 1
Above ground, covered and permanently dry – moisture content less than 20%.
Hazard Class 2
Above ground, covered but with risk of high humidity and occasional wetting – moisture content occasionally over 20%.
Hazard Class 3
Above ground, not covered and frequently wet – moisture content frequently over 20%.
Hazard Class 4
Ground- or fresh water contact, permanently wet – moisture content permanently over 20%.
Hazard Class 5
In salt water, permanently wet – moisture content permanently over 20%.

The level of treatment required to give the necessary performance is classified according to the depth of penetration into the timber and by retention or loading within the defined zone of the timber. The depth of penetration is defined by nine classes (P1 to P9) of increasing zones of preservative retention. As full preservative treatment is not appropriate in all cases, a range of service factors (A to D) define the level of safety and economic considerations appropriate to preservative treatment. These are listed in the British Standard BS 8417: 2003 (Table 4.9).

The British Standard (BS 8417: 2003) recommends the durability classes of timber (BS EN 350–2: 1994) which can be used without preservative treatment in relation to their use and environmental factors. For example, dry roof timbers in an area not affected by house longhorn beetle:

Table 4.9 Service factors for preservative treatment of timber to BS 8417: 2003

Service factors	Need for treatment	Safety and economic considerations
A	Unnecessary	Negligible
B	Optional	Remedial action is easy Preservation is an insurance against cost of repairs
C	Desirable	Remedial action is expensive and difficult
D	Essential	Consequences of structural collapse would be serious

Building component – (e.g. dry roof timbers in a non-*Hylotrupes* area)
Hazard class of wood against biological attack (e.g. 1 for dry roof timbers)
Service factor – safety and economic considerations (e.g. B for dry roof timbers)
Desired Service life – 15, 30 or 60 years (e.g. 60 years for dry roof timbers)
Durability class 5 – non-durable timber is appropriate.

The British Standard (BS 8417: 2003) also gives guidance on the types of preservative treatment appropriate for timbers to be used in more severe situations. For example, occasionally wet roof timbers in a house longhorn beetle affected location:

Building component – roof timbers (with risk of wetting in an *Hylotrupes* area)
Timber species – e.g. European whitewood
Treatment – required for insect hazard and desirable for fungal hazard
Service life required – e.g. 60 years
Preservative type – organic solvent containing active fungicide and insecticide
(The British Standard (BS 8417: 2003) gives guidance on the required depth of penetration and retention of the preservative.)

Fire

Timber is an organic material and therefore combustible. As timber is heated it initially evolves any absorbed water as vapour. By about 230–250°C, decomposition takes place with the production of charcoal, and combustible gases such as carbon monoxide and methane are evolved, which cause the flaming. Finally the charcoal smoulders to carbon dioxide and ash.

However, despite its combustibility timber, particularly in larger sections, performs better in a fire than the equivalent sections of exposed steel or aluminium. Timber has a low thermal conductivity which, combined with the protection afforded by the charred surface material, insulates the interior from rapid rises in temperature and loss of strength. The rate of charring of timber under the standard Fire Resistance Test ranges between 30 mm and 50 mm per hour per surface exposed, according to the timber density (Table 4.10). It is therefore possible to predict the fire resistance of any timber component using the British Standards (BS 5268–4.1: 1978 and 5268–4.2: 1990). Additionally, as all timbers have a low coefficient of expansion, timber beams will not push over masonry walls as sometimes occurs with steel beams and trusses during fires. Solid timber with a minimum density of 350 kg/m^3 and a thickness equal to or greater than 10 mm may be assigned to Euroclass D with respect to fire without testing.

Flame retardants
Within a fire, volatile combustible components are evolved from the surface of the timber and these cause the flaming. The two alternative types of treatment which may be used on timber to reduce this effect are impregnation or the application of surface coatings.

Impregnation involves forcing inorganic chemicals, which on heating evolve non-combustible gases, into the timber under high pressure and then vacuum. Timbers should be machined to their final dimensions before treatment. For interior use, typical compositions include water-borne inorganic salts such as ammonium sulfate or phosphate with sodium borate

or zinc chloride. As these materials are hygroscopic, the timber should not be used in areas of high humidity. For exterior use, a leach-resistant flame-retardant material based on an organo-phosphate is used as this is heat fixed by polymerisation within the timber.

Surface treatments, which cause the evolution of non-combustible gases in fire, include antimony trioxide flame-retardant paints which are suitable for both interior and exterior use. Intumescent coatings, which swell up and char in fire, are suitable for most environments if overcoating is applied. However, the protection afforded by surface treatments may be negated by unsuitable covering or removal by redecoration.

Untreated timber, which is normally Class 3 Spread of Flame to BS 476–7: 1997, can be improved to Class 1 by surface treatments. Class 0 can only be achieved by heavy impregnation, a combination of impregnation and surface coating, or certain very specific surface-coating treatments.

European fire classification of construction materials
The European fire classification of construction products and building elements is defined in BS EN 13501–1: 2002. All construction products except floorings may be classified to one of the following seven classes: A1, A2, B, C, D, E or F based on performance. Class A1 represents products which do not contribute to the fire load even within a fully developed fire. Class A2 products do not significantly contribute to the fire load and growth, while the other classes reflect decreasingly stringent fire performance criteria down to Class E products which can resist a small flame for a short period of time without substantial flame spread. Class F products are outside the other classes or have no determined performance. In addition to fire load factors, the standard includes classifications relating to smoke production (s1, s2 and s3, where s1 is the most stringent criterion) and flaming droplets (d0, d1 and d2, where d0 indicates no flaming droplets or particles). For floorings the seven classes are A1$_{FL}$, A2$_{FL}$, B$_{FL}$, C$_{FL}$, D$_{FL}$, E$_{FL}$, and F$_{FL}$ with sub-classifications for smoke production (s1, s2 and s3). Correlation between the UK and European classes is not exact and materials require testing to the European Standard before assignment to a particular class to BS EN 13501–1: 2002.

The Euroclass fire performance rating under the conditions specified in BS EN 13986: 2004 for 12 mm untreated solid wood panels of minimum density

Table 4.10 Rate of burning for timber from each exposed face

Rate of Burning	Timber	Typical Density (kg/m^3)
30 mm per hour	Hardwoods e.g. keruing, teak	over 650
40 mm per hour	Structural softwoods e.g. European redwood/ European whitewood	450–550
50 mm per hour	Western red cedar	380

400 kg/m^3 is Class D–s2, d0 for non-floor use and Class D_{FL}–s1 for floorings.

TIMBER CONNECTORS

A range of steel timber connectors is commercially available. Trussed rafters, which account for a large market, are usually constructed with galvanised steel nail plates (Fig. 4.21). Plates are hydraulically pressed into both sides of the timbers to be connected at the butt joints. Other types of connector include single- and double-sided circular toothed plates fixed with a central bolt, also joist hangers and roof truss clips. Laminated timber beams are usually fixed with purpose-made bolted shoes or plates.

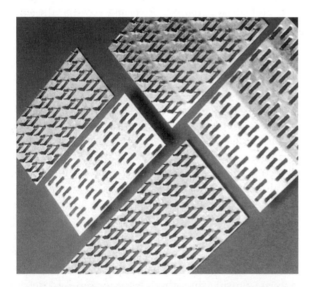

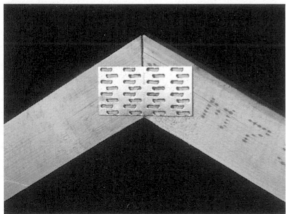

Fig. 4.21 Nail plates

Timber products

A wide range of products is manufactured from wood material ranging in size from small timber sections and thin laminates through chips and shavings down to wood fibres. The physical properties of the materials produced reflect a combination of the sub-division of the wood, the addition of any bonding material and the manufacturing process. The physical properties then determine the products' appropriate uses within the building industry. Many of the products are manufactured from small timber sections or timber by-products from the conversion of solid timber which otherwise would be wasted. Compressed straw slabs and thatch are additionally included in this section.

The product range includes:

- laminated timber;
- structural insulated panels;
- laminated veneer lumber;
- plywood;
- blockboard and laminboard;
- particleboard;
- fibreboard;
- wood wool slabs;
- compressed straw slabs;
- thatch;
- shingles;
- 'Steko' blocks;
- flexible veneers.

Within the European Union, whenever wood-based panels are used in construction, compliance with the Construction Products Directive must be demonstrated. This may be achieved by adherence to European Harmonised Standard for wood-based panels EN 13986 (BS EN 13986: 2004 in the UK). This requires that products used in construction comply with its specifications and also to the additional performance-based criteria within the various EN standards listed for each specific material. Most European countries now use the CE mark on boards and panels to show compliance with this harmonised standard.

LAMINATED TIMBER

Manufacture

Large solid-timber sections are limited by the availability of appropriate lumber; in addition, their calculated strength must be based on the weakest part of the variable material. Laminated timber sections overcome both of these difficulties and offer additional opportunities to the designer. Laminated timber is manufactured by curing within a jig, layers of accurately cut smaller timber sections which are continuously glued together with a resin adhesive. Laminates may be vertically or horizontally orientated. The use of strength-graded timber and the staggering of individual scarf or finger joints ensures uniformity of strength within the product; although, under BS 5268–2: 2002 and BS EN 387: 2001, large finger joints through the whole section of a *glulam* member are permissible. The manufacturing process ensures greater dimensional stability and less visual defects than in comparable solid timber sections. Laminated timber may be homogeneous with all laminates of the same strength class of timber or combined, in which lower strength class laminates are used for the centre of the units. Table 4.11 gives the European strength classes to BS EN 1194: 1999 for the two alternatives. Laminated timber manufactured from spruce or pine and phenolic or urea/melamine formaldehyde resins would normally achieve a Euroclass fire performance rating, under the conditions specified in BS EN 13986:2004 of Class D, subject to testing.

Forms

Sections can be manufactured up to any transportable size, typically 30 m, although spans over 50 m are possible. Standard size straight sections (315 × 65 and 90 mm; 405 × 90 and 115 mm; and 495 × 115 mm) are stock items, but common sizes range from 180 × 65

Table 4.11 Strength classes to BS EN 1194: 1999 for homogeneous and combined glulam

Glulam strength classes	GL24	GL28	GL32	GL36
Homogeneous glulam: Strength class of laminates	C24	C30	C40	
Combined glulam: Strength class of outer laminates	C24	C30	C40	
Strength class of inner laminates	C18	C24	C30	

mm to 1035 × 215 mm. Sections may be manufactured to order, to any uniform or non-uniform linear or curved form. Figure 4.22 illustrates typical laminated-timber arches, columns and portal frames as generators of structural forms. The aesthetic properties of laminated timber can be enhanced by the application of suitable interior or exterior timber finishes. The majority of laminated-timber structures are manufactured from softwoods such as European redwood or whitewood, although the rib members within the roof structure of the Thames Flood Barrier (Fig. 4.23) were manufactured from the West African hardwood, iroko. Steel fixing devices and joints may be visually expressed (Fig. 4.22) or almost unseen by the use of concealed bolted steel plates. Laminated timber performs predictably under fire conditions with a charring rate of 40 mm per hour as defined within BS 5268:4–1: 1978. Preservative treatments are necessary when the material is to be used under conditions in which the moisture content is likely to exceed 20%. The three service classes of glulam structures relate the environmental conditions.

Service classes for glulam:

Service Class 1 Internal conditions with heating and protection from damp (typical moisture content <12%).
Service Class 2 Protected, but unheated conditions (typical moisture content <20%).
Service Class 3 Exposed to the weather (typical moisture content >20%).

STRUCTURAL INSULATED PANELS

Structural insulated panels (SIPs) are prefabricated lightweight building components, used for load-bearing internal and external walls and roofs. Unlike cladding sandwich panels, structural insulated panels can support considerable vertical and horizontal loads without internal studding. They are manufactured from two high-density face layers separated by a lightweight insulating core. The three layers are strongly bonded together to ensure that the composite acts as a single structural unit. The outer layers of oriented strand board (OSB), cement-bonded particleboard or gypsum-based products are typically 8 mm to 15 mm thick. The core is composed of a rigid cellular foam, such as polyurethane (PU), polyisocyanurate (PUR), phenolic foam (PF), expanded (EPS) or extruded (XPS) polystyrene, giving an overall unit thickness

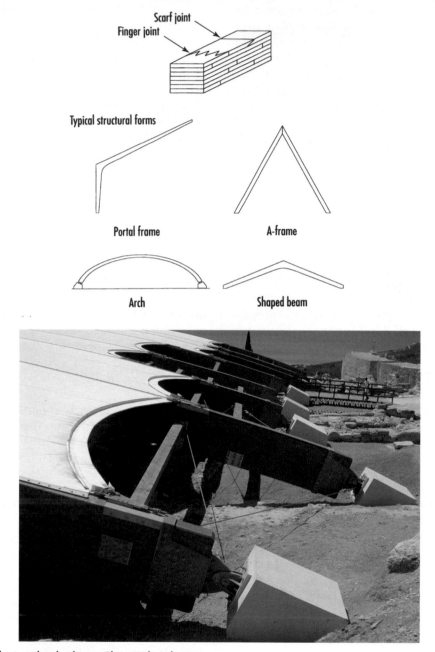

Fig. 4.22 Glued laminated-timber beams. Photograph: Arthur Lyons

between 70 mm and 250 mm. The structural performance is predominantly influenced by the thickness and physical properties of the outer layers, and the thermal performance is largely determined by the width and insulating characteristics of the core material.

Structural insulated panels (typically 1.2 m wide by 2.4 m high) offer a thermally efficient and air-tight form of construction, which is rapidly erected on site.

Jointing between panels is usually some form of tongue and groove system. Sound reduction for separating walls is typically 58 dB depending upon construction details. External cladding may be brickwork, wooden panelling or rendering with plasterboard as the standard internal finish. Table 4.12 gives typical thermal performance data for structural insulated panels.

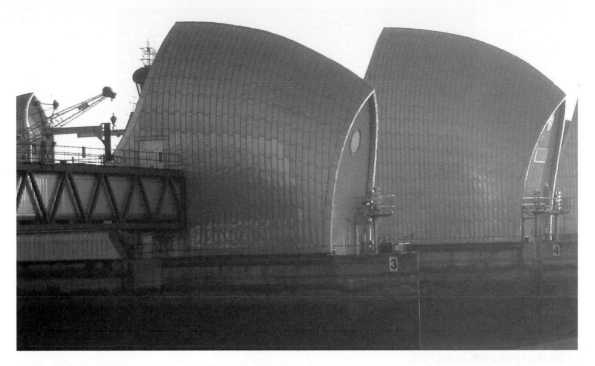

Fig. 4.23 Thames Barrage, London. Photograph: Arthur Lyons

Table 4.12 Thermal performance data for Structural Insulated Panels (SIPs)

Core material	Face material	Panel thickness (mm)	Thermal performance (W/m²K)
Polyurethane foam	cement bonded particleboard	86	0.28
Polyurethane foam	oriented strand board	100	0.23
Polyisocyanurate foam	oriented strand board	140	0.22

The thermal performance data (U-values) are typical for the listed SIPs when constructed with a brick outer leaf and 50 mm clear vented cavity.

LAMINATED VENEER LUMBER

Laminated veneer lumber (LVL) (Fig. 4.24), also known as *microlam*, is more economical than laminated timber as there is little waste in the production process. It is manufactured to three grades by laminating timber strands with polyurethane resin under heat and pressure. In one process, logs are cut into flat timber strands 300 mm long; these are then treated with resin, aligned and hot-pressed into billets of reconstituted wood. In the other processes, 3-mm-thick timber strands or sheets of veneer are coated with waterproof adhesive and bundled together with the grain parallel. The strands or veneers are pressed together and microwave cured to produce structural timber billets or sheets up to 26 m long. The versatile material is suitable for use in columns, beams, purlins and trusses, and can be machined as solid timber (Fig. 4.25). I-section joists, with LVL flanges and web, are suitable for flat and pitched roofs and floor construction. Untreated LVL has a Class 3 surface spread of flame classification (BS 476–7: 1997). Three grades of laminated veneer lumber are classified according to their serviceability in dry and exposed conditions.

Grades of laminated veneer lumber:

Purpose/ Loading	Environmental Conditions	Type
Load-bearing	dry (hazard class 1)	LVL/1
Load-bearing	humid (hazard class 2)	LVL/2
Load-bearing— exterior conditions	exterior (hazard class 3)	LVL/3
(subject to testing or appropriate finish)		

Fig. 4.24 Laminated veneer lumber (LVL)

Fig. 4.25 Laminated veneer lumber (LVL) construction – Finnforest Office, Boston, Lincolnshire. Architects: Arosuo and Vapaavuori Oy. Photograph: Courtesy of Finnforest

Monocoque structures

Interesting and innovative built forms can be created using LVL (and other timber products) to create flat or curved form monocoque structures. These work on the well-established principles from the motor industry, in which the hard body skin acts in concert with any stiffeners to form the structure. Using this technology, structurally efficient and elegant forms, which may be slender, tapered, flat or curved may be produced. LVL is quickly becoming a significant material to complement the more established products such as plywood, OSB and glulam, particularly because of its availability in very large sections.

PLYWOOD

Manufacture

Plywood is manufactured by laminating a series of thin timber layers, or plies, to the required thickness. The timber log is softened by water or steam treatment and rotated against a full-length knife to peel off a veneer or ply of constant thickness (Fig. 4.26). The ply is then cut to size, dried and coated with adhesive prior to laying up to the required number of layers. Not all the plies are of the same thickness; often, thicker plies of lower-grade material are used in the core. However, the sheets must be balanced about the centre to prevent distortions caused by differential movement. Plies are normally built up with adjacent grain directions at right angles to each other to give uniform strength and reduce overall moisture movement, although with even plywoods, the central pair of plies has parallel grains. The laminate of plies and glue is cured in a hot press, sanded and trimmed to standard dimensions for packaging. Decorative veneers of hardwood or plastic laminate may be applied to one or both faces. Most plywood imported into the UK is made from softwood (largely pine and spruce), from North America and Scandinavia. Smaller quantities of plywood produced from temperate hardwoods are imported from Finland (birch) and Germany (beech), while tropical hardwood products are imported from Indonesia, Malaysia, South America and Africa.

The standard sheet size is 2440 × 1220 mm, with some manufactures producing sheet sizes of up to 3050 × 1525 mm or slightly larger. Sheet thicknesses range from 4 mm to 25 mm for normal construction use, although thinner sheets down to 1.5 mm are available for specialist purposes.

Under the European fire classification of construction materials (BS EN 13501–1: 2002), an untreated plywood panel would normally achieve a class D–s2, d0 rating, excluding its use as flooring when the rating is class D_{FL}–s1 (depending on a minimum thickness of 9 mm, a minimum density of 400 kg/m^3 and fixing to a non-combustible substrate [class A1 or A2] without an air gap. The secondary classifications 's' and 'd' relate to smoke production and flaming droplets).

Grades

Plywood is classified according to its general appearance and physical properties (BS EN 313–1: 1996). The key characteristics are the form of construction, durability, and the nature of the surface. The durability of plywood is largely determined by the bonding class of the adhesive used. This ranges from class 1, to the most durable class 3 (BS EN 314–2: 1993), which can be used externally without delamination, providing that the timber itself is durable or suitably protected against deterioration.

Bonding classes for plywood:

Class 1   Dry conditions (suitable for interior use).
Class 2   Humid conditions (protected external applications, e.g. behind cladding or under roof coverings).

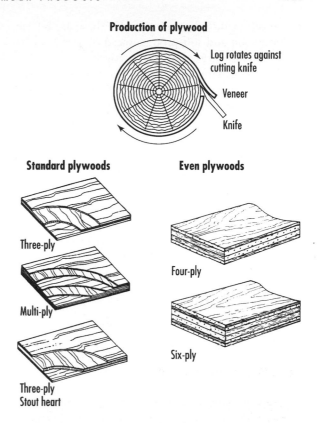

Production of plywood

Log rotates against cutting knife
Veneer
Knife

Standard plywoods

Three-ply

Multi-ply

Three-ply Stout heart

Even plywoods

Four-ply

Six-ply

Fig. 4.26 Manufacture and standard types of plywood

Class 3   Exterior conditions (exposed to weather over sustained periods).

Phenol formaldehyde resins are the most frequently used for the most durable plywoods. Marine plywood (BS 1088–1: 2003) is a combination of a moderately durable timber bonded with phenolic or melamine-formaldehyde resin. The standard class of marine plywood is suitable for regular wetting or permanent exposure to salt or fresh water. The lower grades of plywood are bonded with melamine-urea-formaldehyde or urea-formaldehyde resins. In addition to the grade of adhesive and the durability of the timber itself, the quality of plywood is affected by the number of plies for a particular thickness and the surface condition of the outer plies which range from near perfect, through showing repaired blemishes, to imperfect. Factory-applied treatments to improve timber durability and fire resistance are normally available.

The standard BS EN 635: 1994 describes five classes of allowable defects (E, and I to IV) according to decreasing quality of surface appearance; class E is practically without surface defects. These are related to hardwood and softwood surfaced plywoods in BS EN 635: 1995 Parts 2 and 3 respectively.

The performance specifications for plywood to be used in dry humid or exerior conditions against the criteria of bonding strengthe and durability with respect to biological decay are described in the standard BS EN 636: 2003.

Biological hazard class conditions for the use of plywood:

Class 1   Dry conditions (average moisture content <10%).

Class 2   Humid conditions (average moisture content < 18%).

Class 3   Exterior conditions (average moisture content >18%).

These biological hazard classes correspond to the service classes in BS EN 1995–1–1: 2004.

The standard BS EN 636: 2003 also has a classification system based on the strength and stiffness of plywood based on bending tests. Plywood is assigned to a four-part code specifying bending strength and modulus in both the length and width directions. Plywood sheets should be identified according to their intended application with 'S' for structural and 'G' for general use.

Uses

Considerable quantities of plywood are used by the construction industry because of its strength, versatility and visual properties. The strength of plywood in shear is used in the manufacture of plywood box and I-section beams in which the plywood forms the web. Increased stiffness can be generated by forming the plywood into a sinusoidal web. Plywood box beams can be manufactured to create pitched and arched roof forms as illustrated in Figure 4.21. Stiffened and stressed skin panels, in which plywood and softwood timbers are continuously bonded to act as T- or I-beams, will span greater distances as floor structures than the same depths of traditional softwood joists with nailed boarding. Such structural units can also be used to form pitched roofs or to form folded plate roof structures or barrel vaulting (Fig. 4.27). Plywood of 8–10 mm thickness is frequently used as the sheeting

material in timber-frame construction and for complex roof forms such as domes. The lower-grade material is extensively used as formwork for *in-situ* concrete.

CORE PLYWOOD

The standard core plywood products are blockboard and laminboard. Both are manufactured with a core of usually softwood strips sandwiched between one or two plies (Fig. 4.28). In blockboard the core strips are between 7 mm and 30 mm wide, but in laminboard, the more expensive product, they are below 7 mm in width and continuously glued throughout. As with plywood, the grain directions are perpendicular from layer to layer. Most core plywoods are bonded with urea-formaldehyde adhesives appropriate to interior applications only. The standard sheet size is 2440 × 1220 mm with a thickness range from 12–25 mm, although larger sheets up to 45 mm thick are available. Blockboard may be finished with a wide range of decorative wood, paper or plastic veneers for use in fitted furniture. Variants on the standard products include plywood with phenolic foam, polystyrene or a particleboard core.

PARTICLEBOARDS

Particleboards are defined as panel materials produced under pressure and heat from particles of wood, flax, hemp or other similar lignocellulosic materials. The wood particles may be in the form of flakes, chips, shavings, saw-dust, wafers or strands (BS EN 309: 2005). Boards may be uniform through their thickness or of a muliti-layer structure. Wood particleboard and cement-bonded particleboard are made from wood chips with resin and cement binder respectively. Oriented strand board is manufactured from large wood flakes and is classified in BS EN 300: 1997.

Wood particleboard (chipboard)

Manufacture

Wood particleboard (chipboard) is manufactured from wood waste or forest thinnings, which are converted into wood chips, dried and graded according to size. The chips are coated with adhesive to approximately 8% by weight and then formed into boards

Plywood beams

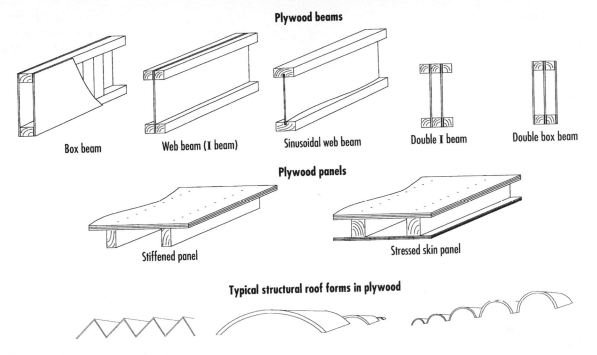

Box beam   Web beam (**I** beam)   Sinusoidal web beam   Double **I** beam   Double box beam

Plywood panels

Stiffened panel   Stressed skin panel

Typical structural roof forms in plywood

Fig. 4.27 Structural uses of plywood

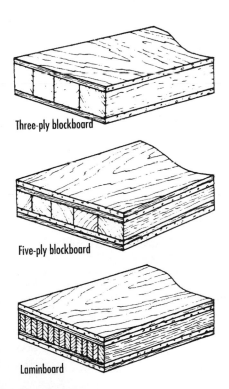

Three-ply blockboard

Five-ply blockboard

Laminboard

Fig. 4.28 Core plywoods

(Fig. 4.29). The woods chips are either formed randomly into boards giving a uniform cross-section or distributed with the coarse material in the centre and the finer chips at the surface to produce a smoother product. The boards are then compressed and cured between the plates of a platen press at 200°C. Boards are finally trimmed, sanded and packed. In the *Mende* process a continuous ribbon of 3–6 mm particleboard is produced by calendering the mix around heated rollers.

The standard sizes are 2440 × 1220 mm, 2750 × 1220 mm, 3050 × 1220 mm and 3660 × 1220 mm, with the most common thicknesses ranging from 12 mm to 38 mm although much larger sheet sizes and thicknesses from 2.5 mm are available.

Extruded particleboard (BS EN 14755: 2005) is manufactured by extruding the mixture of wood chip and resin through a die into a continuous board; however, in this method the wood chips are predominantly orientated at right angles to the board face, thus generating a weaker material. Extruded particleboard is specified within four grades according to its density and whether it is solid or has tube voids.

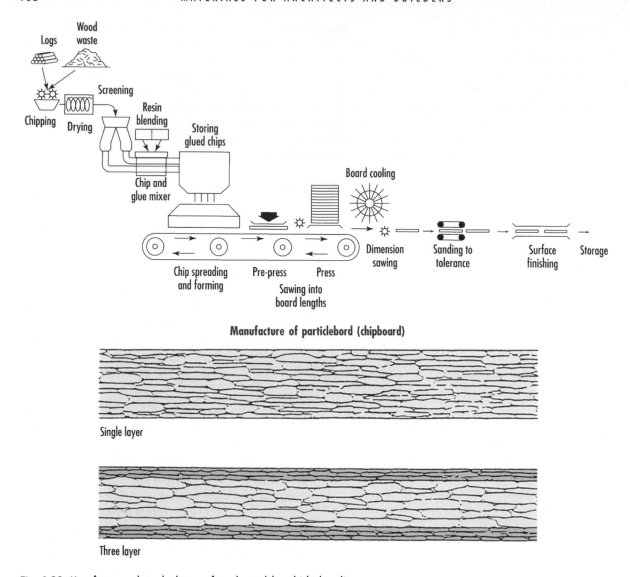

Manufacture of particlebord (chipboard)

Single layer

Three layer

Fig. 4.29 Manufacture and standard types of wood particleboard (chipboard)

Types

The durability of particleboards is dependent upon the resin adhesive. Much UK production uses urea-formaldehyde resin although the moisture-resistant grades are manufactured with melamine-urea-formaldehyde or phenol-formaldehyde resins. Wood chipboards are categorised into seven colour-coded types to BS EN 312: 2003 according to the anticipated loading and environmental conditions. The standard specifies requirements for mechanical and swelling properties and also formaldehyde emissions. The first colour code defines the loading and the second colour the moisture conditions.

Grades of wood particleboard:

Purpose/ Loading	Environmental Conditions	Colour codes		Type
General purpose	dry	white, white	blue	P1
Interior fitments	dry	white	blue	P2
Non-load-bearing	humid	white	green	P3
Load-bearing	dry	yellow, yellow	blue	P4
Load-bearing	humid	yellow, yellow	green	P5
Heavy duty, load-bearing	dry	yellow	blue	P6
Heavy duty, load-bearing	humid	yellow	green	P7

Grades of extruded particleboard:

ES   Extruded Solid board with a minimum density of 550 kg/m³
ET   Extruded Tubes board with a minimum solid density of 550 kg/m³
ESL   Extruded Solid Light board with a density of less than 550 kg/m³
ETL   Extruded Tubes Light board with a minimum solid density of 550 kg/m³
(Grade ET must have at least 5 mm of material over the void spaces.)

Standard particleboards are hygroscopic and respond to changes in humidity. A 10% change in humidity will typically increase the sheet length and breadth by 0.13% and the thickness by 3.5%. Dry grades should not be exposed to moisture even during construction. Humid-tolerant grades are resistant to occasional wetting and relative humidities over 85%. However, no particleboards should be exposed to prolonged wetting, as they are all susceptible to wet rot fungal attack.

All untreated wood particleboards have Class 3 spread of flame (BS 476–7: 1997). However, they can be treated to the requirements of Class 1 by chemical addition in manufacture, by impregnation or the use of intumescent paints. Class 0 may also be achieved. For untreated particleboard with a minimum density of 600 kg/m³ and a minimum thickness of 9 mm, the Euroclass fire performance rating under the conditions specified in BS EN 13986: 2004 is Class D–s2, d0 for non-floor use and Class D_{FL}–s1 for floorings.

A wide range of wood veneer, primed/painted, paper and plastic (PVC, phenolic film or frequently melamine) finishes is available as standard products. Pre-cut sizes are available edged to match. Domestic flooring grade particleboard, usually 18 mm or 22 mm, may be square-edged or tongued and grooved. The industrial flooring grades are typically from 38 mm upwards in thickness.

Uses

Significant quantities of wood particleboard (chipboard) are used in the furniture industry. Much flat-pack DIY furniture is manufactured from painted or veneered particleboard. Particleboard can be effectively jointed by use of double-threaded particleboard wood screws and various specialist fittings. Where high humidity is anticipated the moisture-resistant grades should be used. The domestic housing market uses large quantities of flooring-grade particleboard as it is competitively priced compared to traditional tongued and grooved softwood. Joist centres should be at 450 mm and 610 mm centres

maximum for 18/19 mm and 22 mm particleboard, respectively. Edges should be tongued and grooved or fully supported and the standard panel size is 2400 × 600 mm. For heavy-duty flooring, flat-roof decking and structural work, the moisture-resistant structural grade must be used. Phenolic film-coated particleboard offers a suitable alternative to plywood as formwork to concrete.

Cement-bonded particleboard

Manufacture

Cement-bonded particleboard is manufactured from a mixture of wood particles or filaments (usually softwood) and cement. The boards which are light grey in colour have a uniform cementitious surface. The material has up to 75% cement by weight, with the cement filling all the void spaces, producing a material with a density of 1000–1250 kg/m³, (c.f. 650–690 kg/m³ for standard grade particleboard).

Types and uses

The material based on Portland cement has good resistance to fire, water, fungal attack and frost. The standard (BS EN 634–2: 1997) specifies only one grade, which is suitable for use both internally and externally. It should be colour coded white, white (non load bearing) and brown (suitable for dry, humid and exterior conditions). The standard BS EN 633: 1994 refers to both Portland and magnesium-based cements. Magnesite-bonded particleboard is used as a lining board but it is not frost-resistant and is unsuitable for use in humid conditions.

Boards frequently have a core of coarse wood chips, sandwiched between finer material, producing a good finish, which may be further treated by sanding and priming. Because of its density, cement-bonded particleboard has good sound-insulation properties. Typically, 18 mm board will give sound reduction of 31–3 dB. The material is frequently used for soffits, external sheathing and roofing on both modular and timber-frame buildings, particularly where racking resistance is required. The heavier grades, generally tongued and grooved, are suitable for flooring, due their resistance to moisture, fire, impact and airborne sound.

The material has a Class 0 Surface Spread of Flame to Building Regulations (Class 1 to BS 476:

Part 7: 1997). The Euroclass fire performance rating under the conditions specified in BS EN 13986:2004 for 10 mm cement-bonded particleboard is Class B–s1, d0 for non-floor use and Class B_{FL}–s1 for floorings.

Board sizes are typically 1200 × 2440, 2600 or 3050 mm, with standard thicknesses of 12 and 18 mm although sheets up to 40 mm in thickness are made. However, because of the density of the material, a 1200 × 2440 × 12 mm board weighs approximately 45 kg and should not be lifted by one operative alone.

Gypsum-bonded particleboard

Gypsum-bonded particleboard, available in sheets of 6 mm thickness upwards, is an alternative multipurpose building board. It is not included in the scope of BS EN 633: 1994.

Oriented strand board

Manufacture

Oriented strand board (OSB) is manufactured from 0.5-mm- thick timber flakes tangentially cut and measuring approximately 75 × 35 mm. These are dried and coated with wax and 2.5% of either phenol formaldehyde or melamine-urea-formaldehyde resin. The mix is laid up in three (or occasionally five) layers with the strands running parallel to the sheet on the outer faces and across or randomly within the middle layer. The boards are then cured under heat and pressure, sanded and packaged (Fig. 4.30). Standard panel sizes are 2440 or 3660 × 1220 mm with densities usually in the range 600 to 680 kg/m^3.

Grades and uses

Orientated strand board is graded according to the anticipated loading and environmental conditions (BS EN 300: 1997). Large quantities are used as sheathing in timber-frame housing. The moisture-resistant grade is suitable for roof sarking, whilst the higher specification grade with enhanced strength properties is suitable for flat-roof decking. Thicker panels are used for heavy-duty flooring, and OSB is often used as the web material in timber I-beams. Oriented strand board is manufactured to a thickness range of 6–38 mm, although 9–18 mm sheets predominate. In Europe it is manufactured from Scots pine and spruce, but in North America from aspen and Southern pine.

Grades of oriented strand board:

Grade	Purpose/Loading	Environmental conditions	Colour codes	
OSB 1	General purpose, Interior fitments	dry conditions (hazard class 1)	white	blue
OSB 2	Load-bearing	dry conditions (hazard class 1)	yellow, yellow	blue
OSB 3	Load-bearing	humid conditions (azard class 2)	yellow, yellow	green
OSB 4	Heavy duty, Load-bearing	humid conditions (hazard class 2)	yellow	green

The Euroclass fire performance rating under the conditions specified in BS EN 13986:2004 for 9 mm untreated oriented strand board with a minimum density of 600 kg/m^3 is Class D–s2, d0 for non-floor use and Class D_{FL}–s1 for floorings.

FIBREBOARDS

Fibreboards are manufactured from wood or other plant fibres by the application of heat and/or pressure. They are bonded by the inherent adhesive properties and felting of the fibres or by the addition of a synthetic binder. In the *wet* process used for the manufacture of hardboard, medium board and softboard, no adhesive is added to the wood fibres. In the case of medium density fibreboard (MDF), a resin-bonding agent is incorporated during the production process.

Manufacture

Wet process

Forest thinnings and wood waste are chipped and then softened by steam heating. The chips are ground down into wood fibres and made into a slurry with water. The slurry is fed onto a moving wire-mesh conveyor, where the excess water is removed by suction and light rolling which causes the fibres to felt together. The *wet lap* is then cut to lengths and transferred to a wire mesh for further pressing and heat treatment to remove the remaining water and complete the bonding process. Boards are then conditioned to the correct moisture content and packaged. The range of products primarily arises from the differing degrees of compression applied during the manufacturing process (Fig. 4.31).

Oriented strand board

Wood wool slab

Compressed straw slab

Fig. 4.30 Oriented strand board, wood wool and compressed straw slabs

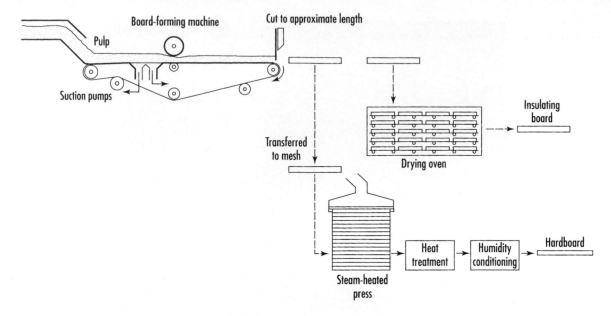

Fig. 4.31 Manufacture of fibreboards

Dry process – MDF

The manufacture of medium-density fibreboard (MDF) involves the addition of adhesive, usually urea-formaldehyde, to the dry wood fibres, which are laid up to an appropriate thickness, slightly compressed to a density of at least 450 kg/m³ and cut to board lengths. The boards are cured under heat and pressure in a press, trimmed to size and sanded. MDF has the advantage of a high-quality machinable finish, and is now used for the production of various mouldings as well as boards. Decorative profiled sheets can be manufactured by laser-cutting of MDF panels to individual client designs. Because of the uniformity of the material, solid sections can be routed to any form. It is therefore widely used for furniture panels as well as internal load-bearing applications.

Four grades are described in BS EN 622–5: 1997 relating to their anticipated loading and environmental conditions. Most MDF is based on urea-formaldehyde resin, but to ensure safety, the quantities of formaldehyde used are strictly controlled by appropriate standards (BS EN 622–1: 2003). Where improved moisture resistance is required, a melamine-urea-formaldehyde resin is used, but this material is not suitable for exterior applications. MDF sheets and mouldings can be finished with a range of coatings including paints, lacquers, stains, plastic laminates, wood veneers and foils.

Moisture-resistant dense MDF (690–800 kg/m³) with all-through colour is available in a range of colours and thicknesses from 8 mm to 30 mm. The material is manufactured from organic-dyed fade-resistant wood fibres and melamine resin with a low residual formaldehyde content. The material can be machined to decorative forms and patterns with a high-quality surface finish which requires only sealing to enhance the colours.

Grades of medium-density fibreboard (MDF):

Grade	Purpose/Loading	Environmental conditions	Colour codes	
MDF	General purpose	dry	white, white	blue
MDF.H	General purpose	humid	white, white	green
MDF.LA	Load-bearing	dry	yellow, yellow	blue
MDF.HLS	Load-bearing	humid	yellow, yellow	green

The following additional grades of MDF are anticipated:

Grade	Purpose/Loading	Environmental conditions
L-MDF	Light MDF material	dry
L-MDF.H	Light MDF material	humid
UL-MDF	Ultra light MDF material	dry
MDF.RWH	Rigid underlays in roofing and walls	

The Euroclass fire performance rating under the conditions specified in BS EN 13986: 2004 for 9 mm untreated MDF with a minimum density of 600 kg/m^3 is Class D–s2, d0 for non-floor use and Class D$_{FL}$–s1 for floorings.

Hardboard

Hardboards are the densest fibreboards, with a minimum density of 900 kg/m^3. The boards range in colour from light to dark brown, usually with one smooth surface and a mesh-textured surface on the underside, although *duo-faced* hardboard – smooth on two faces – is available. Standard sheet sizes are 1220 × 2440 mm to 3600 mm and 1700 × 4880 mm; there are also door sizes. Standard thicknesses range from 3.2 to 6.4 mm, although a wider range is available.

The standard BS EN 622–2: 1997 specifies six grades of hardboard according to load-bearing properties and environmental conditions.

Grades of hardboard:

Grade	Purpose/Loading	Environmental conditions	Colour codes	
HB	General purpose	dry	white, white	blue
HB.H	General purpose	humid	white, white	green
HB.E	General purpose	exterior	white, white	brown
HB.LA	Load-bearing	dry	yellow, yellow	blue
HB.LA1	Load-bearing	humid	yellow, yellow	green
HB.LA2	Heavy duty, load-bearing	humid	yellow	green

Standard hardboard is suitable for internal use, typically panelling, wall and ceiling linings, floor underlays and furniture. A range of perforated, embossed and textured surfaces is available. Applied coatings include primed or painted and various printed woodgrain, PVC or melamine foils.

The Euroclass fire performance rating under the conditions specified in BS EN 13986:2004 for 6 mm untreated hardboard with a minimum density of 900 kg/m^3 is Class D–s2, d0 for non-floor use and Class D$_{FL}$–s1 for floorings.

Tempered hardboard

Tempered hardboards, impregnated with oils during manufacture, are denser and stronger than the standard hardboards, with enhanced water and abrasion resistance. Tempered hardboards are dark brown to black in colour and have a density usually exceeding 960 kg/m^3. Tempered hardboards are suitable for structural and exterior applications. The high shear strength of the material is used within hardboard-web structural box beams and I beams. Typical exterior applications include claddings, fasciae and soffits, where weather resistance is important. The moisture resistance of tempered hardboard makes it suitable for lining concrete formwork.

Mediumboard and softboard

Mediumboard and softboard are manufactured by the *wet* process. Mediumboard (high-density and low-density) and softboard exhibit a range of physical properties which reflects the degree of compression applied during the manufacturing process. High-density mediumboard (density 560–900 kg/m^3) has a dark brown shiny surface like hardboard. Low-density mediumboard (density 400–560 kg/m^3) has a light brown soft finish. Softboard (density 210–400 kg/m^3) is light in colour with a fibrous, slightly textured finish. Softboard impregnated with bitumen offers an increased moisture resistance over the untreated material.

The general-purpose exterior grades (E) may be used for exterior cladding. The higher density grades (H) are used for wall linings, sheathing, partitioning, ceilings and floor underlays. Low-density mediumboard (L) is use for wall linings, panelling, ceilings and notice boards. Softboard is used for its acoustic and thermal insulating properties. Bitumen-impregnated softboard is suitable for use as a floor underlay to chipboard on concrete.

The standard BS EN 622–3: 2004 specifies ten grades of low (L) and high (H) density mediumboard according to load-bearing requirements and environmental conditions.

Grades of mediumboard:

Grade	Purpose/Loading	Environmental conditions	Colour codes	
MB.L	General purpose	dry	white, white	blue
MB.H	General purpose	dry	white, white	blue
MBL.H	General purpose	humid	white, white	green
MBH.H	General purpose	humid	white, white	green
MBL.E	General purpose	exterior	white, white	brown
MBH.E	General purpose	exterior	white, white	brown
MBH.LA1	Load-bearing	dry	yellow, yellow	blue
MBH.LA2	Heavy duty, load-bearing	dry	yellow	blue
MBH.HLS1	Load-bearing	humid	yellow, yellow	green
MBH.HLS2	Heavy duty, load-bearing	humid	yellow	green

The Euroclass fire performance rating under the conditions specified in BS EN 13986:2004 for 9 mm untreated high-density mediumboard of 600 kg/m³ is Class D–s2, d0 for non-floor use and Class D_{FL}–s1 for floorings. For untreated low-density mediumboard of 400 kg/m³ the equivalent rating is Class E, pass for non-floor use and Class E_{FL} for floorings.

The standard BS EN 622–4: 1997 specifies five grades of softboard according to load-bearing properties and environmental conditions.

Grades of softboard:

Grade	Purpose/Loading	Environmental conditions	Colour codes	
SB	General purpose	dry	white, white	blue
SB.H	General purpose	humid	white, white	green
SB.E	General purpose	exterior	white, white	brown
SB.LS	Load-bearing	dry	yellow, yellow	blue
SB.HLS	Load-bearing	humid	yellow, yellow	green

The Euroclass fire performance rating under the conditions specified in BS EN 13986:2004 for 9 mm untreated softboard of 250 kg/m³ is Class E, pass for non-floor use and Class E_{FL}–s1 for floorings.

WOOD WOOL SLABS

Manufacture

Wood wool slabs are manufactured by compressing long strands of chemically stabilised wood fibres coated in Portland cement (Fig. 4.30). The grey product has an open texture which may be left exposed, spray painted or used as an effective substrate for plastering. It is also a suitable material for permanent shuttering for concrete. Slabs are available in a range of thicknesses from 25 to 150 mm, typically 500, 600 or 625 mm wide and up to 3 metres in length. Standard sizes are listed in BS EN 13168: 2001.

Types and uses

Wood wool slabs are available either plain-edged or with interlocking galvanised steel channels to the longitudinal edges. Thicknesses in the range 15 mm to 50 mm are suitable for ceilings, partitions, wall linings and permanent concrete shuttering. The thicker grades from 50 mm to 150 mm may be used for roof decking, with spans up to 3 m according to the anticipated loading. Some products incorporate an additional insulation layer of mineral wool or cellular plastics to enhance thermal properties.

The material is rated as Class 0 with respect to Building Regulations and Class I (BS 476–7: 1997) in terms of surface spread of flame. Classification to the European Standard (BS EN 13501–1: 2002) is subject to manufacturer's testing. The material is resistant to fungal attack and is unaffected by wetting. Wood wool slabs offer good sound-absorption properties due to their open-textured surface. This is largely unaffected by the application of sprayed emulsion paint. The material is therefore appropriate for partitions, internal walls and ceilings where sound absorption is critical. Acoustic insulation for a pre-screeded 50 mm slab is typically 30 dB. The relatively high proportion of void space affords the material good thermal insulating properties with a typical thermal conductivity of 0.077 W/m K at 8% moisture content.

The material is workable, being easily cut and nailed. Where a gypsum plaster or cement/lime/sand external rendering is to be applied all joints should be reinforced with scrim. Wood wool slabs form a suitable substrate for flat roofs finished with built-up bitumen sheet, asphalt or metals.

COMPRESSED STRAW SLABS

Compressed straw slabs are manufactured by forming straw under heat and pressure, followed by encapsulation in a fibreglass mesh and plastering grade paper (Fig. 4.30). Typically used for internal partitioning, the panels are mounted onto a timber sole plate and butt jointed with adhesive or dry-jointed with proprietary sheradised clip fixings. All joints are jute scrimmed and the whole partitioning finished with a 3 mm skim of board plaster. The slabs are 58 mm thick by 1200 mm wide in a range of standard lengths from 2270–400 mm. Service holes may be incorporated into the panels at 300 mm centres for vertical electrical wiring. While normal domestic fixtures can be fitted directly to the panels, heavier loads require coach-bolt fixings through the panels. The product should not be used where it will be subjected to moisture. Compressed straw slabs, when skim plastered, have a 30-minute fire resistance rating, a Class 0 spread of flame, and a sound reduction of typically 35 dB over the range 100–3200 Hz. The standard BS 4046: 1991 describes four types of compressed straw slabs depending upon boron-based insecticide treatment and the provision of continuous longitudinal voids for services. Alternative finishes are plain paper for direct decoration or showerproof paper.

Types of compressed straw slabs:

Type A	untreated	solid core
Type B	insecticide treated	solid core
Type C	untreated	continuous voids in core
Type D	insecticide treated	continuous voids in core

THATCH

Thatch was the roof covering for most buildings until the end of the Middle Ages, and remained the norm in rural areas until the mid-nineteenth century. For most of the twentieth century thatch was only used in conservation work; however, with the new resurgence of interest in the material, partially associated with the reconstruction of the Globe Theatre in London (Fig. 4.32) thatch, once again, has become a current construction material.

Materials

The three standard materials for thatching within the UK are water reed (*Phragmites australis*), long straw (usually wheat) and combed wheat reed (also known as *Devon reed*). Water reed is associated with the Norfolk broads, the fens, south Hampshire and the Tay estuary, but much is imported from Turkey, Poland, Romania and China. Long straw is the standard thatch in the Midlands and Home Counties, while combed wheat reed is more common in Devon and Cornwall. Water reed is the most durable, lasting typically 50 to 60 years, but long straw and combed wheat reed last approximately 20 and 30 years respectively, depending upon location and roof pitch. All thatched roofs will need reridging at 10 to 15 year intervals; in the case of water reed this is often done with saw sedge (*Cladium mariscus*) which is more flexible than the reed itself.

Both long straw and combed wheat reed are often grown and harvested specifically as thatching materials to ensure long undamaged stems. Long straw is threshed winter wheat, whereas combed wheat straw is wheat with any leaves and the grain head removed. An alternative to combed wheat reed is triticale (*Triticale hexaploide*), which is a cross between wheat and rye. Triticale produces a more reliable harvest than other forms of wheat straw and it is indistinguishable from combed wheat reed when used as a thatching material. Water reed for thatching is usually

Fig. 4.32 Thatched roof – Globe Theatre, London. Photograph: Arthur Lyons

between 915 mm and 1830 mm in length. Typical lengths for long straw and combed wheat reed are 760 mm and 915 mm respectively.

Appearance

Long straw roofs show the lengths of the individual straws down the roof surface and are also characterised by the use of split hazel rodding around the eaves and gables to secure the thatch. To prevent attack by birds they are frequently covered in netting. Combed wheat reed and water reed both have a closely packed finish with the straw ends forming the roof surface. A pitch of about 50° is usual for thatch with a minimum of 45°, the steeper pitches being more durable. The ridge, which may be a decorative feature, is produced by either wrapping wheat straw over the apex or butting-up reeds from both sides of the roof. Traditionally hazel twigs are used for fixings although these can be replaced with stainless steel wires. The durability of thatch is significantly affected by the climate. All materials tend to have shorter service lives in warmer locations with high humidity, which encourage the development of fungi. Chemical treatment, consisting of an organic heavy metal compound, may be used, preferably on new thatch, to delay the biological decomposition. Thatch is usually laid to a thickness of between 220 and 400 mm.

Properties

Fire
The fire hazards associated with thatched roofs are evident; however, fire retardants can be used although these may denature the material. In the case of the Globe Theatre in London, a sparge water-spray system has been installed. In other new installations, the granting of planning consent has been facilitated by the location of permanent water drenching systems near the ridge and by the use of fire-resisting board and foil under the thatch to prevent internal fire spread. Electrical wiring and open fire chimneys are the most common causes of thatch fires, although maintenance work on thatched roofs is also a risk if not carefully managed.

Insulation
Thatch offers good insulation, keeping buildings cool in summer and warm in winter, a typical 300 mm of water reed achieving a U-value of 0.35 W/m² K.

SHINGLES

Western red cedar (*Thuja plicata*), as a naturally durable material, is frequently used as shingles or shakes for roofing or cladding. Shingles are cut to shape, whereas shakes are split to the required thickness, usually between 10 and 13 mm. Both shakes and shingles may be tapered or straight. Shingles, typically 400, 450 or 600 mm long and between 75 and 355 mm wide, may be treated with copper/chrome/arsenate (CCA) wood preservative to improve their durability. Additionally they may be treated with flame retardant to satisfy the AA fire rating of BS 476–3: 1975. Shingles should be fixed with corrosion-resistant nails, leaving a spacing of 5–6 mm between adjacent shingles. A minimum pitch of 14° is necessary and three layers are normally required. Whilst the standard laying pattern is straight coursing, staggered patterns and the use of profiled shingles on steeper pitches can create decorative effects. Shingles should normally be laid over a waterproof breather membrane. The standard BS 5534: 2003 suggests a minimum side lap of 40 mm for shakes and shingles.

Figure 4.33 illustrates the typical detailing of cedar shingles as a lightweight cladding material. The natural colour of the material gives an overall warmth to the exterior envelope of the building. In exposed locations the red-brown cedar wood surface gradually weathers to a silver-grey, while in very sheltered locations the shingles become green with lichen.

'STEKO' BLOCKS

Steko is an innovative wall system, which uses very accurately engineered large hollow timber blocks which simply slot together. No fixings or glue is needed to form load-bearing wall construction. The blocks are manufactured from small off-cuts of spruce timber, which can be readily obtained from renewable sources. The blocks (640 × 320 × 160 mm wide) weighing 6.5 kg, are manufactured from two 20-mm-thick panels glued to horizontal battens and separated by vertical studs (Fig. 4.34). Units fit snugly together with their tongued and grooved profiles and vertical dowels. Special units include quarter, half and three-quarter blocks as well as components for corners, lintels, wall closers, base plates, wall plates and solid blocks for point loads. Walls can be up to 20 m long without additional bracing, and 3 m high, but up to 4 or 5 storeys high, if the necessary horizontal bracing is provided by the intermediate floors and roof.

Fig. 4.33 Western red cedar shingles. Photograph: Arthur Lyons

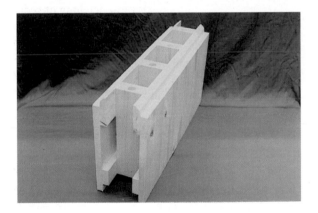

Fig. 4.34 'Steko' block. Photograph: Courtesy of Construction Resources

Where two-storey unrestrained walls are required they can be post-tensioned with threaded steel rods. The blocks are untreated except when borax protection against house longhorn beetle is required. Electrical services can run in the voids, which are finally dry-injected with cellulose insulation. Steko construction must work from a 300 mm upstand to avoid contact with surface water, and should be protected externally with a breather paper to prevent rain damage during construction. Internal finishes may be left exposed or finished with 15 mm gypsum plasterboard, which gives 30 minutes fire resistance. Additional insulation is required externally to achieve the current Building Regulation requirements. Typically, 100 mm of mineral wool and a 20 mm render externally on cellulose filled and internally plastered blocks will achieve a U-value of 0.20 W/m K. The first UK house built with the Steko system, which originated in Switzerland, has been designed and built on a cliff-top site in Downderry, Cornwall.

FLEXIBLE VENEERS

Flexible veneers are sheets of wood veneer which have been backed with paper or fibre reinforcement to allow the material to be handled without splitting. Flexible veneers can be moulded onto profiled components of MDF or plywood by a rolling process. Typical standard components are cornice and pelmet trims for kitchen furniture. The material can be rolled up for storage without damage, unlike traditional veneer. The veneer is fixed to the substrate with polyvinyl acetate (PVA) or urea-formaldehyde (UF) adhesive under pressure or with contact adhesive. The material is increasingly being specified by designers for creating high-quality polished wood finishes to complex curved forms such as reception desks and wall panelling.

References

FURTHER READING

Bedding, B. 2001: *Timber frame construction*. High Wycombe: TRADA Technology.

Billett, M. 2003: *The complete guide to living with thatch*. London: Hale.

BRE. 1972: *Handbook of hardwoods*. London: HMSO.

BRE. 1977: *Handbook of softwoods*. London: HMSO.

Breyer, D.E. 2003: *Design of wood structures*. New York: McGraw Hill Education.

Brunskill, R.W. 2004: *Timber in building*. London: Orion.

Burch, M. 2004: *The complete guide to building log homes*. Sterling: New York.

Constantine, A. 2005: *Know your woods*. Connecticut: The Lyons Press.

Cox. J. and Letts, J 2000: *Thatch: thatching in England 1940–1994*. London: James & James.

Desch, H.E. 1996: *Timber: its structure, properties and utilisation*. 7th ed. London: Macmillan.

Dinwoodie, J.M. 2000: *Timber: its nature and behaviour*. 2nd ed. London: E. & F.N. Spon.

Fearn, J. 2004: *Thatch and thatching*. Princes Risborough: Shire Publications.

Goldstein, E.W. 1998: *Timber construction for architects and builders*. New York: McGraw Hill Education.

Gutdeutsch, G. 1996: *Building in wood*. Basel: Birkhäuser.

Hislop, P. 2001: *External timber cladding*. High Wycombe: TRADA Technology.

Hugues, T. Steiger, L. and Weber, J. 2004: *Timber construction: Details, products, case studies*. Basle: Birkhäuser.

Jayanetti, L. and Follett, P. 2000: *Timber pole construction*. London: ITDG Publishing.

Keyworth, B. and Woodbridge, D. 1992: *Environmental aspects of timber in use in the UK*. Oxford: Timber Tectonics Ltd.

Leffteri, C. 2003: *Wood: Materials for inspirational design*. East Sussex: Rotovision.

McKenzie, W.M.C. 2000: *Design of structural timber*. London: Palgrave.

Mueller, C. 2000: *Laminated timber construction*. Berlin: Birkhäuser.

Newman, R. 2005: *Oak-framed buildings*. Lewes: Guild of Master Craftsman Publications.

Ojeda, O.R. and Pasnik, M. 2003: *Materials: architecture indetail*. Massachusetts USA: Rockport.

Pryce, W. 2005: *Architecture in wood, a world history*. London: Thames & Hudson.

Ridout, B.V. 1999: *Timber decay in buildings: the conservation approach to treatment*. London: E. & F.N. Spon.

Ruske, W. 2004: *Timber construction for trade, industry, administration*. Berlin: Birkhäuser.

Sunley, J. and Bedding, B. (eds.) 1985: *Timber in construction*. London: Batsford/TRADA.

Thelandersson, S. and Larsen, H.J. 2003: *Timber engineering*. Chichester: John Wiley and Sons.

TRADA, 2003: *List of British Standards relating to timber*. High Wycombe: TRADA.

STANDARDS

BS 144: 1997 Specification for coal tar creosote for wood preservation.

BS 373: 1957 Methods for testing small clear specimens of timber.

BS 476 Fire tests on building materials and structures:
    Parts 3, 4, 6, 7, 10–13, 15, 20–24, 31–33

BS 644: 2003 Timber windows. Factory assembled windows of various types.

BS 1088 Marine plywood:
    Part 1: 2003 Requirements.
    Part 2: 2003 Determination of bonding quality.

BS 1186 Timber for and workmanship in joinery:
    Part 2: 1988 Specification for workmanship.
    Part 3: 1990 Specification for wood trim and its fixing.

BS 1187: 1959 Wood blocks for floors.

BS 1203: 2001 Hot-setting phenolic and aminoplastic wood adhesives.

BS 1282: 1999 Wood preservatives – Guidance on choice, use and application.

BS 1297: 1987 Specification for tongued and grooved softwood flooring.

BS 1336: 1971 Knotting.

BS 4046: 1991 Compressed straw building slabs.

BS 4050 Specification for mosaic parquet panels:
    Part 1: 1977 General characteristics.
    Part 2: 1966 Classification and quality requirements.

BS 4072: 1999 Copper/chrome/arsenic preparations for wood preservation.

BS 4787 Internal and external wood doorsets, door leaves and frames:
    Part 1: 1980 Specification for dimensional requirements.

BS 4965: 1999 Specification for decorative laminated plastics sheet veneered boards and panels.

BS 4978: 1996 Specification for visual strength grading of softwood.

BS 5268 Structural use of timber:

Part 2: 2002 Code of practice for permissible stress design, materials and workmanship.

Part 3: 1998 Code of practice for trussed rafter roofs.

Part 4: 1978/90 Fire resistance of timber structures.

Part 5: 1989 Code of practice for the preservative treatment of structural timber.

Part 6: 1996/01 Code of practice for timber frame walls.

Part 7: 1989/90 Recommendations for the calculation basis for span tables.

BS 5277: 1976 Doors. Measurement of defects of general flatness of door leaves.

BS 5278: 1976 Doors. Measurement of dimensions and of defects of squareness of door leaves.

BS 5369: 1987 Methods of testing doors; behaviour under humidity variations of door leaves placed in successive uniform climates.

BS 5395 Stairs, ladders and walkways:

Part 1: 2000 Code of practice for straight stairs.

Part 2: 1984 Code of practice for the design of helical and spiral stairs.

Part 3: 1985 Code of practice for the design of industrial type stairs, permanent ladders and walkways.

BS 5534: 2003 Code of practice for slating and tiling (including shingles).

BS 5589: 1989 Code of practice for preservation of timber.

BS 5666 Methods of analysis of wood preservatives and treated timber:

Parts 2–7

BS 5707: 1997 Specification for preparations of wood preservatives in organic solvents.

BS 5756: 1997 Specification for visual strength grading of hardwood.

BS 6100 Glossary of building and civil engineering terms:

Part 4 Forest products.

BS 6178 Joist hangers:

Part 1: 1990 Specification for joist hangers for building into masonry walls of domestic dwellings.

BS 6446: 1997 Manufacture of glued structural components of timber and wood based panels.

BS 6559: 1985 General introductory documents on European (or CEN) methods of test for wood preservatives.

BS 7331: 1990 Direct surfaced wood chipboard based on thermosetting resins.

BS 7359: 1991 Commercial timbers including sources of supply.

BS 8103 Structural design of low-rise buildings:

Parts 1–4

BS 8201: 1987 Code of practice for flooring of timber, timber products and wood-based panel products.

BS EN 300: 1997 Oriented strand board (OSB) – definitions, classification and specifications.

BS EN 301: 1992 Adhesives, phenolic and aminoplastic for loadbearing timber structures.

BS EN 309: 2005 Wood particleboards – definition and classification.

BS EN 311: 2002 Wood-based panels – surface soundness – test method.

BS EN 312: 2003 Particleboards – specifications.

BS EN 313 Plywood – classification and terminology:

Part 1: 1996 Classification.

Part 2: 2000 Terminology.

BS EN 314 Plywood – bonding quality:

Part 1: 2004 Test methods.

Part 2: 1993 Requirements.

BS EN 315: 2000 Plywood – tolerances for dimensions.

BS EN 316: 1999 Wood fibreboards – definition, classification and symbols.

BS EN 317: 1993 Particleboards and fibreboards – determination of swelling in thickness after immersion in water.

BS EN 318: 2002 Wood-based panels – determination of dimensional changes associated with changes in relative humidity.

BS EN 319: 1993 Particleboards and fibreboards – determination of tensile strength perpendicular to the plane of the board.

BS EN 320: 1993 Fibreboards – determination of resistance to axial withdrawal of screws.

BS EN 321: 2002 Wood-based panels – determination of moisture resistance.

BS EN 322: 1993 Wood-based panels – determination of moisture content.

BS EN 323: 1993 Wood-based panels – determination of density.

BS EN 324 Wood-based panels – determination of dimensions of boards:

Part 1: 1993 Determination of thickness, width and length.

Part 2: 1993 Determination of squareness and edge straightness.

BS EN 325: 1993 Wood-based panels – determination of test pieces.

BS EN 326 Wood-based panels – sampling, cutting and inspection:

Part 1: 1994 Sampling and cutting of test pieces and expression of test results.

Part 2: 2000 Quality control in the factory.

Part 3: 2003 Inspection of a consignment of panels.

BS EN 330: 1993 Wood preservatives – field test method for determining the relative protective effectiveness of a wood preservative for use under a coating.

BS EN 335 Hazard classes of wood and wood-based products against biological attack:

Part 1: 1992 Classification of hazard classes.

Part 2: 1992 Guide to the application of hazard classes to solid wood.

Part 3: 1996 Durability of wood and wood-based products. Definition of hazard classes of biological attack – application to wood-based panels.

BS EN 336: 2003 Structural timber – coniferous and poplar – sizes – permissible deviations.

BS EN 338: 2003 Structural timber – strength classes.

BS EN 350 Durability of wood and wood-based products – natural durability of solid wood:

Part 1: 1994 Guide to the principles of testing and classification of the natural durability of wood.

Part 2: 1994 Guide to the natural durability and treatability of selected wood species of importance in Europe.

BS EN 351 Durability of wood and wood-based products – preservative-treated solid wood:

Part 1: 1996 Classification of preservative penetration and retention.

Part 2: 1996 Guidance on sampling for the analysis of preservative-treated wood.

BS EN 380: 1993 Timber structures – test methods – general principles for static load testing.

BS EN 382 Fibreboards – determination of surface absorption:

Part 1: 1993 Test method for dry process fibreboard.

Part 2: 1994 Test methods for hardboard.

BS EN 383: 1993 Timber structures – test methods – determination of embedded strength and foundation values for dowel type fasteners.

BS EN 384: 2004 Structural timber – determination of characteristic values of mechanical properties and density.

BS EN 385: 2001 Finger jointed structural timber – performance requirements and minimum production requirements.

BS EN 386: 2001 Glued laminated timber – performance requirements and minimum production requirements.

BS EN 387: 2001 Glued laminated timber – production requirements for large finger joints. Performance requirements and minimum production requirements.

BS EN 390: 1995 Glued laminated timber – sizes – permissible deviations.

BS EN 391: 2002 Glued laminated timber – delamination test of glue lines.

BS EN 392: 1995 Glued laminated timber – shear test of glue lines.

BS EN 408: 2003 Timber structures – structural timber and glued laminated timber.

BS EN 409: 1993 Timber structures – test methods – determination of the yield moment of dowel type fasteners – nails.

BS EN 460: 1994 Durability of wood and wood-based products – natural durability of solid wood – guide to the durability requirements for wood to be used in hazard classes.

BS EN 518: 1995 Structural timber – grading – requirements for visual strength grading standards.

BS EN 519: 1995 Structural timber – grading – requirements for machine strength grading timber and grading machines.

BS EN 594: 1996 Timber structures – test methods – racking strength and stiffness of timber frame wall panels.

BS EN 595: 1995 Timber structures – test methods – test of trusses for the determination of strength and deformation behaviour.

BS EN 596: 1995 Timber structures – test methods – soft body impact test of timber framed walls.

BS EN 599 Durability of wood and wood-based products, performance of wood preservatives as determined by biological tests:

Part 1: 1997 Specification according to hazard class.

Part 2: 1997 Classification and labelling.

BS EN 622 Fibreboards – specifications:

Part 1: 2003 General requirements.

Part 2: 1997 Requirements for hardboards.

Part 3: 2004 Requirements for medium boards.

Part 4: 1997 Requirements for softboards.

Part 5: 1997 Requirements for dry process boards (MDF).

BS EN 633: 1994 Cement-bonded particleboards – definition and classification.

BS EN 634 Cement-bonded particleboards – specification:

Part 1: 1995 General requirements.

Part 2: 1997 Requirements for OPC bonded particleboards for use in dry, humid and exterior conditions.

BS EN 635 Plywood – classification by surface appearance:

Part 1: 1995 General.

Part 2: 1995 Hardwood.

Part 3: 1995 Softwood.

Part 5: 1999 Methods for measuring and expressing characteristics and defects.

BS EN 636: 2003 Plywood – specifications.

BS EN 789: 2004 Timber structures – test methods – determination of mechanical properties of wood-based panels.

BS EN 844–1: 1995 Round and sawn timber – terminology.

BS EN 912: 2000 Timber fasteners – specifications for connectors for timber.

BS EN 942: 1996 Timber in joinery – general classification of timber quality.

BS EN 975 Sawn timber – appearance grading of hardwoods:

Part 1: 1996 Oak and Beech.

BS EN 1014 Wood preservatives:

Part 1: 1995 Procedure for sampling creosote.

Part 2: 1996 Procedure for obtaining a sample of creosote from creosoted timber.

Part 3: 1998 Determination of the benzoypyrene content of creosote.

Part 4: 1996 Determination of the water extractable phenols content of creosote.

BS EN 1026: 2000 Windows and doors – Air permeability.

BS EN 1027: 2000 Windows and doors – Watertightness.

BS EN 1072: 1995 Plywood – description of bending properties for structural plywood.

BS EN 1087 Particleboards – determination of moisture resistance:

Part 1: 1995 Boil test.

BS EN 1128: 1996 Cement-bonded particleboards – determination of hard body impact resistance.

BS EN 1193: 1998 Timber structures – structural timber and glued laminated timber – determination of shear strength and mechanical properties.

BS EN 1194: 1999 Timber structures – glued laminated timber – strength classes and determination of characteristic values.

BS EN 1195: 1998 Timber structures – performance of structural floor decking.

BS EN 1294: 2000 Door leaves. Determination of the behaviour under humidity variations.

BS EN 1309 Round and sawn timber – method of measurement of dimensions:

Part 1: 1997 Sawn timber.

BS EN 1313 Round and sawn timber – permitted deviations and preferred sizes:

Part 1: 1997 Softwood sawn timber.

Part 2: 1999 Hardwood sawn timber.

BS EN 1611–1:2000 Sawn timber – appearance grading of softwoods.

BS EN 1912: 2004 Structural timber – strength classes assignment of visual grades and species.

BS EN 1995 Eurocode 5: Design of timber structures:

Part 1.1: 2004 Common rules and rules for buildings.

Part 1.2: 2004 Structural fire design.

BS EN 12211: 2000 Windows and doors. Resistance to wind.

BS EN 12369 Wood-based panels – characteristic values for structural design:

Part 1: 2001 OSB, particleboards and fibreboards.

Part 2: 2004 Plywood.

BS EN 12436: 2002 Adhesives for load-bearing timber structures – casein adhesives.

BS EN 12512: 2001 Timber structures – cyclic testing of joints made with mechanical fasteners.

BS EN 12871: 2001 Wood-based panels. Performance specifications and requirements for load bearing boards.

BS EN 13017 Solid wood panels classified by surface appearance:

Part 1: 2001 Softwood.

Part 2: 2001 Hardwood.

BS EN 13168: 2001 Thermal insulation products for building – factory made wood wool products.

BS EN 13271: 2002 Timber fasteners – characteristic load-carrying capacities.

BS EN 13501 Fire classification of construction products and building elements:

Part 1: 2002 Classification using test data from reaction to fire tests.

Part 2: 2003 Classification using data from fire resistance tests.

BS EN 13986: 2004 Wood-based panels for use in construction. Characteristics, evaluation of conformity and marking.

BS EN 14080: 2005 Timber structures – glued laminated timber – requirements.

BS EN 14081 Timber structures – strength graded timber with rectangular cross-section:

Part 1: 2005 General requirements.

Part 2: 2005 Machine grading – additional requirements for initial type testing.

Part 3: 2005 Machine grading – additional requirements for production.

Part 4: 2005 Machine grading – grading machine settings.

pr EN 14220: 2004 Timber and wood-based materials in external windows, external door leaves and external doorframes.

pr EN 14221: 2004 Timber and wood-based materials in internal windows, door leaves and internal doorframes.

BS EN 14279: 2004 Laminated veneer lumber – definition, classification and specifications.

BS EN 14298: 2004 Sawn timber – assessment of drying quality.

BS EN 14342: 2005 Wood flooring – characteristics, evaluation of conformity and marking.

BS EN 14351-1: 2006 Windows and doors, product standard, performance characteristics.

BS EN 14519: 2005 Solid softwood panelling and cladding.

pr EN 14544: 2002 Timber structures – strength graded structural timber with round cross-section – requirements.

pr EN 14545: 2002 Timber structures – connectors– requirements.

pr EN 14732: 2003 Timber structures – prefabricated wall, floor and roof elements.

BS EN 14755: 2005 Extruded particleboard – specifications.

BS EN 14761: 2006 Wood flooring – solid wood parquet.

BS EN 14762: 2006 Wood flooring – sampling procedures for evaluation of conformity.

BS EN 26891: 1991 Timber structures – joints made with mechanical fasteners.

BS EN 28970: 1991 Timber structures – testing of joints made with mechanical fasteners.

DD ENV 839: 2002 Wood preservatives – determination of the protective effectiveness against wood destroying basidiomycetes.

DD ENV 1250 Wood preservatives – method of measuring loss of active ingredients:

Part 1: 1995 Losses by evaporation to air.

Part 2: 1995 Losses by leaching into water.

DD ENV 1390: 1995 Wood preservatives – determination of the eradication action against *Hylotrupes bajulus* larvae.

REGULATIONS

Control of Pesticides Regulations 1986.
Control of Substances Hazardous to Health 1988.
Wildlife & Countryside Act 1981.

BUILDING RESEARCH ESTABLISHMENT PUBLICATIONS

BRE Special digest

SD2: 2002 Timber frame dwellings: U-values and building regulations.

BRE Digests

BRE Digest 208: 1988 Increasing the fire resistance of existing timber floors.

BRE Digest 299: 1993 Dry rot: its recognition and control.

BRE Digest 301: 1985 Corrosion of metals by wood.

BRE Digest 307: 1992 Identifying damage by wood-boring insects.

BRE Digest 327: 1993 Insecticidal treatments against wood boring insects.

BRE Digest 340: 1989 Choosing wood adhesives.

BRE Digest 345: 1989 Wet rots: recognition and control.

BRE Digest 351: 1990 Re-covering old timber roofs.

BRE Digest 364: 1991 Design of timber floors to prevent decay.

BRE Digest 375: 1992 Wood-based panel products, their contribution to the conservation of forest resources.

BRE Digest 393: 1994 Specifying preservative treatments: the new European approach.

BRE Digest 407: 1995 Timber for joinery.

BRE Digest 416: 1996 Specifying structural timber.

BRE Digest 417: 1996 Hardwoods for construction and joinery.

BRE Digest 423: 1997 The structural use of wood-based panels.

BRE Digest 429: 1998 Timbers: their natural durability and resistance to preservative treatment.

BRE Digest 431: 1998 Hardwoods for joinery an construction (Parts 1, 2 and 3)

BRE Digest 435: 1998 Medium-density fibreboard.
BRE Digest 443: 1999 Termites and UK buildings (Parts 1 and 2).
BRE Digest 445: 2000 Advances in timber grading.
BRE Digest 470: 2002 Life cycle impacts of timber.
BRE Digest 476: 2003 Guide to machine strength grading of timber.
BRE Digest 477 Part 1: 2003 Wood-based panels: Oriented Strand Board (OSB).
BRE Digest 477 Part 2: 2003 Wood-based panels: particleboard (chipboard).
BRE Digest 477 Part 3: 2003 Wood-based panels: cement-bonded particleboard.
BRE Digest 477 Part 4: 2004 Wood-based panels: plywood.
BRE Digest 477 Part 5: 2004 Wood-based panels: medium density fibreboard (MDF).
BRE Digest 477 Part 6: 2004 Wood-based panels: hardboard, medium board and softboard.
BRE Digest 477 Part 7: 2004 Wood-based panels: selection.
BRE Digest 479: 2003 Timber piles and foundations.
BRE Digest 487 Part 4: 2004. Structural fire engineering design: materials behaviour – timber.
BRE Digest 492: 2005 Timber grading and scanning.
BRE Digest 494: 2005 Using UK-grown Douglas fir and larch for external cladding.
BRE Digest 496: 2005 Timber-frame building.

BRE Defect action sheets

BRE DAS 74: 1986 Suspended timber ground floors: repairing rotted joists.
BRE DAS 103: 1987 Wood floors: reducing risk of recurrent dry rot.

BRE Good building guide

BRE GBG 21: 1996 Joist hangers.
BRE GBG 32: 1999 Ventilating thatched roofs.
BRE GBG 60: 2004 Timber frame construction: an introduction.

BRE Information papers

BRE IP 7/88 The design and manufacture of ply-web beams.
BRE IP 5/90 Preservation of hem-fir timber.
BRE IP 5/91 Exterior wood stains.
BRE IP 7/91 Serviceability design of ply-web roof beams.

BRE IP 9/91 Blue staining of timber in service: its cause, prevention and treatment.
BRE IP 10/91 The selection of timber for exterior joinery from the genus *Shorea*.
BRE IP 12/91 Fibre building board: types and uses.
BRE IP 14/91 *In-situ* treatment of exterior joinery using boron-based implants.
BRE IP 11/92 Schedules for the preservation of hem-fir timber.
BRE IP 14/92 Cement-bonded particleboard.
BRE IP 19/92 Wood-based panel products: moisture effects and assessing the risk of decay.
BRE IP 9/93 Perspectives on European Standards for wood-based panels.
BRE IP 8/94 House longhorn beetle: geographical distribution and pest status in the UK.
BRE IP 2/96 Assessment of exterior medium density fibreboard.
BRE IP 8/96 Moisture resistance of laminated veneer lumber (LVL).
BRE IP 9/96 Preservative-treated timber for exterior joinery – European standards.
BRE IP 4/97 Preservative-treated timber for exterior joinery – applying European standards.
BRE IP 8/98 Reducing kiln-drying twist of Sitka spruce.
BRE IP 6/99 Preservative-treated timber – ensuring conformity with European standards.
BRE IP 2/01 Evaluating joinery preservatives.
BRE IP 13/01 Preservative-treated timber – the UK's code of best practice.
BRE IP 14/01 Durability of timber in ground contact.
BRE IP 1/03 European Standards for wood preservatives and treated wood.
BRE IP 9/03 Best practice of timber waste management.
BRE IP 13/04 An introduction to building with structural insulated panels.
BRE IP 10/05 Green gluing of timber: a feasibility study.
BRE IP 13/05 Incising UK-grown Sitka spruce.

BRE Reports

BR 76: 1986 Timber drying manual – second edition.
BR 226: 1992 A review of tropical hardwood consumption.
BR 229: 1992 Wood preservation in Europe: Development of standards for preservatives and treated wood.
BR 232: 1992 Recognising wood rot and insect damage in buildings.

BR 241: 1992 The strength properties of timber.

BR 249: 1990 Long-term field trials on preserved timber out of ground contact.

BR 256: 1994 Remedial treatment of wood rot and insect attack in buildings.

BR 276: 1995 Long-term field trials on preserved timber in ground contact.

TRADA PUBLICATIONS

Wood information sheets

WIS 1–6: 2003 Glued laminated timber.

WIS 1–17: 2003 Structural use of hardwoods.

WIS 1–25: 2003 Structural use of timber – an introduction to BS 5268–2 2002.

WIS 1–37: 2000 Eurocode 5 – an introduction.

WIS 1–45: 2001 Structural use of wood-based panels.

WIS 1–46: 2004 Decorative timber flooring.

WIS 2/3–1: 2005 Finishes for external timber.

WIS 2/3–3: 2003 Flame retardant treatments for timber.

WIS 2/3–10: 1999 Timbers – their properties and uses.

WIS 2/3–11: 1999 Specification and treatment of exterior plywood.

WIS 2/3–23: 1999 Introduction to wood-based panel products.

WIS 2/3–28: 2003 Introducing wood.

WIS 2/3–31: 2003 Adhesively-bonded timber connections.

WIS 2/3–32: 2004 Timber: fungi and insect pests.

WIS 2/3–33: 2005 Wood preservation – chemicals and processes.

WIS 2/3–37: 2005 Softwood sizes.

WIS 2/3–38: 1995 Durability and preservative treatment of wood – European standards.

WIS 2/3–42: 1997 Particleboards – European standards.

WIS 2/3–46: 1997 Fibreboards – European standards.

WIS 2/3–49: 1997 Plywood – European standards.

WIS 2/3–51: 2003 Timber engineering hardware and connectors.

WIS 2/3–54: 1999 Exterior coatings on 'alternative' hardwoods.

WIS 2/3–55: 2001 UK-grown birch – suitable end-uses.

WIS 4–7: 2003 Timber strength grading and strength classes.

WIS 4–11: 1991 Timber and wood-based sheet materials in fire.

WIS 4–14: 1999 Moisture in timber.

WIS 4–16: 2002 Timber in joinery.

WIS 4–25: 1997 Fire tests for building materials – European standards.

WIS 4–28: 1998 Durability by design.

WIS 4–29: 2002 Dry-graded structural softwood.

ADVISORY ORGANISATIONS

British Wood Preserving & Damp Proofing Association, 1 Gleneagles House, Vernon Gate, South Street, Derby DE1 1UP (01332 225100).

British Woodworking Federation, 55 Tufton Street, London SW13QL (0870 458 6939).

English Nature, Northminster House, Peterborough, Cambridgeshire PE1 1UA (01733 455100).

Finnish Plywood International, Stags End House, Gaddesden Row, Hemel Hempstead HP2 6HN (01532 794661).

Glued Laminated Timber Association, Chiltern House, Stocking Lane, Hughenden Valley, High Wycombe, Bucks. HP14 4ND (01494 565180).

Thatching Advisory Services Ltd., The Old Stables, Redenham Park Farm, Redenham, Andover, Hampshire SP11 9AQ (01264 773820).

Timber Research and Development Association, Stocking Lane, Hughenden Valley, High Wycombe, Bucks. HP14 4ND (01494 569600).

UK Timber Frame Association Ltd., 14 Kinnerton Place South, London SW1X 8EH (020 7235 3364).

Wood Panel Industries Federation, 28 Market Place, Grantham, Lincolnshire NG31 6LR (01467 563707).

5

FERROUS AND NON-FERROUS METALS

Introduction

A wide range of ferrous and non-ferrous metals and their alloys are used within construction, but iron, steel, aluminium, copper, lead and zinc predominate. Over the last decade titanium has featured significantly in construction, having previously been used mainly in the chemical process industry and for military purposes. Recent trends have been towards the development of more durable alloys and the use of coatings both to protect and give visual diversity to the product ranges. Generally the metals require a large energy input for their production from raw materials; however, this high embodied energy is partially offset by the long life and recycling of most metals. The recovery rate of steel from demolition sites is 94%, with 10% being reused and 84% being recycled. Approximately 50% of current steel production is from scrap, and steel can be recycled any number of times without any degradation of the material.

Ferrous metals

Ferrous metals are defined as those in which the element iron predominates. The earliest use of the metal was for the manufacture of implements and weapons in the Iron Age commencing in Europe *circa* 1200 BC. Significant developments were the use by Wren in 1675 of a wrought-iron chain in tension to restrain the outward thrust from the dome of St Paul's Cathedral; the use of cast iron in compression for the Ironbridge at Coalbrookdale in 1779; and by Paxton in the prefabricated sections of the Crystal Palace in 1851. Steel is a relatively recent material, only being available in quantity after the development of the Bessemer converter in the late nineteenth century. The first steel-frame high-rise building of 10 storeys was built in 1885 in Chicago by William le Baron Jenney.

The platform level of the Waterloo International Terminal in London (Fig. 5.1) is covered by curved and tapered 3-pin steel arches, which are designed to accommodate the flexing inevitably caused by the movement of trains at this level. The steel arches, each consisting of two prismatic bow-string trusses connected by a knuckle joint, are asymmetrical to allow for the tight curvature of the site. The tops of the longer trusses are covered with toughened glass providing views towards old London, with profiled stainless steel spanning between. The area spanned by the shorter trusses is fully glazed. The structure is designed for a minimum lifetime of 100 years.

MANUFACTURE OF STEEL

The production of steel involves a sequence of operations which are closely inter-related in order to ensure maximum efficiency of a highly energy-intensive process. The key stages in the production process are the making of pig iron, its conversion into steel, the casting of the molten steel and its formation into sections or strip. Finally, coils of steel strip are cold rolled into thin sections and profiled sheet.

Manufacture of pig iron

The raw materials for the production of iron are iron ore, coke and limestone. Most iron ore is imported from America, Australia and Scandinavia, where the iron content of the ore is high. Coke is produced from coking coal, mainly imported from Europe, in batteries

Fig. 5.1 Structural steelwork — Waterloo International Terminal, London. Architects: Nicholas Grimshaw and Partners. Photographs: Courtesy of Jo Reid & John Peck, Peter Strobel

of coking ovens. Some of this coke is then sintered with iron ore prior to the iron-making process.

Iron ore, coke, sinter and limestone are charged into the top of the blastfurnace (Fig. 5.2). A hot air blast, sometimes enriched with oxygen, is fed through the tuyères into the base of the furnace. This heats the furnace to white heat, converting the coke into carbon monoxide which then reduces the iron oxide to iron. The molten metal collects at the bottom of the furnace. The limestone forms a liquid slag, floating on the surface of the molten iron. Purification occurs as impurities within the molten iron are preferentially absorbed into the slag layer.

$$2C + O_2 \longrightarrow 2CO$$
carbon (coke)   oxygen   carbon monoxide

$$Fe_2O_3 + 3CO \longrightarrow 2Fe + 3CO_2$$
iron ore   carbon   iron   carbon dioxide
(haematite)   monoxide

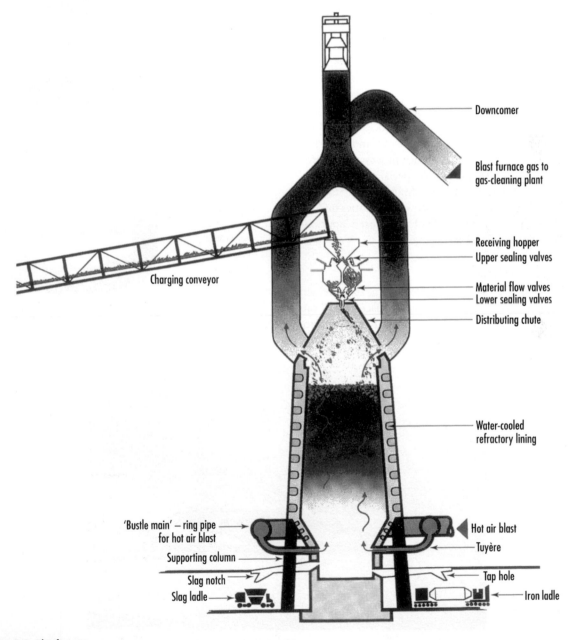

Downcomer

Blast furnace gas to gas-cleaning plant

Receiving hopper
Upper sealing valves

Material flow valves
Lower sealing valves

Distributing chute

Water-cooled refractory lining

Charging conveyor

'Bustle main' – ring pipe for hot air blast

Supporting column

Slag notch

Slag ladle

Hot air blast

Tuyère

Tap hole

Iron ladle

Fig. 5.2 Blastfurnace

The whole process is continuous, as relining the blastfurnace with the special refractory bricks is expensive and time-consuming. From time to time as the molten slag level rises, excess is tapped off for subsequent disposal as a by-product of the steel-making industry. When *hot metal* is required for the subsequent steel-making process it is tapped off into huge ladles for transportation direct to the steel converter. At this stage the iron is only 90–95% pure with sulfur, phosphorus, manganese and silicon as impurities and a carbon content of 4–5%. Waste gases from the blastfurnace are cleaned and recycled as fuel within the plant. A blastfurnace will typically operate non-stop for ten years producing 40 000 tonnes per week.

Steelmaking

There are two standard processes used within the UK for making steel. The basic oxygen process is used for the manufacture of bulk quantities of standard-grade steels and the electric arc furnace process is used for the production of high-quality special steels and particularly stainless steel. The Manchester Stadium (Fig. 5.3) built for the Commonwealth Games in 2002 and Manchester City Football Club used approximately 2000 tonnes of structural steel.

Basic oxygen process
Bulk quantities of steel are produced by the basic oxygen process in a refractory lined steel furnace which can be tilted for charging and tapping. A typical furnace (Fig. 5.4) will take a charge of 350 tonnes and convert it into steel within 30 minutes. Initially scrap metal, accounting for one quarter of the charge, is loaded into the tilted furnace, followed by the remainder of the charge as hot metal direct from the blastfurnace. A water-cooled lance is then lowered to blow high-pressure oxygen into

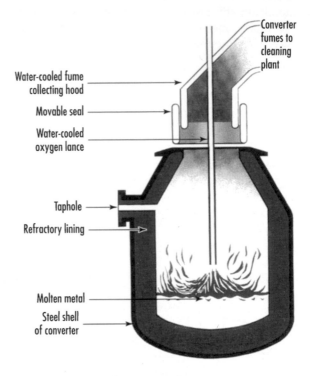

Fig. 5.4 Basic oxygen furnace

Fig. 5.3 Structural steelwork – Manchester City Stadium. Architects: Arup Associates. Photograph: Courtesy of Arup Associates

the converter. This burns off impurities and reduces the excess carbon content while raising the temperature. Argon and a small quantity of nitrogen are introduced at the bottom of the furnace. Lime is added to form a floating slag to remove further impurities and alloying components are added to adjust the steel composition, prior to tapping. Finally, the furnace is inverted to run out any remaining slag prior to the next cycle.

Electric arc process
The electric arc furnace (Fig. 5.5) consists of a refractory lined hearth, covered by a removable roof, through which graphite electrodes can be raised and lowered. With the roof swung open, scrap metal is charged into the furnace, the roof is closed and the electrodes lowered to near the surface of the metal. A powerful electric arc is struck between the electrodes and the metal, which heats it up to melting point. Lime and fluorspar are added to form a slag, and oxygen is blown into the furnace to complete the purification process. When the temperature and chemical analysis are correct, the furnace is tilted to tap off the metal, to which appropriate alloying components may then be added. A typical furnace will produce 150 tonnes of high-grade or stainless steel within 90 minutes.

Casting

Traditionally the molten steel was cast into ingots, prior to hot rolling into slabs and then sheet. However, most steel is now directly poured, or teemed, and cast into continuous billets or slabs, which are then cut to appropriate lengths for subsequent processing. Continuous casting (Fig. 5.6), which saves on reheating, is not only more energy efficient than processing through the ingot stage but also produces a better surface finish to the steel. However, components such as the nodes for rectangular and circular hollow-section constructions and large pin-joint units, are manufactured directly as individual castings. They may then be welded to the standard milled steel sections to give continuity of structure.

Hot-rolled steel

Sheet steel is produced by passing 25-tonne hot slabs at approximately 1250°C through a series of computer-controlled rollers which reduce the thickness to typically between 1.5 and 20 mm prior to water cooling and coiling. A 25-tonne slab would produce 1 km coil of 2 mm sheet. Steel sections such as universal beams and columns, channels and angle (Fig. 5.7) are rolled from hot billets through a series of *stands* to the appropriate section.

Cold-rolled steel

Sheet steel may be further reduced by cold rolling, which gives a good surface finish and increases its tensile strength. Light round sections may be processed into steel for concrete reinforcement, whilst coiled sheet may be converted into profiled sheet or light steel sections (Fig. 5.7). Cold-reduced steel for construction is frequently factory finished with zinc, alloys including terne (lead and tin) or plastic coating.

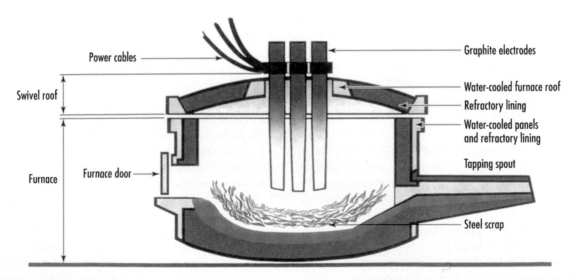

Fig. 5.5 Electric arc furnace

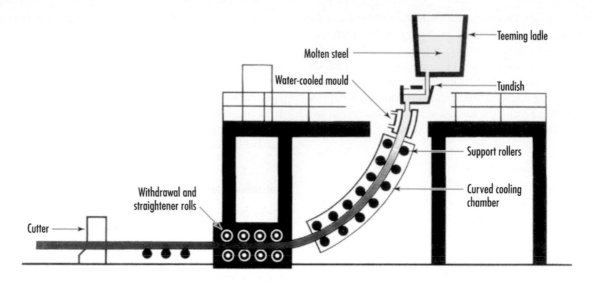

Fig. 5.6 Continuous casting

CARBON CONTENT OF FERROUS METALS

The quantity of carbon alloyed with iron has a profound influence on the physical properties of the metal due to its significant effect on the microscopic crystal structure (Fig. 5.8). At ambient temperature a series of crystal forms (ferrite, pearlite and cementite) associated with different proportions of iron and carbon are stable. However, on increasing the temperature, crystal forms that were stable under ambient conditions, become unstable and are recrystallised into the high temperature form (austenite). This latter crystal structure can be trapped at room temperature by the rapid quenching of red-hot steel, thus partially or completely preventing the natural recrystallisation processes which otherwise would occur on slow cooling. These effects are exploited within the various heat treatments that are applied to steels in order to widen the available range of physical properties.

Wrought iron

Wrought iron contains only about 0.02% of carbon. It was traditionally made by re-melting and oxidising pig iron in a reverberatory furnace. The process was continued until virtually all the high carbon content of the pig iron had been burnt off to produce a pasty wrought iron which was withdrawn from the furnace and then hammered out. Wrought iron is fibrous in character due to the incidental incorporation within the metal of slag residues and impurities such as magnesium sulfide, which are formed into long veins by the hammering

process. Wrought iron has a high melting point, approaching 1540°C, depending upon its purity. It was traditionally used for components in tension due to its tensile strength of about 350 N/mm². It is ductile and easily worked or forged when red hot, thus eminently suitable for crafting into ornamental ironwork, an appropriate use because of its greater resistance to corrosion than steel. Because of its high melting point, wrought iron cannot be welded or cast. Production ceased in the UK in 1973 and modern wrought iron is either recycled old material or, more frequently, low-carbon steel, with its attendant corrosion problems.

Cast iron

Cast iron contains in excess of 2% carbon in iron. It is manufactured by the carbonising of pig iron and scrap with coke in a furnace. The low melting point of around 1130°C and its high fluidity when molten, give rise to its excellent casting properties but, unlike wrought iron, it cannot be hot worked and is generally a brittle material. The corrosion resistance of cast iron has been exploited in its use for boiler castings, street furniture and traditional rainwater goods. Modern foundries manufacture castings to new designs and as reproduction Victorian and Edwardian components.

Differing grades of cast iron are associated with different microscopic crystal structures. The common grey cast iron contains flakes of graphite, which cause the characteristic brittleness and impart the grey colour to fractured surfaces. White cast iron contains the carbon as crystals of cementite (iron

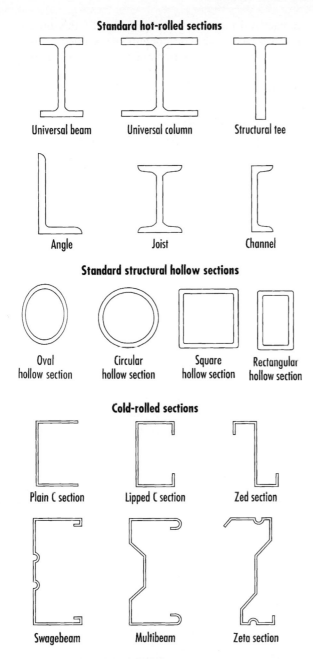

Fig. 5.7 Hot-rolled and cold-rolled sections (after Trebilcock, P.J. 1994: *Building design using cold formed steel sections: an architect's guide.* Steel Construction Institute)

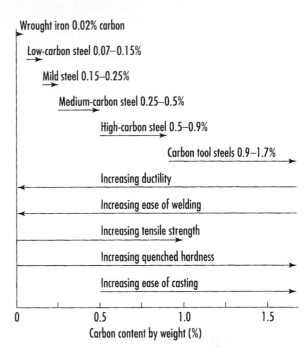

Fig. 5.8 Effect of carbon content on the properties of wrought iron and steels

carbide, Fe_3C) formed by rapid cooling of the melt. This material may be annealed to reduce its brittle character. A more ductile cast iron (spheroidal cast iron) is produced by the addition of magnesium and ferrosilicon and annealing which causes the carbon to crystallise into graphite nodules. This material has an increased tensile strength and significantly greater impact resistance. All cast irons are strong in compression.

Road iron goods, such as manhole covers, made from largely recycled grey cast iron are heavy but brittle. Where increased impact resistance is required for public roads, lighter and stronger ductile iron components are used. Traditional sand cast rainwater goods are usually manufactured from grey cast iron, while cast iron drainage systems are manufactured from both grey and ductile iron. Unlike steel, cast iron does not soften prematurely in a fire, but may crack if cooled too quickly with water from a fire hose. Cast iron drainage systems in both grey and spheroidal cast iron are covered by the standard BE EN 877: 1999. Cast iron drainage systems are particularly appropriate in heritage and conservation areas.

Steels

A wide range of steels are commercially available reflecting the differing properties associated with carbon content, the various heat treatments and the addition of alloying components.

Carbon contents of steels range typically between 0.07% and 1.7% and this alone is reflected in a wide spectrum of physical properties. The low-carbon (0.17–0.15%) and mild steels (0.15–0.25%) are relatively soft and can be subjected to extensive cold working. Medium-carbon steels (0.25–0.5%), which are often heat treated, are hard wearing. High-carbon steels (0.5–0.9%) and carbon tool steels (0.9–1.7%) exhibit increasing strength and wear resistance with increasing carbon content.

HEAT TREATMENT OF STEELS

The physical properties of steels can be modified by various heat treatments which involve heating to a particular temperature followed by cooling under controlled conditions.

Hardening

Rapidly quenched steel, cooled quickly from a high temperature in oil or water, thus retaining the high temperature crystalline form, is hard and brittle. This effect becomes more pronounced for the higher carbon content steels, which are mostly unsuitable for engineering purposes in this state.

Annealing and normalising

These processes involve the softening of the hard steel, by recrystallisation, which relieves internal stresses within the material and produces a more uniform grain structure. For annealing, the steel is reheated and soaked at a temperature of over 700°C, then cooled slowly at a controlled rate within a furnace or cooling pit. This produces the softest steel for a given composition. With normalising, the steel is reheated to a similar temperature for a shorter period and then allowed to cool more rapidly in air. This facilitates subsequent cold working and machining processes.

Tempering

Reheating the steel to a moderate temperature (400–600°C), followed by cooling in air, reduces the brittleness by allowing some recrystallisation of the metal. The magnitude of the effect is directly related to the tempering temperature, with ductility increasing and tensile strength reducing for the higher process temperatures.

Carburising

Components may be case hardened to produce a higher carbon content on the outer surface, whilst leaving the core relatively soft; thus giving a hard wearing surface without embrittlement and loss of

impact resistance to the centre. Usually this process involves heating the components surrounded by charcoal or other carbon-based material to approximately 900°C for several hours. The components are then heat treated to fully develop the surface hardness.

SPECIFICATION OF STEELS

Steels within the European Union are designated by a series of European Standards, BS EN 10025: 2004.

Hot-rolled structural steels:

BS EN 10025–1: 2004	General technical data
BS EN 10025–2: 2004	Non-alloy structural steels
BS EN 10025–3: 2004	Weldable fine-grained structural steels
BS EN 10025–4: 2004	Rolled weldable fine-grained structural steels
BS EN 10025–5: 2004	Steels with improved atmospheric corrosion resistance
BS EN 10025–6: 2004	High yield strength structural steels

In addition, BS EN 10210–1: 2006 and BS EN 10219–1: 2006 relate to hot- and cold-formed structural hollow sections respectively. The standard grades and their associated characteristic strengths are illustrated in Tables 5.1, 5.2, 5.3, 5.4 and 5.5. In the standards, S refers to structural steel and the subsequent coding numbers relate to the minimum yield strength. The sub-grade letters refer to impact resistance and other production conditions and compositions, such as W for weather-resistant steel. Steel numbers for each grade of steel are defined by BS EN 10027–2: 1992.

The following example illustrates the two coding systems for one standard grade of steel:

S275JR (BS EN 10027–1: 2005)
1.0044 (BS EN 10027–2: 1992)

S275JR	S refers to structural steel. The yield strength is 275 MPa. J is the lower impact strength at room temperature R.
1.0044	The first digit is the material group number with steel 1. The second pair of digits is the steel group number with 00 referring to a non-alloy base steel. The final digits refer to the particular grade of non-alloy steel.

Table 5.1 Steel designations for standard grades to BS EN 10025–2: 2004 (Hot-rolled products of non-alloy structural steels)

Designation		Properties	
BS EN 10027–1: 2005 & BS EN 10027–2: 1992		BS EN 10025–2: 2004 limits	
Grade	Number	Ultimate tensile strength (MPa)	Minimum yield strength (MPa)
S185	1.0035	290–510	185
S235JR	1.0038	360–510	235
S235J0	1.0114	360–510	235
S235J2	1.0117	360–510	235
S275JR	1.0044	410–560	275
S275J0	1.0143	410–560	275
S275J2	1.0145	410–560	275
S355JR	1.0045	470–630	355
S355J0	1.0553	470–630	355
S355J2	1.0577	470–630	355
S355K2	1.0596	470–630	355

Notes:

Sub-grades JR, J0, J2 and K2 indicate increasing impact resistance as measured by the Charpy V-notch test. K has a higher impact energy than J, the symbols R, 0 and 2 refer to the impact test at room temperature, $0°C$ and $-20°C$ respectively.

Data is for thicknesses of 16 mm or less.

Table 5.2 Steel designations for higher grade structural steels to BS EN 10025–3: 2004 (Hot-rolled products in weldable fine grain structural steels)

Designation		Properties	
BS EN 10027–1: 2005 & BS EN 10027–2: 1992		BS EN 10025–3: 2004 limits	
Grade	Number	Ultimate tensile strength (MPa)	Minimum yield strength (MPa)
S275N	1.0490	370–510	275
S275NL	1.0491	370–510	275
S355N	1.0545	470–630	355
S355NL	1.0546	470–630	355
S420N	1.8902	520–680	420
S420NL	1.8912	520–680	420
S460N	1.8901	550–720	460
S460NL	1.8903	550–720	460

Notes:

Sub-grade N (normalised or normalised rolled) relates to the physical state of the steel and L (low temperature impact) to high impact resistance.

Data is for thicknesses of 16 mm or less.

Table 5.3   Steel designations for higher grades to BS EN 10025–4: 2004 (Hot-rolled products in thermomechanical-rolled weldable fine-grain structural steels)

Designation		Properties	
BS EN 10027–1: 2005 & BS EN 10027–2: 1992		BS EN 10025–4: 2004 limits	
Grade	Number	Ultimate tensile strength (MPa)	Minimum yield strength (MPa)
S275M	1.8818	370–530	275
S275ML	1.8819	370–530	275
S355M	1.8823	470–630	355
S355ML	1.8834	470–630	355
S420M	1.8825	520–680	420
S420ML	1.8836	520–680	420
S460M	1.8827	540–720	460
S460ML	1.8838	540–720	460

Notes:

Sub-grade M (thermomechanical rolled) relates to the physical state of the steel and L (low temperature impact) to high impact resistance.

Data is for thicknesses of 16 mm or less.

Table 5.4   Steel designations for weather-resistant grades to BS EN 10025–5: 2004

Designation		Properties	
BS EN 10027–1: 2005 & BS EN 10027–2: 1992		BS EN 10025–5: 2004 limits	
Grade	Number	Ultimate tensile strength (MPa)	Minimum yield strength (MPa)
S235J0W	1.8958	360–510	235
S235J2W	1.8961	360–510	235
S355J0WP	1.8945	470–630	355
S355J2WP	1.8946	470–630	355
S355J0W	1.8959	470–630	355
S355J2W	1.8965	470–630	355
S355K2W	1.8966	470–630	355

Notes:

Sub-grades J0, J2 and K2 respectively indicate increasing impact resistance.

Sub-grade W refers to weather resistant steel.

P indicates a high phosphorus grade.

Data is for thicknesses of 16 mm or less.

STRUCTURAL STEELS

Weldable structural steels, as used in the Wembley Stadium, London (Fig. 5.9), have a carbon content within the range 0.16–0.25%. Structural steels are usually normalised by natural cooling in air after hot rolling. The considerable size effect which causes the larger sections to cool more slowly than the thinner sections gives rise to significant differences in physical properties, thus an 80 mm section can typically have a 10% lower yield strength than a 16 mm section of the same steel. Whilst grade S275 had

Table 5.5   Steel designations for high-yield strength quenched and tempered steels to BS EN 10025–6: 2004

Designation		Properties	
BS EN 10027–1: 2005 & BS EN 10027–2: 1992		BS EN 10025–6: 2004 limits	
Grade	Number	Ultimate tensile strength (MPa)	Minimum yield strength (MPa)
S460Q	1.8908	550–720	460
S500Q	1.8924	590–770	500
S550Q	1.8904	640–820	550
S620Q	1.8914	700–890	620
S690Q	1.8931	770–940	690
S890Q	1.8940	940–1100	890
S960Q	1.8941	980–1150	960

Notes:
Q indicates quenched steel.
Data is for thicknesses between 3 mm and 50 mm.

Fig. 5.9 Structural steelwork – Wembley Stadium, London. Architects: Foster and Partners. Photograph: Arthur Lyons

previously been considered to be the standard grade structural steel and is still used for most small beams, flats and angles, the higher grade S355 is increasingly being used for larger beams, columns and hollow sections.

Hollow sections

Circular, oval, square and rectangular hollow sections are usually made from flat sections which are progressively bent until almost round. They are then passed through a high-frequency induction coil to raise the edges to fusion temperature, when they are forced together to complete the tube. Excess metal is removed from the surface. The whole tube may then be reheated to normalising temperature (850–950°C), and hot-rolled into circular, oval, rectangular or square sections. For smaller sizes, the tube is heated to 950–1050°C and stretch reduced to appropriate dimensions. The standard steel grades to BS EN 10210–1: 1994 are S275J2H and S355J2H (Table 5.6). Cold-formed hollow sections differ in material characteristics from the hot-finished sections and conform to BS EN 10219: 1997. The lowest grade S235, with a minimum yield strength of 235 MPa is imported, but the standard non-alloy grades are S275 and S355. Grades S420 and S460 are designated as alloy special steels (Table 5.7).

Bending of structural sections

Castellated beams, rolled, hollow and other sections can be bent into curved forms by specialist metal bending companies. The minimum radius achievable depends upon the metallurgical properties, thickness and the cross-section. Generally, smaller sections can be curved to smaller radii than the larger sections, although for a given cross-section size the heavier-gauge sections can be bent to smaller radii than the thinner-gauge sections. Normally universal sections can be bent to tighter radii than hollow sections of the same dimensions. Elegant

Table 5.6   Steel designations for hot-finished structural hollow sections to BS EN 10210: 2006 (Hot-finished structural sections of non-alloy and fine grain structural steels)

Designation		Properties	
BS EN 10027–1: 2005 & BS EN 10027–2: 1992		BS EN 10210–1: 2006 limits	
Grade	Number	Ultimate tensile strength (MPa)	Minimum yield strength (MPa)
S235JRH	1.0039	360–510	235
S275J0H	1.0149	410–560	275
S275J2H	1.0138	410–560	275
S355J0H	1.0547	470–630	355
S355J2H	1.0576	470–630	355
S355K2H	1.0512	470–630	355
S275NH	1.0493	370–510	275
S275NLH	1.0497	370–510	275
S355NH	1.0539	470–630	355
S355NLH	1.0549	470–630	355
S420NH	1.8750	520–680	420
S420NLH	1.8751	520–680	420
S460NH	1.8953	540–720	460
S460NLH	1.8956	540–720	460

Notes:
H refers to hollow sections.
Sub-grades JR, J0 and J2 indicate impact resistance at room temperature, 0°C and −20°C respectively.
K2 refers to higher impact energy than J2.
Sub-grade N (normalised or normalised rolled) relates to the physical state of the steel and L (low temperature impact) to high impact resistance.
The standard UK production grades are the S275J2H and S355J2H designations.
Data is for thicknesses between 3 mm and 16 mm.

Table 5.7 Steel designations for cold-formed structural hollow sections to BS EN 10219–1: 2006 (Cold-formed welded structural hollow sections of non-alloy and fine grain steels)

Designation		Properties	
BS EN 10027–1: 2005 & BS EN 10027–2: 1992		BS EN 10219–1: 2006 limits	
Grade	Number	Ultimate tensile strength (MPa)	Minimum yield strength (MPa)
S235JRH	1.0039	360–510	235
S275J0H	1.0149	410–560	275
S275J2H	1.0138	410–560	275
S355J0H	1.0547	470–630	355
S355J2H	1.0576	470–630	355
S355K2H	1.0512	470–630	355
S275NH	1.0493	370–510	275
S275NLH	1.0497	370–510	275
S355NH	1.0539	470–630	355
S355NLH	1.0549	470–630	355
S460NH	1.8953	540–720	460
S460NLH	1.8956	540–720	460
S275MH	1.8843	360–510	275
S275MLH	1.8844	360–510	275
S355MH	1.8845	450–610	355
S355MLH	1.8846	450–610	355
S420MH	1.8847	500–660	420
S420MLH	1.8848	500–660	420
S460MH	1.8849	530–720	460
S460MLH	1.8850	530–720	460

Notes:

H refers to hollow sections.

Sub-grades JR, J0 and J2 indicate impact resistance at room temperature, 0°C and −20°C respectively.

K2 refers to higher impact energy than J2.

Sub-grades M (thermomechanical rolled) and N (normalised or normalised rolled) relate to the physical state of the steel and L (low temperature impact) to high impact resistance.

The standard UK production grades are the S275J2H and S355J2H designations.

Data is for thicknesses between 3 mm and 16 mm.

structures, such as Merchants Bridge, Manchester (Fig. 5.10) can be produced with curved standard sections and also curved tapered beams. The cold bending process work hardens the steel, but without significant loss of performance within the elastic range appropriate to structural steelwork. Tolerances on units can be as low as ± 2 mm with multiple bends, reverse curvatures and bends into three dimensions all possible. Increasingly cold bending is replacing induction or hot bending which require subsequent heat treatment to regain the initial steel properties.

BI-STEEL

Bi-steel panels consist of two steel plates held apart by an array of welded steel bar connectors (Fig. 5.11).

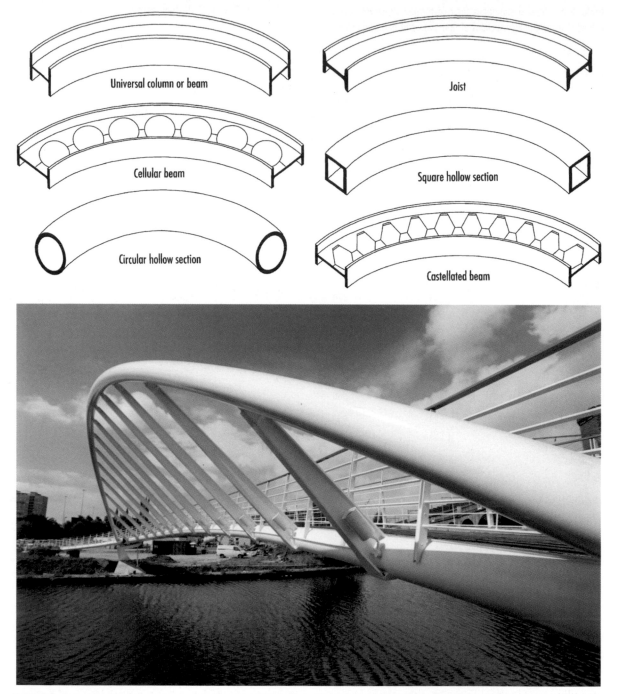

Fig. 5.10 Curved steel sections – Merchants Bridge, Manchester. Photograph: Courtesy of The Angle Ring Company Ltd

The panels are usually assembled into larger modules for delivery to site, where they are erected and the void space filled with concrete. The combination of permanent steel formwork and concrete fill acts as reinforced concrete, with the steel providing resistance to in-plane and bending forces and the concrete offering resistance to compression and shear. Units are manufactured up to 2 m wide and 18 m long in S275 or S355 steel to thicknesses between 200 mm and 700 mm and may be flat or curved. Adjacent panels may be bolted or fixed with proprietary connectors, giving fast erection times on site.

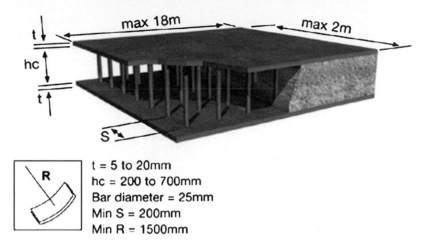

t = 5 to 20mm
hc = 200 to 700mm
Bar diameter = 25mm
Min S = 200mm
Min R = 1500mm

Fig. 5.11 Bi-steel unit. Photograph: Courtesy of Corus

FIRE PROTECTION OF STRUCTURAL STEEL

The fire protection of structural steel may be approached either by the traditional method involving the application of insulation materials with standard fire resistance periods (Fig. 5.12) or by a structural fire engineering method, which predicts the potential rate of rise of temperature of exposed steel members in each situation, based on the calculated fire load and particular exposure of the steel.

Applied protection to structural steel

Intumescent coatings

Thin-film intumescent coatings, which do not seriously affect the aesthetic of exposed structural steelwork, offer up to 120 minutes' fire protection. A full colour range for application by spray, brush or roller can be used on steel and also for remedial work on old cast-iron or wrought-iron structures.

Sprayed coatings

Sprayed coatings based on either vermiculite cement or mineral-fibre cement may be applied directly to steel to give up to 240 minutes' fire protection. The process is particularly appropriate for structural steel in ceiling voids, where the over-spray onto other materials is less critical. The finish, which can be adjusted to the required thickness, is heavily textured, and the products are relatively cheap.

Boarded systems

Lightweight boards boxed around steel sections offer between 30 and 240 minutes' fire protection according

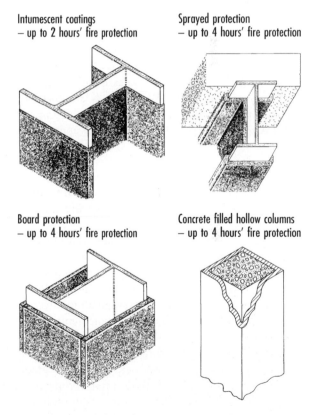

Intumescent coatings
– up to 2 hours' fire protection

Sprayed protection
– up to 4 hours' fire protection

Board protection
– up to 4 hours' fire protection

Concrete filled hollow columns
– up to 4 hours' fire protection

Fig. 5.12 Structural steelwork, typical fire-protection systems

to their thicknesses. Products are generally based on vermiculite or mineral fibres within cement, calcium silicate or gypsum binders. Boarded systems are screwed either directly to the structural steel, to light-gauge steel fixings or to a box configuration. Precoated products are available, or the standard systems may be subsequently decorated.

Preformed casings

Preformed sheet-steel casings which encase light-weight vermiculite plaster give a high-quality appearance and up to 240 minutes' fire resistance. The calculated fire resistance is based solely on the thickness of insulation and does not take into account any additional protection afforded by the sheet steel.

Masonry and concrete

Structural steel may be fully encased with masonry or suitably reinforced lightweight concrete in which non-spalling aggregates should be used. Hollow steel columns may be filled with plain fibre-reinforced or bar-reinforced concrete to give up to 120 minutes' fire resistance. For plain or fibre-reinforced concrete a minimum section of 140 mm × 140 mm or 100 mm × 200 mm is required and 200 mm × 200 mm or 150 mm × 250 mm for bar-reinforced concrete filling.

Water-filled systems

Interconnecting hollow steel sections can be given fire protection by being filled with water as part of a gravity feed or pumped system. Water loss is automatically replaced from a tank, where corrosion inhibitor and anti-freeze agents are added to the system as appropriate.

Fire engineering

The heating rate of a structural steel section within a fire depends upon the severity of the fire and the degree of exposure of the steel. Where a steel section has a low surface/cross-sectional area (Hp/A) ratio (Fig. 5.13), its temperature will rise at a slower rate than a section with a high Hp/A ratio. Fire-engineered solutions calculate the severity of a potential fire based on the enclosure fire loads, ventilation rates and thermal characteristics, and then predict temperature rises within the structural steel based on exposure. The stability of the structural member can therefore be predicted, taking into consideration its steel grade, loading and any structural restraint. From these calculations it can be determined whether additional fire protection is required and at what level to give the required fire resistance period.

Depending upon the particular circumstances, a fully loaded unprotected column with a section factor (Hp/A) of less than 50 m^{-1} may offer 30 minutes' structural fire resistance; similarly lighter columns with lightweight concrete blocks in the web can achieve 30 minutes' fire resistance. Shelf angle floors of suitable section, and in which a high proportion of

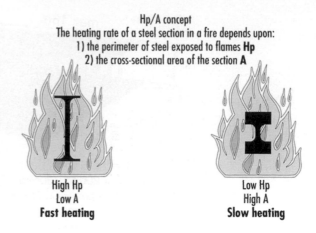

Hp/A concept
The heating rate of a steel section in a fire depends upon:
1) the perimeter of steel exposed to flames **Hp**
2) the cross-sectional area of the section **A**

High Hp
Low A
Fast heating

Low Hp
High A
Slow heating

Fig. 5.13 Hp/A ratios and rates of heating in fire

the steel is encased by the concrete floor construction, can achieve 60 minutes' fire resistance (Fig. 5.14).

PROFILED STEEL SHEETING

The majority of profiled sheet steel is produced by shaping the precoated strip through a set of rolls which gradually produce the desired section without damage to the applied coating. The continuous profiled sheet is then cut and packaged to customer requirements. The standard sections have a regular trapezoidal profile, with the depth of the section dependent on the loading and required span (Fig. 5.15). In cases where there is the risk of buckling, stiffeners are incorporated into the profile. Curved profiled sheets for eaves and soffits are manufactured by brake-pressing from the same coated strip. Trapezoidal profiles may be crimped in this process, although sinusoidal sheets and shallow trapezoidal sections can be curved without this effect. The rigidity of curved sections reduces their flexibility and thus the tolerances of these components. Proprietary spring-clip fixings may be used when concealed fixings are required for certain profiled sheet sections.

STEEL CABLES

Steel cables are manufactured by drawing annealed thin steel rod through a series of lubricated and tapered tungsten carbide dies, producing up to a ten-fold elongation. The drawing process increases the strength and reduces the ductility of the steel; thus the higher carbon steels, required for the production of high tensile wires, need special heat treatment before they are sufficiently ductile for the sequence of drawing processes.

Shelf angle floor
Up to 60 minutes' fire resistance

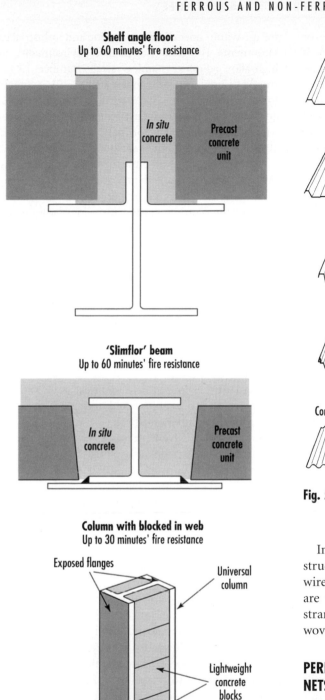

'Slimflor' beam
Up to 60 minutes' fire resistance

Column with blocked in web
Up to 30 minutes' fire resistance

Fig. 5.14 Fire resistance of structural-steel systems

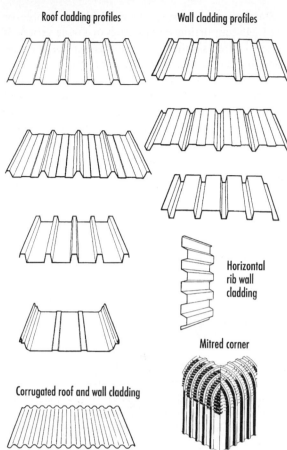

Fig. 5.15 Typical profiles for sheet-steel roofing and cladding

In order to manufacture steel cables for suspended structures or prestressed concrete, a set of individual wires are twisted into a strand, then a series of strands are woven around a central core of steel or fibre strand to produce a rope. A series of ropes are then woven to produce cable to the required specifications.

PERFORATED STEEL SHEETS, STEEL MESHES AND NETS

Perforated steel sheets are manufactured in mild steel, galvanised steel and stainless steel for use in architectural features, sunshades, balustrades, also wall and ceiling panels. Perforated sheets, also available in aluminium, copper, brass and bronze, may have round, square or slotted holes within a wide range of sizes and spacings to produce the desired aesthetic effect. Metal sheets are either punched or plasma profile cut.

Stainless steel meshes are available as flexible or rigid self-supporting weaves, each in a wide variety of patterns offering choice for use as external facades and sun-screening, also internal space dividers, balustrades, wall coverings and suspended ceilings. Patterns range from traditional weave and expanded metal to chain mail with a wide variation in texture and transparency. Some patterns are available in mild steel and also in non-ferrous metals. Stainless steel netting can be used creatively to form open tent and canopy structures.

Ferrous alloys

WEATHERING STEEL

Weathering steels are structural steels which have been alloyed with small proportions of copper, usually between 0.25 and 0.55%, together with silicon, manganese, chromium and either vanadium or phosphorus as minor constituents (BS 7668: 2004). The alloying has the effect of making the naturally formed brown rust coating adhere tenaciously to the surface thus preventing further loss by spalling. The use of weathering steels is not appropriate within marine environments, and all weathering steel must be carefully detailed to ensure that the rainwater run-off does not impinge on other materials, particularly concrete or glass where it will cause severe staining during the first few years of exposure to the elements. *Cor-Ten* is the commercial name for weathering steels. Table 5.4 gives the steel specification and steel number to the European Standards.

STAINLESS STEELS

Stainless steels are a range of alloys containing at least 10.5% chromium. The corrosion resistance of the material is due to the natural passive film of chromium oxide which immediately forms over the material in the presence of oxygen; thus if the surface is subsequently scratched or damaged the protective film naturally reforms. The corrosion resistance is increased by the inclusion of nickel and molybdenum as additional alloying components. The standard (austenitic) grades used within construction are 18% chromium, 10% nickel (1.4301) and 17% chromium, 12% nickel, 2.5% molybdenum (1.4401). The 18/10 alloy is suitable for use in rural and lightly polluted urban sites, while the 17/12/2.5 higher specification alloy is more appropriate for use within normal urban, marine and industrial environments. For certain aggressive environments, the high-alloy (duplex) stainless steel (number 1.4462) should be used. Ferritic stainless steel (1.4016) containing only chromium, with a reduced corrosion resistance, is appropriate for internal building use where corrosion is a less critical factor. Standard grades to BS EN 10088–1: 1995 for stainless steels are given in Table 5.8.

Stainless steel is manufactured by a three-stage process. Scrap is melted in an electric arc furnace, then refined in an argon–oxygen decarburizer and alloyed to the required composition in a ladle furnace by the addition of the minor constituents. Most molten metal is continuously cast into billets or slabs for subsequent forming. Stainless steel is hot rolled into plate, bar and sheet, while thin sections may be cold rolled. Heavy universal sections are made up from plates. Stainless steel may be cast or welded and is readily formed into small components such as fixings and architectural ironmongery. Polished, brushed, matt, patterned and profiled finishes are available; additionally, the natural oxide film may be permanently coloured by chemical and cathodic treatment to bronze, blue, gold, red, purple or green according to its final thickness.

Stainless steel is available in square, rectangular and circular hollow sections as well as the standard sections for structural work. Its durability is illustrated by the Lloyd's Building (Fig. 5.16), which maintains its high quality finish within the urban context of the City of London. Stainless steel is widely used for roofing, cladding, interior and exterior trim owing to its combined strength and low maintenance. The corrosion resistance of stainless steel also makes it eminently suitable for masonry fixings such as corbels, anchor bolts, cavity wall ties and for concrete reinforcement. Austenitic stainless steels are used for the manufacture of pipework, catering and drainage products where durability and corrosion resistance are critical. Exposed exterior stainless steel should be washed regularly to retain its surface characteristics. Pitting corrosion causing surface pin-point attack, crevice corrosion under tight-fitting washers and stress corrosion cracking, where the material is under high tensile load, may occur where inappropriate grades are used in aggressive environments.

High performance (superduplex) stainless steel (1.4507) wire ropes which have 50% more tensile strength than standard (austenitic) stainless steel (1.4401) are appropriate for architectural support and restraint systems. These alloys containing around

1.6% copper have a lower elastic stretch than standard stainless steels and a higher fatigue resistance, which makes them appropriate for architectural tensile elements including those within marine and swimming pool environments.

HEAT-TREATED STEELS

The size effect, which causes a reduced yield strength in large sections due to their slower cooling rates than the equivalent thin sections, can be ameliorated by the addition of small quantities of alloying elements such as chromium, manganese, molybdenum and nickel.

Coated steels

To inhibit corrosion, steel may be coated with metallic or organic finishes. Metallic finishes are typically zinc (Z), zinc-iron alloy (ZF), zinc-aluminium alloy (ZA), aluminium-zinc alloy (AZ) and aluminium-silicon-alloy (AS) all of which may be applied by hot-dipping of steel into the molten metal. The degree of corrosion protection is directly related to the thickness of the metallic coating (BS EN 10326: 2004). Organic coatings (BS EN 10169: 2003) may be divided into liquid paints, powder coatings and films. Certain products are suitable only for interior applications.

ZINC-COATED STEEL

The zinc coating of steel has for many years been a standard method for its protection against corrosion. The zinc coating may be applied by hot-dipping or spraying with the molten metal, sheradizing in heated

Fig. 5.16 Stainless steel construction – Lloyd's Building, London. Architects: Richard Rogers. Photograph: Arthur Lyons

Table 5.8 Stainless steel compositions and grades to BS EN 10088–1: 2005 for different environmental conditions

Designation			Suitable environments
Type	Name (indicating composition of alloying components)	Number	
Austenitic	X5CrNi18–10	1.4301	rural and clean urban
	X5CrNiMo17–12–2	1.4401	urban, industrial and marine
Ferritic	X6Cr17	1.4016	interior
Duplex	X2CrNiMoN22–5–3	1.4462	severe industrial and marine

Notes:
Cr, Ni, Mo and N refer to chromium, nickel, molybdenum and nitrogen respectively.
X2, X5 and X6 refer to the carbon contents of 0.02, 0.05 and 0.06% respectively.

zinc powder or electrodeposition. In hot-dip galvanising the steel is cleaned by pickling in acid followed by immersion in molten zinc or iron-zinc alloy. The zinc coating protects the steel by acting as a physical barrier between the steel and its environment, and also by sacrificially protecting the steel where it is exposed by cutting or surface damage. The iron-zinc alloy coating gives a better surface for painting or welding.

The durability of the coated steel is dependent upon the thickness of the coating (standard 275 g/m^2, i.e. 137.5 g/m^2 per face) and the environment. Coastal situations and industrial environments with high concentrations of salt and sulfur dioxide respectively may cause rapid deterioration. The alkalis in wet cement, mortar and plaster etch zinc coatings, but once dry, corrosion is slow; however, calcium chloride used as an accelerator in plaster is aggressive and should only be used sparingly. Fixings for zinc-coated sheet should be carefully chosen to avoid the formation of bimetallic couples, which can cause accelerated corrosion. In particular, no copper or brass should make contact with either zinc or iron-zinc alloy coated steel. Other metals such as lead, aluminium and stainless steel have less serious effects in clean atmospheres, but generally all fixings should be sealed and insulated by rubber-faced washers. Where zinc-coated steel is to be fixed to unseasoned timber or timber impregnated with copper-based preservatives, the wood should be coated with bitumen paint. Where damaged in cutting, fixing or welding, the zinc coating should be repaired with the application of zinc-rich paint.

Zinc-coated steel may be painted for decoration or improved corrosion resistance. However, the *normal spangle* zinc finish will show through paint and the *minimised spangle* or iron-zinc alloy finish is more appropriate for subsequent painting.

ALUMINIUM-ZINC ALLOY COATED STEEL

Steel coated with an alloy of aluminium (55%), zinc (43.5%) and silicon (1.5%) is more durable than that coated with an equivalent thickness of pure zinc, and may be used without further protection in non-aggressive environments. It is also used as the substrate for certain organic coatings.

TERNE-COATED AND LEAD-CLAD STEEL

Lead and terne, an alloy of lead (80–90%) and tin (20–10%), are used as finishes to steel and stainless steel for cladding and roofing units. Terne can be applied to sheet stainless steel as a 20 micron layer by immersion in the molten alloy. Terne-coated stainless steel does not suffer from bimetallic corrosion and can normally be used in contact with lead, copper, aluminium or zinc. Thermal movement is similar to stainless steel, allowing for units up to 9 m in length to be used for roofing and cladding. The composite material, lead-clad steel, is produced by cold-roll bonding 0.75 mm lead to 1.0 mm terne-plated steel or to 0.8 mm terne-plated stainless steel. Lead-clad steel is suitable for cladding and roofing systems and has the appearance and corrosion resistance of milled lead. Because of the support afforded by the steel substrate, lead-clad steel or stainless steel can be used for self-supported fasciae, soffits, gutters and curved sections. Joints can be lead burned and soldered as traditional lead. Cut ends should be protected by soldering in the case of lead-clad steel, although the stainless steel version requires no protective treatment. Unlike traditional lead, the material is virtually theft-proof and does not suffer significantly from creep. Patination oil should be applied to the lead surface after installation to prevent staining effects.

ORGANIC COATED STEEL

Since the 1960s, a range of heat-bonded organic coatings for steel has been developed including PVC plastisol *(Colorcoat)*, polyvinylidene fluoride (PVDF), polyesters and PVC film *(Stelvetite)*. Within this product range the PVC plastisol currently has the largest market share for cladding and roofing within the UK.

PVC plastisol coating

PVC plastisol is applied to zinc or aluminium/zinc-coated steel to a thickness of 0.2 mm. It has a tough leather grain finish and is available in a wide range of colours, although the pastel shades are recommended for roofing applications. The reverse side is usually coated with a grey corrosion-resistant primer and polyester finish, although PVC plastisol may be specified for unusually aggressive internal environments. Careful site storage and handling is required to prevent physical damage to the surface. For non-marine environments the most durable colours will give a period to first maintenance of greater than 20 years. Very deep colours, and the non-pastel shades in coastal locations, will have reduced periods to first repainting.

Polyvinylidene fluoride coating

Polyvinylidene fluoride (PVDF), an inert fluorocarbon, when applied as a 0.027 mm coating to zinc coated steel, has good colour stability at temperatures up to 120°C, making it suitable for worldwide use and for buildings which are likely to be extended at a later date. The finish is smooth and self-cleaning, although considerable care is required on site to prevent handling damage. A period to first maintenance of 15 years is typical within the UK for non-coastal locations. The wide colour range includes metallic silver. Polyvinylidene fluoride finished zinc/tin (terne)-coated steel or stainless steel is also available in a range of colour finishes which include copper, copper patina and stainless steel. The material can be used for cladding and roofing; the solar reflective paint finish reduces excessive solar gain.

Polyester coating

Polyester and silicone polyester-coated galvanised steels are economic products, but offer only medium-term life in non-aggressive environments. Externally the period to first maintenance will be typically 10 years in unpolluted inland locations, but they are suitable for internal use. Silicone polyester should not be used in marine or hot humid environments. Polyester and silicone polyester coatings are smooth and typically 0.025 mm in thickness.

Enamel coating

Organic enamel-coated steels offering good light reflectance are suitable for internal use as wall and roof linings. Coatings, usually 0.022 mm thick, are typically applied to hot-dip zinc/aluminium alloy-coated steel and are easily cleaned. The standard colour is brilliant white, but a range of light colours is also available.

PVC film coating

PVC film (0.02 mm) in a range of colours, decorative patterns and textured finishes is calendered to zinc-coated steel strip. The product is suitable only for internal applications.

Steel tiles and slates

Lightweight steel tile and slate units, manufactured from galvanised steel coated with acrylic resin and a granular finish, give the appearance of traditional slate or pantile roofs. The products have the advantage, particularly for refurbishment work, of lightness in comparison to the traditional materials. Units can

typically be used for roof pitches between 12 and 90°. A span of 1200 mm allows for wider spacing of roof-trussed rafters. Units in a range of traditional material colours are available with appropriate edge and ventilation accessories.

Aluminium

Aluminium has only been available as a construction material for about a hundred years. Possibly the most well known early use of the metal was for the cast statue of Eros which has stood in Piccadilly Circus, London since 1893. Because of its durability, it is widely in construction, particularly for secondary components as illustrated in the permanent shading devices on the Faculty of Divinity building of the University of Cambridge (Fig. 5.17).

MANUFACTURE

Aluminium, the most common metallic element in the earth's crust, is extracted from the ore bauxite, an impure form of aluminium oxide or alumina. The bauxite is dissolved in caustic soda, filtered, reprecipitated to remove impurities and dried. The pure alumina is then dissolved in fused cryolite (sodium aluminium fluoride) within a carbon-lined electrolytic cell. Electrolysis of the aluminium oxide produces oxygen and the pure aluminium, which is tapped off periodically and cast. The process is highly energy intensive, and typically the production of 1 tonne of aluminium requires 14 000 kWh of electrical energy. In the western world, 60% of this energy is from renewable hydroelectric power. Currently, 63% of new aluminium used in the UK is from recycled sources, and recycling requires only 5% of the energy input compared to primary production. Cast ingots or slabs are hot rolled at 500°C into 5 mm coiled sheet which subsequently can be cold rolled into thinner sheet or foil. Due to the ductility of aluminium, the metal can be extruded into complex shapes or drawn into wire. Forming and machining processes are generally easier than with steel. Aluminium components may also be formed by casting.

PROPERTIES

Aluminium is one of the lightest metals with a density of 2700 kg/m^3 compared to steel 7900 kg/m^3. Standard-grade aluminium (99% pure) has a tensile strength between 70 and 140 MPa, depending on

Fig. 5.17 Aluminium shading devices – Faculty of Divinity, University of Cambridge. Architects: Edward Cullinan Architects. Photograph: Arthur Lyons

temper; however, certain structural aluminium alloys (e.g. alloy 5083) achieve 345 MPa comparable to the 410–560 MPa for S275 steel. This compares favourably on a strength-to-weight basis, but the modulus of elasticity for aluminium is only one third that of steel, so deflections will be greater unless deeper sections are used. For an aluminium section to have the same stiffness as an equivalent steel member, the aluminium section must be enlarged to approximately half the weight of the steel section.

DURABILITY

The durability of aluminium as a construction material is due to the protection afforded by natural oxide film, which is always present on the surface of the metal. The aluminium oxide film, which is immediately produced when the surface of the metal is cut or scratched, is naturally only 0.01 micron thick, but may be thickened by the process of anodisation.

FIRE

The strength of aluminium is halved from its ambient value at a temperature of 200°C, and for many of the alloys is minimal by 300°C.

CONTACT WITH OTHER BUILDING MATERIALS

Whilst dry cement-based materials do not attack aluminium, the alkalinity of wet cement, concrete and mortar causes rapid corrosion. Thus, where these materials make contact during the construction process, the metal should be protected by a coating of bitumen paint. Furthermore, anodised and particularly coloured sections, such as glazing units, can be permanently damaged by droplets of wet cement products, and should be protected on site by a removable lacquer or plastic film. Under dry conditions aluminium is unaffected by contact with timber; however, certain timber preservatives, particularly those containing copper compounds, may cause corrosion under conditions of high humidity. Where this risk is present the metal should be protected with a coating of bitumen.

Although aluminium is highly resistant to corrosion in isolation, it can be seriously affected by corrosion when in contact with other metals. The most serious effects occur with copper and copper-based alloys, and rainwater must not flow from a copper roof or copper pipes into contact with aluminium. Except in marine and industrial environments it is safe to use stainless steel fixings or lead with aluminium, although zinc and zinc-coated steel fixings are more durable. Unprotected mild steel should not be in electrical contact with aluminium.

ALUMINIUM ALLOYS

Aluminium alloys fall into two major categories: cast or wrought. Additionally, the wrought alloys may be subjected to heat treatment. The majority of aluminium used in the construction industry is wrought, the content and degree of alloying components being directly related to the physical properties required, with the pure metal being the most malleable. BS EN 573–1: 2004 designates aluminium alloys into categories according to their major alloying components (Table 5.9).

The following example illustrates the coding system for aluminium and its alloys:

Structural aluminium is alloy EN AW–6082

Where EN refers to the European Norm, A for aluminium and W for wrought products. 6082 is within the 6000 series of magnesium and silicon alloys and the final three digits refer to the exact chemical composition as listed in BS EN 573–3: 2003.

For flashings where on-site work is necessary, 99.8% pure aluminium (alloy EN AW–1080A) or 99.5% (EN AW–1050A) offer the greatest malleability, although the standard commercial grade 99% pure aluminium (alloy EN AW–1200) is suitable for insulating foils and for continuously supported sheet roofing.

Profiled aluminium for roofing and cladding, requiring additional strength and durability is alloyed with 1.25% manganese (alloy EN AW–3103). It is produced from the sheet by roll-forming, and can be manufactured into curved sections to increase design flexibility. Preformed rigid flashings to match the profile sheet are manufactured from the same alloy and finish. The alloy with 2% magnesium (alloy EN AW–5251) is more resistant to marine environments. Aluminium rainscreen cladding panels up to 2.8×1.5 m in size, may also be shaped using the superplastic forming (SPF) process, which relies on the high extensibility of the alloy EN AW–5083SPF. Sheet alloy, typically 2 mm in thickness, is heated to 380–500°C, and forced by air pressure into the three-dimensional form of the mould. Horizontal or vertical ribs are frequently manufactured to give enhanced rigidity, but cladding panels may be formed to individual designs including curvature in two directions. Coloured finishes are usually polyester powder or polyvinylidene fluoride coatings.

Extruded sections for curtain walling, doors and windows require the additional strength imparted by alloying the aluminium with magnesium and silicon (alloy EN AW–6063). Thermal insulation within such extruded sections is achieved by a hidden thermal break or by an internal plastic or timber insulating cladding.

Structural aluminium for load-bearing sections and space frames typically contains magnesium, silicon and manganese (alloy EN AW–6082). Tempering increases the tensile strength to the range 270–310 MPa, which is more comparable to the standard grade of structural steel S275 (minimum tensile strength 410 MPa).

FINISHES FOR ALUMINIUM

Anodising

The process of anodising thickens the natural aluminium oxide film to typically 10–25 micron. The component is immersed in sulfuric acid and electrolytically made anodic, which converts the surface metal into a porous aluminium oxide film, which is then sealed by boiling in water. The anodising process increases durability and can be used for trapping dyes within the surface to produce a wide range of coloured products. Some dyes fade with exposure to sunlight, the most durable colours being gold, blue,

Table 5.9 Broad classification of aluminium alloys to BS EN 573–1: 2004

Alloy series	Major alloying components
1000	greater than 99% aluminium
2000	copper alloys
3000	manganese alloys
4000	silicon alloys
5000	magnesium alloys
6000	magnesium and silicon alloys
7000	zinc alloys
8000	other elements

(In many cases minor alloying components are also present.)

red and black. Exact colour matching for replacement or extensions to existing buildings may be difficult, and manufacturers will normally produce components within an agreed band of colour variation. If inorganic salts of tin are incorporated into the surface during the anodising process then colour-fast bronzes are produced. Depending upon the period of exposure to the electrolytic anodisation process, a range of colours from pale bronze to black may be produced. Different aluminium alloys respond differently to the anodising treatment. Pure aluminium produces a silver mirror finish, whereas the aluminium–silicon alloys (e.g. alloy EN AW–6063) produce a grey finish.

Surface textures

A range of surface textures is achieved by mechanical and chemical processing. Finishes include bright polished, matt, etched and pattern-rolled according to the pretreatments applied, usually before anodising and also the particular alloy used. The aluminium discs on the facade of the Selfridges building in Birmingham give an innovative decorative finish to the steel-frame store, forming an elegant contrast to the blue-painted rendered surface (Fig. 5.18).

Coatings

Zinc-coated aluminium is a foldable cladding material combining the durability of aluminium with the appearance of pre-weathered zinc. Matching rainwater goods including half-round gutters and down pipes are available. An equivalent pre-patinated or pre-oxidised titanium-coated aluminium with the appearance of bright steel is under development.

Polyester coatings, predominantly white, but with a wide range of colour options, are used for double-glazing systems, cladding panels and rainwater goods. Electrostatically-applied polyester powder is heat cured to a smooth self-cleaning finish. PVC simulated wood-grain and other pattern finishes may also be applied to aluminium extrusions and curtain-wall systems.

Paint

Where aluminium is painted for decorative purposes it is important that the appropriate primer is used. The aluminium should be abraded or etched to give a good key to the paint system, although cast aluminium normally has a sufficiently rough surface. Oxide primers are appropriate but red lead should be avoided.

Maintenance of finished aluminium

For long-term durability, all external aluminium finishes should be washed regularly, at intervals not normally exceeding three months, with a mild detergent solution. Damaged paint coatings may be touched-up on site, but remedial work does not have the durability of the factory-applied finishes.

ALUMINIUM IN BUILDING

Typical applications for aluminium and its alloys in building include roofing and cladding, curtain wall and structural glazing systems, flashings, rainwater goods, vapour barriers and – internally – ceilings, panelling, luminaires, ducting, architectural hardware and walkways.

Monocoque construction

The Lord's Cricket Ground Media Centre (Fig. 5.19) was the world's first semi-monocoque building in aluminium. The media centre is a streamlined pod raised 14 m off the ground on two concrete support towers, giving journalists and commentators an uninterrupted view over the cricket ground. The structure consists of a curved 6 and 12 mm aluminium-plate skin welded to a series of ribs. Thus acting together, the skin and the ribs provide both the shape and the structural stability, a system typically used in the boat-building and aircraft industries. The building was made in 26 sections and transported to the site for assembly.

Thermal breaks in aluminium

In order to overcome thermal bridging effects, where aluminium extrusions are used for double-glazing systems, thermal breaks are inserted between the aluminium in contact with the interior and exterior spaces. These may be manufactured from preformed polyamide strips or alternatively the appropriate extrusions are filled with uncured polymer, then the bridging aluminium is milled out after the plastic has set.

Jointing methods

Aluminium components may be joined mechanically with aluminium bolts or rivets; non-magnetic stainless steel bolts are also appropriate. If aluminium is to be electric-arc welded, the use of an inert-gas shield, usually argon, is necessary to prevent oxidation of the metal surface. A filler rod, compatible with the alloy

Fig. 5.18 Aluminium discs – Selfridges Store Birmingham. Architects: Foster and Partners. Photographs: Arthur Lyons

Fig. 5.19 Aluminium semi-monocoque construction – Lord's Media Centre, London. Architects: Future Systems. Photograph: Courtesy of Richard Davies

to be welded, supplies the additional material to make up the joint. Strong adhesive bonding of aluminium components is possible, providing that the surfaces are suitably prepared.

Copper

Copper was probably one of the first metals used by man, and evidence of early workings suggests that the metal was smelted as early as 7000 BC. Later it was discovered that the addition of tin to copper improved the strength of the material and by 3000 BC the Bronze Age had arrived. The Romans made extensive use of copper and bronze for weapons, utensils and ornaments. Brass from the alloying of copper and zinc emerged from Egypt during the first century BC. By the mid-eighteenth century South Wales was producing 90% of the world's output of copper, with the ore from Cornwall, but now the main sources are the United States, Chile and Europe. The traditional visual effect of copper is illustrated in a modern context by the millennium

project, Swan Bells in Perth, Australia (Fig. 5.20). The copper was initially clear-coated to prevent gradual oxidation and patinisation within the marine environment of the harbour.

MANUFACTURE

The principal copper ores are the sulfides (e.g. chalcocite), and sulfides in association with iron (e.g. chalcopyrite). Ores typically contain no more than 1% copper and therefore require concentration by flotation techniques before the copper is extracted through a series of furnace processes. The ores are roasted, then smelted to reduce the sulfur content, and produce *matte*, which contains the copper and a controlled proportion of iron sulfide. The molten matte is refined in a converter by a stream of oxygen. This initially oxidises the iron which concentrates into the slag and is discarded; sulfur is then burnt off to sulfur dioxide, leaving 99% pure metal, which on casting evolves the remaining dissolved gases and solidifies to *blister* copper. The blister copper is further refined in a furnace to remove remaining sulfur with air and then

Fig. 5.20 Copper cladding — Swan Bells Tower, Perth Australia. Architects: Hames–Sharley Architects. Photographs: Arthur Lyons

oxygen with methane or propane. Finally, electrolytic purification produces 99.9% pure metal. Approximately 40% of copper and the majority of brass and bronze used within the UK are recycled from scrap. Recycling requires approximately 25% of the energy used in the primary production of copper depending upon the level of impurities present.

GRADES OF COPPER

Grades of copper and copper alloys are designated by both a symbol and a number system to the current standards BS EN 1172: 1997 and BS EN 1412: 1996.

Thus, for example, phosphorus deoxidised non-arsenical copper, typically used for roofing and cladding would be defined as:

Symbol system:
Cu-DHP with very good welding, brazing and soldering properties.

Number system:
CW024A–R240
(C refers to copper, W refers to wrought products, 024A identifies the unique composition – Table 5.10). The subsequent letters and numbers define a specific requirement such as tensile strength or hardness; in this case, a minimum tensile strength of 240 MPa, which is half-hard temper.)

Only four of the numerous grades of copper are commonly used within the construction industry.

Table 5.10   Broad classification of copper alloys

Number series	Letters	Materials
000–099	A or B	copper
100–199	C or D	copper alloys (less than 5% alloying elements)
200–299	E or F	copper alloys (more than 5% alloying elements)
300–349	G	copper – aluminium alloys
350–399	H	copper – nickel alloys
400–449	J	copper – nickel – zinc alloys
450–499	K	copper – tin alloys
500–599	L or M	copper – zinc alloys
600–699	N or P	copper – zinc – lead alloys
700–799	R or S	copper – zinc alloys, complex

Notes:
The three-digit number designates each material and the letter indicates copper or the alloy group.

Electrolytic tough pitch high-conductivity copper (Cu-ETP or CW004A)

Electrolytic tough pitch high-conductivity copper is used mainly for electrical purposes; however, the sheet material is also used for fully supported traditional and long strip copper roofing. It contains approximately 0.05% dissolved oxygen which is evolved as steam if the copper is heated to 400°C in a reducing flame, thus rendering the metal unsuitable for welding or brazing.

Fire refined tough pitch copper (Cu-FRHC or CW005A)

Fire refined tough pitch copper has a similar specification to Cu-ETP, but with marginally more impurities.

Tough pitch non-arsenical copper (Cu-FRTP or CW006A)

Tough pitch non-arsenical copper is used for general building applications. It is suitable for sheet roofing.

Phosphorus deoxidised non-arsenical copper (Cu-DHP or CW024A)

Phosphorus deoxidised non-arsenical copper is the standard grade for most building applications including roofing, but not for electrical installations. The addition of 0.05% phosphorus to refined tough pitch copper isolates the oxygen rendering the metal suitable for welding and brazing. It is therefore used for plumbing applications where soldering is inappropriate.

COPPER FORMS AND SIZES

Copper is available as wire, rod, tube, foil, sheet and plate. Typical roofing grades are 0.45, 0.6 and 0.7 mm. The metal is supplied dead soft (fully annealed), one eighth or one quarter hard, half-hard or full-hard. It rapidly work hardens on bending, but this can be recovered by annealing at red heat. Copper can be worked at any temperature since, unlike zinc, it is not brittle when cold. The standard grade of copper used for roofs, pipes and domestic water-storage cylinders is phosphorus deoxidised non-arsenical copper CW024A, although the other tough pitch grades CW004A, CW005A and CW006A may also be used for roofs. Copper for pipework is supplied in annealed coils for mini/microbore systems, in 6 m lengths half-hard and hard for general plumbing work. The hard temper pipes cannot be bent. Plastic-coated tubes, colour coded to identify the service (e.g. yellow – gas), are available.

PATINA

The green patina of basic copper sulfate or carbonate on exposed copper gradually develops according to the environmental conditions. On roofs within a marine or industrial environment the green patina develops within five years; under heavy pollution it may eventually turn dark brown or black. Within a town environment, the patina on roofs will typically develop over a period of ten years. However, vertical copper cladding will normally remain a deep brown, due to the fast rainwater run-off, except in marine environments when the green colour will develop. On-site treatment to accelerate the patinisation process is unreliable, but pre-patinised copper sheet is available if the effect is required immediately. Green pre-patinised copper sheet should not be welded, brazed or soldered as heat treatment causes discolouration of the patina. The factory-generated bright green patina will weather according to the local environmental conditions, often turning quickly to a blue-green. Tinned copper, which is grey in colour, quickly weathers to a matt surface with the appearance of zinc or lead, but has the durability and workability of copper. The Urbis Centre in Manchester illustrates well the visual effect of a large feature patinised copper roof (Fig. 5.21).

CORROSION

Generally, copper itself is resistant to corrosion; however, rainwater run-off may cause staining on adjacent materials and severe corrosion to other metals. Zinc, galvanised steel and non-anodised aluminium should not be used under copper, although in this respect lead, stainless steel and brass are unaffected. Copper may cause corrosion to steel or anodised aluminium in direct contact, if moisture is present. Specifically, copper should not be installed below exposed bitumen, bitumen paint, or cedarwood shingles where leaching action producing acid solutions can cause localised attack on the metal. Additionally, some corrosion may arise from the acid produced by algae on tiled roofs. The accidental splashing of lime or cement mortar onto copper causes a blue-green discolouration; however, this can readily be removed with a soft brass brush. Some corrosion of copper pipework may be caused by soft water, particularly if high levels of dissolved carbon dioxide are present; hard waters generally produce a protective film of calcium compounds which inhibits corrosion. Pitting corrosion has been reported in rare cases associated with either hard deep-well waters or hot soft waters with a significant manganese content. Additionally, excessive acidic flux residues not removed by flushing the system may cause corrosion. Within heating systems in which oxygen in the primary circulating water is constantly being replenished through malfunction or poor design, bimetallic corrosion will occur between steel radiators and copper pipework. This will result in the build-up of iron oxide residues at the bottom of the radiators. The use of appropriate inhibitors will reduce this effect.

Fig. 5.21 Copper roof construction – Urbis Centre, Manchester. Architects: Ian Simpson Architects. Photograph: Courtesy of Chris Hodson

PROTECTIVE COATINGS FOR COPPER

A range of clear coatings is available for external and internal application to copper where the original colour and surface finish is to be retained. Air-drying acrylic thermosetting resins and soluble fluoropolymers are suitable for exterior use. The acrylic resins, which incorporate benzotriazole to prevent corrosion if the coating is damaged, have a life expectancy in excess of 10 years, whilst the recently developed fluropolymers should last for 20 years. For internal use polyurethane, vinyls, epoxy and alkyd resins are appropriate. Silicones are necessary for high temperature applications.

COPPER ROOFING SYSTEMS

Traditional and long-strip systems

Copper roofing systems may be categorised as traditional or long strip. The latter has the advantage that bays between 8.1 m and 14.6 m depending upon the pitch may be constructed without the necessary cross welts on sloping roofs or drips on flat roofs appropriate to the traditional system. This has significant cost benefits in terms of installation costs. The long-strip copper roof system (Fig. 5.22) with bays up to approximately 600 mm wide may be laid on roofs with pitches between 3° and 90° and uses one quarter or half-hard temper 0.6 mm or 0.7 mm copper strip.

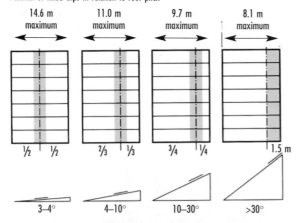

Position of fixed clips in relation to roof pitch

Recommended widths and standing seam centres for long-strip copper roofing

Width of strip (mm)	400	450	500	600	670
Standing seam centres (mm)	325	375	425	525	595

(using 0.6/0.7 mm copper strip at ¼ or ½ hard temper with a fixed zone of 1.5 m)

Fig. 5.22 Long-strip copper roofing

The system requires specified areas of the roof to be fixed with conventional welted joint clips, and the remainder with expansion clips, which allow for the longitudinal expansion of the bays, but ensure a secure fixing to the substructure. Lateral thermal movement is accommodated by a space at the base of the standing seams. Long-strip copper is laid on a breather membrane which allows free movement between the metal and the structure, whilst isolating the copper from any ferrous fixings in the structure and providing some sound reduction from the effects of wind and rain. All fixings should be made from the same copper as the roof. Nails with minimum 6 mm diameter heads should be copper or brass.

Within the traditional copper roofs (Fig 5.23), standing seams or batten-roll jointing systems are used depending on the pitch and appearance required. For pitches of 5° or less, batten rolls are appropriate, as standing seams are vulnerable to accidental flattening and subsequent failure by capillary action. Cross welts may be continuous across roofs where batten rolls are used, but should be staggered where standing seams are used. Bays should not exceed approximately 1700 mm in length. Either soft or one-quarter-hard temper copper is normally used. The substructure, breather membrane and fixed clips are as used in long-strip roofing. These differences in articulation within the traditional roofing systems, and particularly by contrast to the smooth line of long-strip system, offer alternative visual effects to the designer of copper roofs. Copper rainwater systems are available with a range of standard components. Copper shingles offer an additional aesthetic to traditional copper roofing and cladding systems.

Bonded copper systems

Proprietary systems offer similar visual effects to traditional copper, aluminium, stainless and terne-coated stainless steel roofing systems, by using the metal bonded to either particleboard or roofing sheet. (The latter is referred to in Chapter 6 on Bitumen and Flat Roofing Materials). Copper bonded to 18 mm high-density moisture-resistant chipboard, offers a smoother finish than that achieved by traditional roofing and cladding methods, whilst still showing the articulation of standing or flat seams.

COPPER ALLOYS

Copper may be alloyed with zinc, tin, aluminium, nickel or silicon to produce a range of brasses and bronzes.

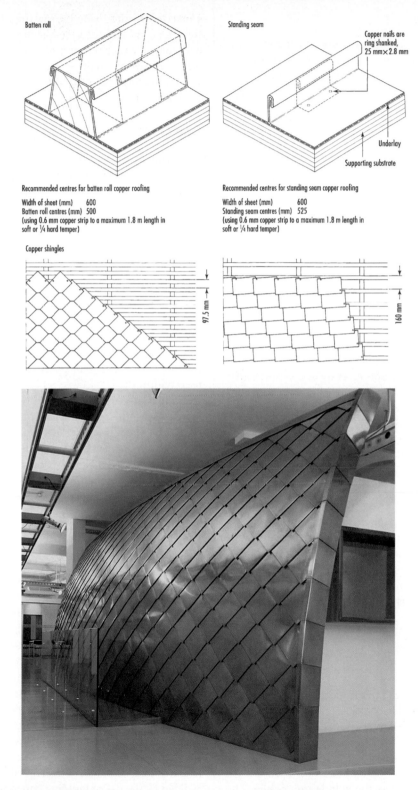

Batten roll

Standing seam

Copper nails are ring shanked, 25 mm × 2.8 mm

Underlay

Supporting substrate

Recommended centres for batten roll copper roofing

Width of sheet (mm)   600
Batten roll centres (mm)   500
(using 0.6 mm copper strip to a maximum 1.8 m length in soft or ¼ hard temper)

Recommended centres for standing seam copper roofing

Width of sheet (mm)   600
Standing seam centres (mm)   525
(using 0.6 mm copper strip to a maximum 1.8 m length in soft or ¼ hard temper)

Copper shingles

97.5 mm

160 mm

Fig. 5.23 Traditional copper roofing and copper shingles. Photograph: Courtesy of Copper Development Association

Brass

Brass is an alloy of copper and zinc, most commonly with a zinc content between 10 and 45%. It is used for small components such as architectural ironmongery, door and window furniture, handrails and balustrades. It may be lacquered to prevent deterioration of the polished finish, although externally and in humid environments the lacquer fails, requiring the brass to be cleaned with metal polish to remove the tarnish. Brass plumbing fittings manufactured from 60/40 brass may corrode in soft high-chloride content waters by dezincification. The process produces insoluble zinc corrosion products and ultimately porous metal fittings which may cause failure of the system. In situations where this problem is likely, dezincification-resistant (DZR) fittings made from alloy CW602N containing 2% lead should be used. Such components are marked with the 'CR' dezincification-resistant symbol.

Bronze

Bronze is an alloy of copper and tin, used for high-quality door furniture and recently as a woven fabric cladding material for the Theatre Royal rehearsal centre, Plymouth. Bronzes are usually harder and more durable than the equivalent brasses and exhibit a greater resistance to corrosion. Phosphor bronze contains up to 0.5% phosphorus in an 8% tin bronze. Because of its load-bearing properties and durability, it is frequently used as corbel plates and fixings for stone, and precast concrete cladding panels. Aluminium bronze (copper and aluminium), silicon bronze (copper and silicon) and gunmetals (copper–tin–zinc alloys) are also used for masonry fixings and cast components by virtue of their strength and durability. Nickel bronze alloys (copper, nickel and zinc) can be manufactured to highly polished *silver* finishes, particularly appropriate for interior fittings.

Lead

The Egyptians used lead in the glazing of their pottery and for making solder by 5000 BC. It was mined in Spain by the Phoenicians around 2000 BC.

MANUFACTURE

Lead occurs naturally as the sulfide ore, *galena*. The manufacturing process involves the concentration of the ore by grinding and flotation. The sulfide is converted to the oxide by roasting, then reduced to the metal in a blastfurnace charged with limestone and coke. Further refining removes impurities which otherwise would reduce the softness of the metal. Approximately 75% of lead used within the UK is recycled material.

LEAD SHEET

The majority of lead sheet for roofing, cladding, flashings and gutter linings is produced by milling thick sheet down to the required thickness. Continuous machine-cast lead, which accounts for approximately 10% of the UK market, is manufactured by immersing a rotating water-cooled metal drum in a bath of molten lead at constant temperature. The lead solidifies on the surface of the drum and is peeled off as it emerges from the melt. The thickness can be adjusted by altering the speed of rotation of the drum. The sheet produced is without the anisotropic directional grain structure associated with the standard rolling process.

The standard BS EN 12588: 1999 defines lead sheet by thickness rather than by the code system formerly used within the UK (Table 5.11). Sand-cast lead sheet is still manufactured by the traditional method, which involves pouring molten lead onto a prepared bed of sand. The sheet thickness is controlled by drawing a piece of timber across the molten metal surface to remove the excess material. Sand-cast lead is normally only used for conservation work on key historic buildings, when much of the old lead may be recycled in the process. A typical cast lead sheet size is 6 m × 1.5 m.

CORROSION

Freshly cut lead has a bright finish, but it rapidly tarnishes in the air with the formation of a blue-grey film of lead carbonate and lead sulfate. In damp conditions a white deposit of lead carbonate is produced, and in cladding this can both be aesthetically unacceptable and cause some staining of the adjacent materials. The effect can be prevented by the application of patination oil after the lead has been fixed. Lead is generally resistant to corrosion due to the protection afforded by the insoluble film; however, it is corroded by organic acids. Acidic rainwater run-off from mosses and lichens may cause corrosion and contact with damp timbers, particularly oak, teak and western red cedar should be avoided by the use of building paper or bitumen paint. Trapped condensation under sheet lead may cause significant corrosion, so consideration must be given to

Table 5.11 Lead sheet colour codes to BS EN 12588: 1999 and typical applications

	green	yellow	blue	red	black	white	orange
European Designation							
Colour code	green	yellow	blue	red	black	white	orange
Thickness range (mm)	1.25–1.50	1.50–1.75	1.75–2.00	2.00–2.50	2.50–3.00	3.00–3.50	3.50–6.00
UK Designation							
Lead codes	3		4	5	6	7	8
Nominal thickness (mm)	1.32		1.80	2.24	2.65	3.15	3.55
Nominal weight (kg/m²)	15.0		20.4	25.4	30.1	35.7	40.3
Typical Application:							
Flat roofing				✓	✓	✓	✓
Pitched roofing				✓	✓	✓	✓
Vertical cladding			✓	✓			
Soakers	✓		✓				
Hip and ridge flashings			✓	✓			
Parapets, box/tapered valley gutters					✓	✓	✓
Pitched valley gutters			✓	✓			
Weatherings to parapets			✓	✓			
Apron and cover flashings			✓	✓			
Chimney flashings			✓	✓			

the provision of adequate ventilation underneath the decking which supports the lead. Dew points must be checked to ensure that condensation will not occur and be trapped under the lead sheet in either new work or renovation. Generally lead is stable in most soils; however, it is attacked by the acids within peat and ash residues. Electrolytic corrosion rarely occurs when lead is in contact with other metals, although within marine environments aluminium should not be used in association with lead. Corrosion does occur between wet Portland cement or lime products and lead during the curing process; thus in circumstances where the drying out will be slow, the lead should be isolated from the concrete with a coat of bitumen paint.

FATIGUE AND CREEP

In order to prevent fatigue failure due to thermal cycling or creep, that is, the extension of the metal under its own weight over extended periods of time, it is necessary to ensure that sheet sizes, thicknesses and fixings are in accordance with the advice given by the Lead Sheet Association in their technical manuals. The metal must be relatively free to move with temperature changes, so that alternating stresses are not focussed in small areas leading to eventual fatigue fracture. A geotextile separating underlay may be used. The addition

of 0.06% copper to 99.9% pure lead refines the crystal structure giving increased fatigue resistance without significant loss of malleability. The composition of lead sheet is strictly controlled by BS EN 12588: 1999.

LEAD ROOFING

Lead roofing requires a smooth, continuous substrate. Generally, the bay sizes depend upon the roof geometry and the thickness of lead to be used (Fig. 5.24). For flat roofs (from 1 in 80 [approx. 1°] to 10°), joints are generally wood-cored rolls down the fall and drips across. For pitched roofs (10° to 80°) joints in the direction of the fall may be wood-cored or hollow rolls, with laps across the fall, unless for aesthetic reasons the bays are to be divided by drips. For steep pitches welts are used and, over 80°, standing seams are appropriate (Fig. 5.25). Fixings are copper or stainless steel nails and clips within the rolls, welts or standing seams. Lead as a highly malleable material can be formed or *bossed* into shape with the specialist tools including the bossing stick and bossing mallet. Welding or lead burning involves the joining of lead to lead using additional material to make the joint thicker by one-third than the adjacent material.

The David Mellor Cutlery Factory, Hathersage, Derbyshire (Fig. 9.10, Chapter 9) illustrates a

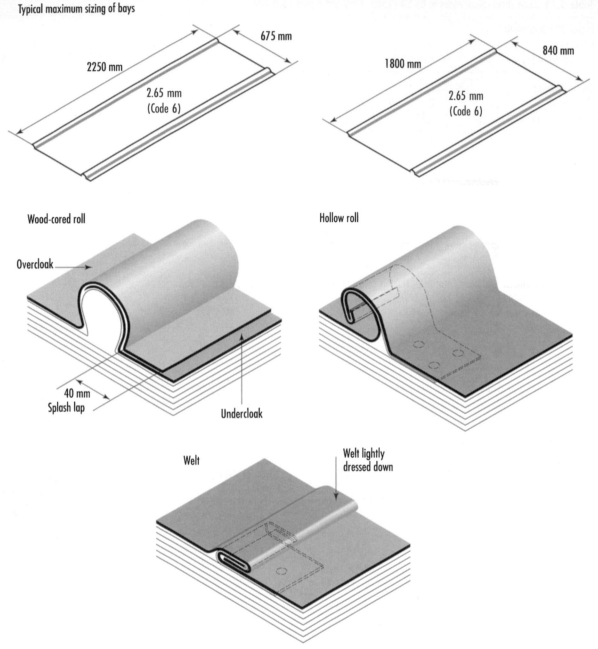

Fig. 5.24 Traditional lead roofing

traditionally detailed lead roof. The wood-cored roll-jointed lead is supported on a stepped deck manufactured from prefabricated stressed-skin insulated ply-wood boxes, tapered to fit the radial design. These units are supported on a series of lightweight steel trusses, tied at the perimeter by a steel tension ring and at the centre lantern by a ring-truss. Around the perimeter, the lead is burnt to ensure a vertical seal. A steeper roof without the need for drips is illustrated in Figure 5.25.

LEAD SHEET CLADDING

For cladding, the thickness of lead to be used dictates the maximum spacing between vertical joints and the

Fig. 5.25 Traditional lead roof. Contractor: Norfolk Sheet Lead. Photograph: Courtesy of Lead Contractors Association

distance between laps. Vertical joints may be wood-cored rolls or welts and occasionally standing seams or hollow rolls, where the risk of physical damage from ladders is negligible. The lead is hung by nailing at the head, with allowance for up to 6 mm thermal movement to occur within the lap joints.

An alternative form of lead cladding is the use of preformed lead-faced cladding panels, which are then fixed to the building facade (Fig. 5.26). Typically, 25 mm exterior-grade plywood covered with 1.80 or 2.24 mm (Code 4 or Code 5) lead is used. The panels are set against a lead-faced timber structural support leaving 25-mm-joints for thermal movement. Standard details are illustrated in the relevant Code of Practice BS 6915: 2001.

LEAD TILES

An innovative use of lead as a roofing material is illustrated by the Haberdashers' Hall in London (Fig. 5.27). The two-storey building, constructed around a courtyard, features a roof clad with diamond-shaped lead tiles. The individual units are formed from 1 × 1.5 m marine plywood diamonds each dressed with lead sheet, and incorporating a flashing on two edges to seal under the panels above.

FLASHINGS

Lead, because of its malleability and durability, is an ideal material from which to form gutters and gutter linings, ridge and hip rolls, and the full range of standard and specialist flashings, including ornamental work to enhance design features. For most flashing applications, lead sheet sof 1.32, 1.80 or 2.24 mm (Codes 3, 4 and 5) are used, fixed with copper or stainless steel and occasionally lead itself.

ACRYLIC-COATED LEAD SHEET

Acrylic-coated lead sheet is produced to order and in the standard colours, white, slate grey, terracotta and dark brown for use in colour-co-ordinated flashings. The colour-coated 1.80 mm (Code 4) milled lead is produced in widths of 250, 300, 450 and 600 mm. The material is moulded by bossing as for standard lead sheet and where welded the exposed grey metal may be touched-up if necessary.

Zinc

Zinc was known to the Romans as the alloy *brass*, but it was not produced industrially until the mid-eighteenth century, and was not in common use on buildings until the nineteenth century. The cut surface tarnishes quickly to a light grey due to the formation of a patina of basic zinc carbonate. The metal is hard at ambient temperatures and brittle when cold. It should therefore not be worked at metal temperatures below 10°C without prior warming, and heavy impacts should not

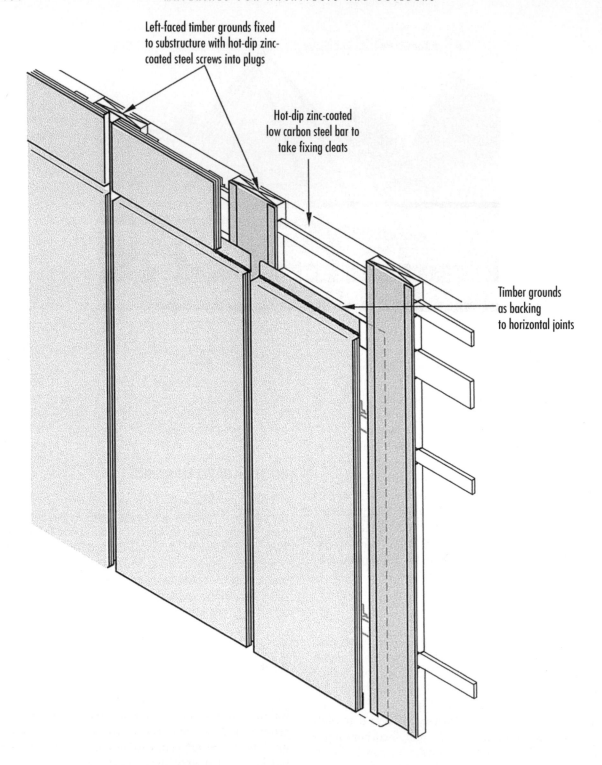

Left-faced timber grounds fixed to substructure with hot-dip zinc-coated steel screws into plugs

Hot-dip zinc-coated low carbon steel bar to take fixing cleats

Timber grounds as backing to horizontal joints

Fig. 5.26 Lead-faced timber cladding panels. Permission to reproduce extracts from BS 6915: 2001 is granted by the British Standards Institute

Fig. 5.27 Lead tile roofing – Haberdashers' Hall, London. Architects: Hopkins Architects. Photograph: Courtesy of Lead Contractors Association

be used within the forming processes of bending and folding.

MANUFACTURE

Zinc occurs naturally as the sulfide ore *zinc blende*. The ore is first concentrated and then roasted to produce zinc oxide. The addition of coal reduces zinc oxide to the metal, which is evolved as the vapour and then condensed. High-grade zinc is produced by the electrolysis of a purified zinc sulfate solution. Zinc is classified according to its purity as specified in the standard BS EN 1179: 2003 (Table 5.12). Approximately 3610 kW hours of energy are consumed in the primary production of one tonne of sheet zinc; however, a large proportion of the metal is recycled.

ZINC SHEET

Zinc sheet is manufactured by continuous casting and rolling in a range of thicknesses (Table 5.13) to a maximum coil width of 1000 mm. The two standard products are the pure metal (99.995% zinc) and its alloy with small additions of titanium and copper (e.g. 0.07% and 0.08% minima respectively). The rolling process modifies the grain structure, particularly in the pure metal; however, this does not affect the working of the sheets. The alloy has improved

Table 5.12 Colour codes for grades of zinc

Grade classification	Z1	Z2	Z3	Z4	Z5
Zinc content (%)	99.995	99.99	99.95	99.5	98.5
Colour code	white	yellow	green	blue	black

Table 5.13 Titanium zinc sheet thicknesses and weights

Nominal thickness (mm)	0.6	0.7	0.8	1.0	1.20	1.50
Nominal weight (kg/m^2)	4.3	5.0	5.8	7.2	8.6	10.8

performance with respect to strength and creep resistance but also a reduced coefficient of thermal expansion which enables the construction of roof bays up to 10 m, or in certain cases up to 16 m in length, depending upon design considerations including bay width. Titanium/copper alloy (BS EN 988: 1997) may be folded or curved to produce interlocking cladding panels for vertical, horizontal or diagonal installation. Both the pure metal and the titanium alloy can be worked by hand at room temperature and do not work harden. The titanium/copper alloy is used to add unity to the suite of elegant buildings forming the University of Cambridge, Centre for Mathematical Sciences, by Edward Cullinan Architects (Fig. 5.28).

Fig. 5.28 Zinc alloy roofing – Centre for Mathematical Sciences, University of Cambridge. Architects: Edward Cullinan Architects. Photograph: Arthur Lyons

PATINA

Bright zinc tarnishes in the air with the production of a thin oxide film, which is rapidly converted into basic zinc carbonate by the action of water and carbon dioxide. The patina then prevents further degradation of the surface. Ordinary zinc has a lighter blue-grey patina than the alloyed sheet, so the two materials should not be mixed within the same construction. Pre-weathered alloys are available if light or slate grey patinated surfaces are required immediately. The lifetime of zinc depends directly upon the thickness. A 0.8 mm roof should last for 40 years in urban conditions, whereas the same sheet as cladding, washed clean by rain, could last for 60 years. The titanium alloy with considerably improved durability has a predicted life of up to 100 years in a rural environment depending upon the pitch of the application.

LACQUERED ZINC SHEET

A factory-applied 25-micron heat-treated polyester lacquer finish to zinc gives a range of colour options through white, brown, terracotta, green grey and blue. Alternative organic coatings include acrylic, silicon-polyester and polyvinylidene fluoride paints or PVC plastisol to BS EN 506: 2000.

CORROSION

Zinc should not be used in contact with copper or where rainwater draining from copper or copper alloys would discharge onto zinc. It may, however, be used in association with aluminium or lead. In contact with steel or stainless steel, the zinc must be the major component to prevent significant corrosion effects. Unprotected cut edges of galvanised steel located above zinc can cause unsightly rust stains and should be avoided. If the underside of zinc sheet remains damp due to condensation for extended periods of time then pitting corrosion will occur, causing eventual failure. It is therefore necessary to ensure that the substructure is designed appropriately with vapour barrier, insulation and ventilation to prevent interstitial condensation. Sulfur dioxide within polluted atmosphere prevents the formation of the protective carbonate film and causes corrosion.

Zinc is not affected by Portland cement mortars or concrete, although it should be coated with an acrylic resin paint where it will be in contact with soluble salts from masonry or cement additives. Zinc may be laid directly onto seasoned softwoods, unless impregnated with copper-salt preservatives, which have a slight corrosion-promoting effect. However, zinc should be not be used on acidic timbers such as oak, chestnut and western red cedar. Furthermore, zinc should not be used in association with western red cedar shingles which generate an acidic discharge. The acidic products from the effect of ultraviolet radiation on bitumen can cause corrosion in zinc. If the bitumen is not protected from direct sunlight by reflective chippings, then any zinc must be separated from the bitumen with an impermeable material.

FIXINGS

Fixings for zinc should be of galvanised or stainless steel. Clips are made of zinc, cut along the rolled direction of the sheet and folded across the grain. Watertight joints may be made by soldering, using tin/lead solder in conjunction with zinc chloride flux.

ROOFING AND CLADDING

Both the roll-cap and standing-seam systems (Fig. 5.29) are appropriate for fully supported zinc and zinc alloy roofing (BS EN 501: 1994). Welted joints are standard practice across the bays at pitches steeper than 15°; below 15° drips are necessary. A minimum fall of 3° is recommended, although a pitch in excess of 7° will ensure self-cleaning, preventing the accumulation of dirt which reduces service life. Where the bay length is greater than 3 m, a section 1.3 m in length is fixed rigidly, whilst the remaining area is secured to the substructure with sliding clips which accommodate the thermal movement. Timber roof boarding, oriented strand board or plywood forms

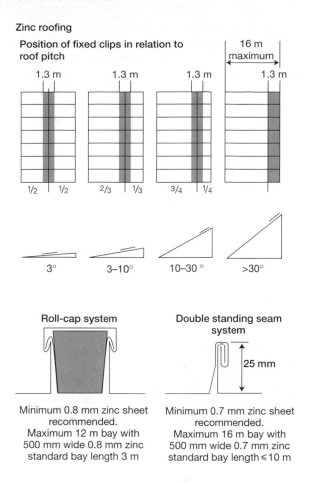

Zinc roofing

Position of fixed clips in relation to roof pitch

Roll-cap system
Minimum 0.8 mm zinc sheet recommended.
Maximum 12 m bay with 500 mm wide 0.8 mm zinc standard bay length 3 m

Double standing seam system
Minimum 0.7 mm zinc sheet recommended.
Maximum 16 m bay with 500 mm wide 0.7 mm zinc standard bay length ≤ 10 m

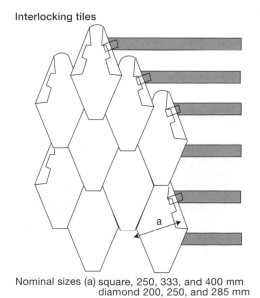

Interlocking tiles

Nominal sizes (a) square, 250, 333, and 400 mm
diamond 200, 250, and 285 mm

Fig. 5.29 Zinc roofing and interlocking tiles

the ideal substructures for zinc, but chipboard is inappropriate except in the case of cement-bonded particleboard for cladding. Where concrete is used it must be sealed against trapped moisture. For cladding, the vertical joints may be welted, standing seam or roll cap with the horizontal joints welted. The maximum bay length for cladding is 6 m, although 3 m is more practical on site. Titanium zinc rainwater systems are available with an appropriate range of standard components.

Titanium zinc interlocking square and diamond tiles are appropriate for vertical hanging and roof pitches down to 25°. They are fixed with soldered and sliding clips to timber battens. A range of sizes is available to give a choice of scale. Tiles are either pre-weathered or bright and manufactured in 0.7 or 0.8 mm alloy. To muffle the sound of rain, a full-surface substructure is advisable.

Titanium

Titanium ore is abundant in the earth's crust, with reserves well exceeding currently anticipated demands. The main producing countries are Russia, USA, Australia and Japan, although the ores, *rutile* (titanium oxide) and *ilmenite* (iron-titanium oxide), are also found in Europe, China and South America. Originally isolated in 1887, it was developed for use in the aerospace industry in the 1950s and has been used as a building cladding material in Japan for over thirty years. The Glasgow Science Centre (Fig. 5.30) illustrates titanium's eye-catching appearance as a modern construction material.

MANUFACTURE

The ore is treated with chlorine to produce titanium tetrachloride, which is then purified to remove other unwanted elements. Treatment with metallic magnesium or sodium reduces the titanium tetrachloride to a sponge of titanium metal, which is then melted under vacuum to produce solid ingots. Ingots are then forged into slabs and rolled out into sheet. Where required, an embossed finish can be applied during the final rolling process. Other sections and forms can be produced by hot rolling or cold forming as for steel. Titanium has a high-embodied energy; however, this is to some extent balanced against its life-cycle costing and ultimate full recycling.

PROPERTIES AND USES

Titanium is an appropriate material for construction due to its corrosion resistance. It is resistant to acids and alkalis, industrial and marine environments. Titanium has a density of $4510 \ kg/m^3$, intermediate between aluminium ($2700 \ kg/m^3$) and steel ($7900 \ kg/m^3$), giving it the advantage of a good strength to weight ratio. It is less ductile than steel so hot forming is required for severe bending. The metal has a modulus of elasticity half that of steel. Titanium has a low coefficient of expansion (8.9×10^{-6}), half that of stainless steel and copper and one-third that for aluminium. This reduces the risk of thermal stress, and enables titanium sheet roofing to be laid in longer lengths than other metals. The use of relatively thin roofing and cladding panels (0.3–0.4 mm) minimises both the dead load and the supporting structural system. Titanium with its very high melting point of 1670°C can withstand fire tests at 1100°C and has been certified as a 'non-combustible material' in Japan for roofing and cladding. Further applications include fascias, panelling, protective cladding for piers and columns and three-dimensional artwork.

DURABILITY

The corrosion resistance of titanium arises from its self-healing and tenacious protective oxide film. However, rainwater run-off from zinc, lead or copper roofs should be avoided. The Guggenheim Museum in Bilbao, clad in 32 000 square metres of commercially pure 0.3–0.4 mm titanium sheet panels, shows some staining due to lack of protection during the construction process and also rainwater run-off. Although initially expensive, on a life-cycle basis, due to its low maintenance costs, titanium may prove to be a highly competitive cladding and roofing material. Already one manufacturer is offering a 100-year guarantee against corrosion failure in roofing applications. Titanium can cause the corrosion of contact aluminium, steel or zinc, but austenitic stainless steel (grade 1.4401) is not affected.

Finishes

The normal oxide film can be thickened by heat treatment or anodising, giving permanent colours ranging from blue and mauve to cream and straw. Control is necessary to ensure the absence of colour variations

Fig. 5.30 Titanium cladding – The Glasgow Science Centre. Architects: BDP – Building Design Partnership. Photograph: Courtesy of Don Clements

within a project. Surface finishes range from reflective bright to soft matte and embossed, as used on the Glasgow Science Centre buildings (Fig. 5.30). In this case, the rolling grain direction was maintained over the building facades to ensure no visible variation of the embossed stipple effect.

Welding titanium

Titanium may be arc welded, but this requires the exclusion of air, usually by the use of argon gas shielding. Other welding technologies, such as plasma arc and laser or electron beam, are used for more specialist applications.

TITANIUM ALLOYS

Titanium is available as a wide range of alloys classified according to increased corrosion resistance, higher strength or higher temperature resistance. However, their current use is mainly confined to aerospace, industrial and medical applications. The standard architectural cladding material is 99% pure titanium (grade 1 or grade 2).

Process of metallic corrosion

Corrosion is an electro-chemical process, which can only occur in the presence of an electrolyte, that is moisture containing some dissolved salts. The process may be understood by considering the action of a simple Daniell cell as shown in Figure 5.31.

When the cell operates, two key processes occur. At the anode, the zinc gradually dissolves, generating zinc ions in solution and electrons which flow along the wire and light up the lamp as they move through its filament. At the copper cathode, the electrons are received at the surface of the metal and combine with copper ions in solution to plate out new shiny metal on the inside of the copper container.

Anode

$$Zn \longrightarrow Zn^{++} + 2e^-$$

zinc   zinc ions   electrons

Cathode

$$Cu^{++} + 2e^- \longrightarrow Cu$$

copper ions   electrons   copper

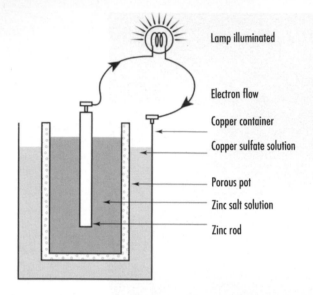

Fig. 5.31 Daniell cell

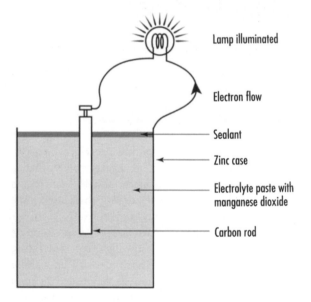

Fig. 5.32 Dry Leclanché cell

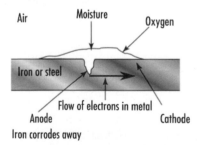

Fig. 5.33 Corrosion of iron

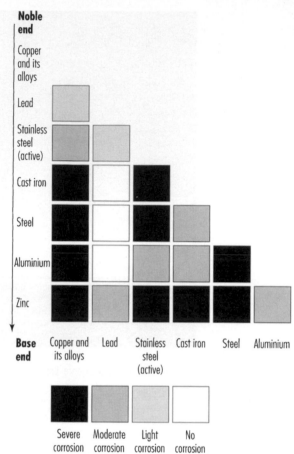

Indication of typical corrosion rates between pairs of metals in contact in the presence of moisture. The metal lower down the series copper (noble end) to zinc (base end) corrodes. The rate of corrosion will be generally increased within more aggressive environments.

Fig. 5.34 Bimetallic corrosion between pairs of metals in building applications

An equivalent process takes place in the dry Leclanché cell – the standard torch battery (Fig. 5.32). However, in this case the central carbon rod replaces the copper and the liquid is replaced by an aqueous paste. The anode process is the same as in the Daniell cell with the gradual dissolution of the zinc container.

Anode

$$Zn \longrightarrow Zn^{++} + 2e^-$$
zinc   zinc ions   electrons

At the cathode the carbon rod is surrounded by manganese dioxide which oxidises the hydrogen gas, which otherwise would have been produced there by the reaction between water and the electrons.

Fig. 5.35 Bimetallic corrosion between aluminium and steel

Cathode

$$H_2O + O + 2e^- \longrightarrow 2OH^-$$

water   oxygen   electrons   hydroxyl ions
(from manganese dioxide)

This sequence is similar to that seen in the corrosion of iron (Fig. 5.33). In this case the presence of both an electrolyte and oxygen are necessary for corrosion to occur.

Anode

$$Fe \longrightarrow Fe^{++} + 2e^-$$

iron   iron ions   electrons

Cathode

$$2H_2O + O_2 + 4e^- \longrightarrow 4OH^-$$

water   oxygen   electrons   hydroxyl ions
     (from the air)

Rust formation

$$Fe^{++} + 2OH^- \longrightarrow Fe(OH)_2 \longrightarrow Fe_2O_3.H_2O$$

iron ions hydroxyl ions   iron hydroxide   rust

Overall summary

$$4Fe + 3O_2 + 2H_2O \longrightarrow 2Fe_2O_3.H_2O$$

iron   oxygen   water   rust

FACTORS AFFECTING THE RATE OF CORROSION

The key factors which accelerate the rate of corrosion are the presence of two dissimilar metals in mutual contact and the degree of pollution within any moisture surrounding the metals. If the more base metal is small in quantity compared to the more noble metal, then rapid corrosion of the more base metal will occur. Figure 5.34 shows which pairs of metals commonly used in construction should not generally be allowed into contact. Within a single metal the following may all cause accelerated corrosion: minor surface variations, such as crystal grain boundaries; the effects of cold working or welding; the presence of impurities or alloying components within the metal; variable cleanliness or access to aerial oxygen. Figure 5.35 illustrates the effect of corrosion between an extruded aluminium gutter and a steel rainwater pipe. The aluminium corroded producing a white deposit near the point of contact between the two metals; this was followed by rapid corrosion of the steel.

References

FURTHER READING

Baddoo, N.R., Burgan, R. and Ogden, R.G. 1997: *Architect's guide to stainless steel*. Ascot: Steel Construction Institute.

Baldwin, K.R. and Wilcox, G.D. 2003: *Corrosion of zinc alloy coatings*. Abington: Woodhead Publishing.

Blanc, A., McEvoy, M. and Plank, R. (ed.) 1993: *Architecture and construction in steel*. London: E. & F.N. Spon.

Byers, M. 2003: *Design in steel*. London: Lawrence King.

Copper Development Association. 2002: *Copper in architecture, copper roofing in detail*. Hemel Hempstead: CDA.

Davis, J.R. 2001: *Copper and its alloys*. Materials Park OH. ASM International.

Eggen, A.P. and Sandaker, B.N. 1995: *Steel, structure and architecture: A survey of the material and its applications*. New York: Whitney Library of Design.

Fröhlich, B. and Schulenburg, S. 2003: *Metal architecture: Design and construction*. Basel: Birkhäuser.

Hind, P. 2001: *Aluminium structures in the entertainment industry*. Royston: Entertainment Technology Press.

International Copper Association, 2001: *European architecture with copper*. New York: International Copper Association.

Jackson, N.N. 1996: *The modern steel house*. London: E. & F.N. Spon.

Lead Sheet Association. 1990: *The lead sheet manual: A guide to good practice. Vol. 1: Lead sheet flashings*. London: Lead Sheet Association.

Lead Sheet Association. 1992: *The lead sheet manual: A guide to good practice. Vol. 2: Lead sheet roofing and cladding*. Tunbridge Wells: Lead Sheet Association.

LeCuyer, A. 2003: *Steel and beyond; new strategies for metals in architecture*. Basel: Birkhäuser.

Leyens, C., Peters, M. and Kumpfert, J. 2003: *Titanium and titanium alloys, fundamentals and applications*. Germany: Wiley.

Mazzolani, F.M. 2003 *Aluminium structural design*. Berlin: Springer-Verlag.

Miettinin, E and Taivalantti, K. 2002: *Stainless steel in architecture*. Finland: Rakennustieto.

National Federation of Roofing Contractors. 1999: *Profiled sheet roofing and cladding– A guide to good practice*, 3rd ed., London: E. & F.N. Spon.

Rheinzink. 2002: *Rheinzink: Applications in architecture*. Datteln: Rheinzink.

Schulitz, H.C. 2000: *Steel construction manual*. Basel: Birkhäuser.

Steel Construction Institute. 1997: *Architects' guide to stainless steel*. Ascot: The Steel Construction Institute.

Steel Construction Institute. 1998: *Structural fire safety: A handbook for architects and engineers*. Ascot: The Steel Construction Institute.

Steel Construction Institute. 1999: *The role of steel in environmentally responsible buildings*. Ascot: The Steel Construction Institute.

Tiainen, J. 2004: *Cor-ten facades*. Manchester: Cornerhouse Publications.

Trebilcock, P. and Lawson, M. 2004: *Architectural design in steel*. London: Spon Press.

Vargel, C. 2004: *Corrosion of aluminium*. Oxford: Elsevier.

Wilquin, H. 2001: *Aluminium architecture: construction and details*. Basel: Birkhäuser.

Zahner, L.W. 2005: *Architectural metal surfaces*. New Jersey: John Wiley and Sons.

Zinc Development Association: *Zinc in building design*. London: Zinc Development Association.

STANDARDS

BS 4 Structural steel sections:
    Part 1: 1993 Specification for hot-rolled sections.
BS 405: 1987 Specification for uncoated expanded metal carbon steel sheets for general purposes.
BS 416 Discharge and ventilating pipes and fittings sand-cast or spun in cast-iron:
    Part 1: 1990 Specification for spigot and socket systems.
BS 417 Galvanised mild steel cisterns and covers, tanks and cylinders:
    Part 2: 1987 Metric units.
BS 437: 1978 Specification for cast-iron spigot and socket drain pipes and fittings.
BS 449 Specification for the use of structural steel in building:
    Part 2: 1969 Metric units.
BS 460: 2002 Cast-iron rainwater goods.
BS 493: 1995 Airbricks and gratings for wall ventilation.
BS 779: 1989 Cast-iron boilers for central heating and indirect water supply (rated output 44kW and above).
BS 1091: 1963 Pressed steel gutters, rainwater pipes, fittings and accessories.
BS 1161: 1977 Specification for aluminium alloy sections for structural purposes.
BS 1189: 1986 Specification for baths made from porcelain enamelled cast iron.
BS 1202 Nails:
    Part 1: 2002 Steel nails.
    Part 2: 1974 Copper nails.
    Part 3: 1974 Aluminium nails.
BS 1210: 1963 Wood screws.
BS 1245: 1975 Metal door frames (steel).

BS 1329: 1974 Metal hand rinse basins.

BS 1390: 1990 Baths made from vitreous enamelled sheet steel.

BS 1449 Steel plate, sheet and strip:

Part 1.1: 1991 Specification for carbon and carbon/manganese plate, sheet and strip.

BS 1494–1: 1964 Specificatiion for fixing accessories for building purposes. Fixings for sheet, roof and wall coverings.

BS 1566 Copper indirect cylinders for domestic purposes:

Part 1: 2002 Open ventilated copper cylinders.

Part 2: 1984 Specification for single feed indirect cylinders.

BS 2997: 1958 Aluminium rainwater goods.

BS 3083: 1988 Specification for hot-dip zinc-coated and hot-dip aluminium/zinc-coated corrugated steel sheets for general purposes.

BS 3198: 1981 Specification for copper hot water storage combination units for domestic purposes.

BS 3830: 1973 Vitreous enamelled steel building components.

BS 3987: 1991 Anodic oxide coatings on wrought aluminium for external architectural applications.

BS 4449: 1997 Specification for carbon steel bars for the reinforcement of concrete.

BS 4513: 1969 Lead bricks for radiation shielding.

BS 4604 The use of high strength friction grip bolts in structural steelwork. Metric series:

Part 1: 1970 General grade.

Part 2: 1970 Higher grade (parallel shank).

BS 4842: 1984 Specification for liquid organic coatings for application to aluminium alloy extrusions, sheet and preformed sections for external architectural purposes.

BS 4868: 1972 Profiled aluminium sheet for building.

BS 4873: 2004 Aluminium alloy windows.

BS 4921: 1988 Specification for sheradized coatings on iron and steel.

BS ISO 4998: 2005 Continuous hot dip zinc-coated carbon steel sheet of structural quality.

BS 5427: 1996 Code of practice for performance and loading criteria for profiled sheeting in building.

BS 5950 Structural use of steelwork in building:

Part 1: 2000 Code of practice for design: rolled and welded sections.

Part 2: 2001 Specification for materials, fabrication and erection: rolled and welded sections.

Part 3: 1990 Design in composite construction.

Part 4: 1994 Code of practice for design of floors with profiled steel sheeting.

Part 5: 1988 Code of practice for design of cold formed thin-gauge sections.

Part 6: 1995 Code of practice for design of light-gauge profiled steel sheeting.

Part 7: 1992 Cold formed sections.

Part 8: 2003 Code of practice for fire-resistant design.

Part 9: 1994 Code of practice for stressed skin design.

BS 5977 Lintels:

Part 1: 1981 Method for assessment of load.

BS 6496: 1984 Specification for powder organic coatings for application and stoving to aluminium alloy extrusions, sheet and preformed sections for external architectural purposes.

BS 6497: 1984 Specification for powder organic coatings for application and stoving to hot-dip galvanised hot-rolled steel sections and preformed steel sheet.

BS 6510: 2005 Specification for steel windows, sills, window boards and doors.

BS 6582: 2000 Specification for continuously hot-dip lead alloy (terne)-coated cold reduced carbon steel flat-rolled products.

BS 6744: 2001 Specification for austenitic stainless steel bars for the reinforcement of concrete.

BS 6915: 2001 Design and construction of fully supported lead sheet roof and wall coverings.

BS 7364: 1990 Galvanised steel studs and channels for stud and sheet partitions and linings using screw fixed gypsum wallboards.

BS 7543: 2003 Guide to durability of buildings and building elements, products and components.

BS 7668: 2004 Specification for weldable structural steels. Hot-finished structural hollow sections in weather-resistant steels.

BS 8118 Structural use of aluminium:

Part 1: 1991 Code of practice for design.

Part 2: 1991 Specification for materials, workmanship and protection.

BS 8202 Coatings for fire protection of building elements:

Part 1: 1995 Code of practice for the selection and installation of sprayed mineral coatings.

Part 2: 1992 Code of practice for the use of intumescent coating systems to metallic substrates for providing fire resistance.

BS EN 124: 1994 Gully tops and manholes – design requirements, type, testing, marking, quality control.

BS EN 485 Aluminium and aluminium alloys – sheet, strip and plate:

Part 1: 1994 Technical conditions for inspection and delivery.

Part 2: 2004 Mechanical properties.

Part 3: 2003 Tolerances on shape and dimensions for hot-rolled products.

Part 4: 1994 Tolerances on shape and dimensions for cold-rolled products.

BS EN 486: 1994 Aluminium and aluminium alloys – extrusion ingots – specifications.

BS EN 487: 1994 Aluminium and aluminium alloys – rolling ingots – specifications.

BS EN 501: 1994 Roofing products from metal sheet – specification for fully supported roofing products of zinc sheet.

BS EN 502: 2000 Roofing products from metal sheet – specification for fully supported products of stainless steel sheet.

BS EN 504: 2000 Roofing products from metal sheet – specification for fully supported roofing products of copper sheet.

BS EN 505: 2000 Roofing products from metal sheet – specification for fully supported products of steel sheet.

BS EN 506: 2000 Roofing products from metal sheet – specification for self-supporting products of copper or zinc sheet.

BS EN 507: 2000 Roofing products from metal sheet – specification for fully supported products of aluminium sheet.

BS EN 508 Roofing products from metal sheet – specification for self-supported products of steel, aluminium or stainless steel sheet:

Part 1: 2000 Steel.

Part 2: 2000 Aluminium.

Part 3: 2000 Stainless Steel.

BS EN 515: 1993 Aluminium and aluminium alloys – wrought products – temper designations.

BS EN 545: 2002 Ductile iron pipes, fittings, accessories and their joints for water pipelines.

BS EN 573 Aluminium and aluminium alloys – chemical composition and form of wrought products:

Part 1: 2004 Numerical designation system.

Part 2: 1995 Chemical symbol-based designation system.

Part 3: 2003 Chemical composition.

Part 4: 2004 Form of products.

BS EN 586 Aluminium and aluminium alloys – forgings:

Part 1: 1998 Technical conditions.

Part 2: 1994 Mechanical properties.

Part 3: 2001 Tolerances on dimensions and form.

BS EN 598: 1995 Ductile iron pipes, fittings, accessories and their joints for sewerage applications – requirements and test methods.

BS EN 754 Aluminium and aluminium alloys:

Parts 1–8 Cold drawn rod/bar and tube.

BS EN 755 Aluminium and aluminium alloys:

Parts 1–9 Extruded drawn rod/bar, tube and profiles.

BS EN 845 Specification for ancillary components for masonry:

Part 1: 2003 Ties, tension straps, hangers and brackets.

Part 2: 2003 Lintels.

Part 3: 2003 Bed-joint reinforcement of steel meshwork.

BS EN 877: 1999 Cast iron pipes and fittings, their joints and accessories.

BS EN 969: 1996 Specification for ductile iron pipes, fittings, accessories and their joints for gas applications – requirements and test methods.

BS EN 988: 1997 Zinc and zinc alloys – specification for rolled flat products for building.

BS EN 1057: 1996 Copper and copper alloys. Seamless round copper tubes for water and gas in sanitary and heating applications.

BS EN: 1172: 1997 Copper and copper alloys – sheet and strip for building purposes.

BS EN 1173: 1996 Copper and copper alloys – material condition or temper designation.

BS EN 1179: 2003 Zinc and zinc alloys – primary zinc.

BS EN 1412: 1996 Copper and copper alloys – European numbering system.

pr EN 1254 Copper and copper alloys – Plumbing fittings:

Part 6: 2004 Fitting with push-fit ends.

Part 7: 2004 Fittings with press ends for metallic tubes.

BS EN: 1993 Eurocode 3: Design of steel structures:

Part 1.1: 2005 General rules and rules for buildings.

Part 1.2: 2005 Structural fire design.

Part 1.8: 2005 Design of joints.

Part 1.9: 2005 Fatigue.

Part 1.10: 2005 Material toughness.

BS EN: 1994 Eurocode 4: Design of composite steel and concrete structures:

Part 1.1: 2004 General rules and rules for buildings.

Part 1.2: 2005 Structural fire design.

BS EN ISO 7441: 1995 Corrosion of metals and alloys – determination of bimetallic corrosion in outdoor exposure corrosion tests.

BS EN 10020: 2000 Definition and classification of grades of steel.

BS EN 10025 Designation system for steel:

Part 1: 2004 General technical delivery conditions.

Part 2: 2004 Non-alloy structural steels.

Part 3: 2004 Normalised/normalised-rolled weldable fine-grain structural steels.

Part 4: 2004 Thermo-mechanical-rolled weldable fine-grain structural steels.

Part 5: 2004 Structural steels with improved atmospheric corrosion resistance.

Part 6: 2004 High yield strength structural steels.

BS EN 10027 Designation systems for steels:

Part 1: 2005 Steel names.

Part 2: 1992 Steel numbers.

BS EN 10034: 1993 Structural steel I and H sections – tolerances on shape and dimensions.

BS EN 10051: 1992 Specification for continuously hot-rolled uncoated plate, sheet and strip of non-alloy and alloy steels – tolerances on dimensions and shape.

BS EN 10056 Specification for structural steel equal and unequal leg angles:

Part 1: 1999 Dimensions.

Part 2: 1993 Tolerances, shape and dimensions.

BS EN 10088 Stainless steels:

Part 1: 2005 List of stainless steels.

Part 2: 2005 Technical delivery conditions for sheet, plate and strip for general purposes.

Part 3: 2005 Technical delivery conditions for semi-finished products, bars, rods and sections for general purposes.

BS EN 10095: 1999 Heat-resisting steels and nickel alloys.

BS EN 10130: 1999 Specification for cold-rolled low carbon steel flat products for cold forming: Technical delivery conditions.

BS EN 10131: 1991 Cold-rolled uncoated low carbon and high yield strength steel flat products for cold forming – tolerances on dimensions and shape.

BS EN 10149 Hot-rolled products made of high yield strength steels for cold forming.

Part1: 1996 General delivery conditions.

Part 2: 1996 Delivery conditions for thermo-mechanically-rolled steels.

Part 3: 1996 Delivery conditions for normalised or normalised-rolled steels.

BS EN 10152: 2003 Specification for electrolytically zinc-coated cold-rolled steel flat products. Technical delivery conditions.

BS EN 10169 Continuously organic-coated (coil-coated) steel flat products:

Part 1: 2003 General information.

Part 2: 2006 Products for building exterior applications.

Part 3: 2003 Products for building interior applications.

BS EN 10210 Hot-finished structural hollow sections of non-alloy and fine-grain structural steels:

Part 1: 2006 Technical conditions.

Part 2: 2006 Tolerances, dimensions and sectional properties.

BS EN 10219 Cold-formed welded structural hollow sections of non-alloy and fine grain steels:

Part 1: 2006 Technical delivery conditions.

Part 2: 2006 Tolerances, dimensions and sectional properties.

BS EN 10242: 1995 Threaded pipe fittings in malleable cast iron.

BS EN 10250 Open steel die forgings for general engineering purposes:

Part 1: 1999 General requirements.

Part 2: 2000 Non-alloy quality and special steels.

Part 3: 2000 Alloy special steels.

Part 4: 2000 Stainless steels.

BS EN 10258: 1997 Cold-rolled stainless steel narrow strip and cut lengths.

BS EN 10259: 1997 Cold-rolled stainless steel wide strip and plate/sheet.

BS EN 10326: 2004 Continuously hot-dip-coated strip and sheet of structural steels.

BS EN 10327: 2004 Continuously hot-dip coated strip and sheet of low carbon steels for cold forming.

BS EN 12588: 1999 Lead and lead alloys – rolled lead sheet for building purposes.

BS EN 13501 Fire classification of construction products and building elements:

Part 1: 2002 Classification using test data from reaction to fire tests.

Part 2: 2003 Classification using data from fire resistance tests.

BS EN 14782: 2006 Self-supporting metal sheet for roofing, external cladding and internal lining.

BS EN 15088: 2005 Aluminium and aluminium alloys – structural products for construction works.

BS ISO 16020: 2005 steel for the reinforcement and prestressing of concrete – vocabulary.

CP 118: 1969 The structural use of aluminium.

CP 143 Sheet roof and wall coverings:

Part 1: 1958 Aluminium, corrugated and troughed.

Part 5: 1964 Zinc.

Part 10: 1973 Galvanised corrugated steel. Metric units.

Part 12: 1970 Copper. Metric units.

Part 15: 1973 Aluminium. Metric units.

BUILDING RESEARCH ESTABLISHMENT PUBLICATIONS

BRE Digests

BRE Digest 301: 1985 Corrosion of metals by wood.
BRE Digest 305: 1986 Zinc-coated steel.
BRE Digest 349:1990 Stainless steel as a building material.
BRE Digest 444: 2000 Corrosion of steel in concrete (Parts 1, 2 and 3).
BRE Digest 455: 2001 Corrosion of steel in concrete – service life design and prediction.
BRE Digest 461: 2001 Corrosion of metal components in walls.
BRE Digest 462: 2001 Steel structures supporting composite floor slabs – design for fire.
BRE Digest 487 Part 2: 2004 Structural fire engineering design: materials behaviour – steel.

Good building guide

GBG 21: 1996 Joist hangers.

Information papers

IP 16/88 Ties for cavities – new developments.
IP 17/88 Ties for masonry cladding.
IP 12/90 Corrosion of steel wall ties – history of occurrence, background and treatment.
IP 13/90 Corrosion of steel wall ties – recognition and inspection.
IP 5/98 Metal cladding – assessing thermal performance.
IP 11/00 Ties for masonry walls: a decade of development.
IP 10/02 Metal cladding – assessing thermal performance of built up systems with use of Z spacers.

BRE Report

BR 142: 1989 The use of light-gauge cold-formed steelwork in construction: developments in research and design.

CORUS PUBLICATIONS

Fire resistance of steel framed buildings (2001).
Fire engineering in sports stands (1997).

ADVISORY ORGANISATIONS

Aluminium Federation Ltd, Broadway House, Calthorpe Road, Five Ways, Birmingham, West Midlands B15 1TN (0121 456 1103).

Aluminium Rolled Products Manufacturers Association, Broadway House, Calthorpe Road, Five Ways, Birmingham, West Midlands B15 1TN (0121 456 1103).

British Constructional Steelwork Association Ltd., 4 Whitehall Court, Westminster, London SW1A 2ES, (020 7839 8566).

British Non-Ferrous Metals Federation, Broadway House, Calthorpe Road, Five Ways, Birmingham, West Midlands B15 1TN (0121 456 6110).

British Stainless Steel Association, Broomgrove, 59 Clarkehouse Road, Sheffield, South Yorkshire S10 2LE (0114 267 1280).

Cast Iron Drainage Development Association, 72 Francis Road, Edgbaston, Birmingham, W. Midlands B16 8SP (0121 693 9909).

Cold Rolled Sections Association, National Metalforming Centre, Birmingham Road, West Bromwich, West Midlands B70 6PY (0121 601 6350).

Copper Development Association, 5 Grovelands Business Centre, Boundary Way, Hemel Hempstead, Herts. HP2 7TE (01442 275700).

Corus Research, Development and Technology, Swindon Technical Centre, Moorgate, Rotherham, South Yorkshire S60 3AR (01709 825 335).

Council for Aluminium in Building, River View House, Bond's Mill, Stonehouse, Gloucestershire GL10 3RF (01453 828851).

Lead Sheet Association, Hawkwell Business Centre, Maidstone Road, Pembury, Tunbridge Wells, Kent TN2 4AH (01892 822773).

Metal Cladding and Roofing Manufacturers Association, 18 Mere Farm Road, Prenton, Birkenhead CH43 9TT (0151 652 3846).

Stainless Steel Advisory Centre, Broomgrove, 59 Clarkehouse Road, Sheffield, South Yorkshire S10 2LE (0114 267 1265).

Steel Construction Institute, Silwood Park, Ascot, Berks. SL5 7QN (01344 23345).

Zinc Information Centre, 6 Wrens Court, 56 Victoria Road, Sutton Coldfield B72 1SY (0121 362 1201).

6

BITUMEN AND FLAT ROOFING MATERIALS

Introduction

The flat roofing materials, which form an impermeable water barrier, include built-up bitumen sheet systems, mastic asphalt, single-ply plastic membranes and liquid coatings. All require continuous support on an appropriate roof decking system. Green roofs are considered as an extension of the standard roofing systems. Metal roofing systems are described in Chapter 5.

FIRE EXPOSURE OF ROOFS

All materials used as finishes for roofs, both pitched and flat, are classified with respect to external fire exposure. The classification system (BS 476–3: 2004) indicates whether the roof is flat or pitched followed by a two-letter coding on fire performance.

Roof system	EXT. F. (flat) or EXT. S. (sloping)
Fire penetration (first letter)	A, B, C or D
	(A = no penetration in one hour, B = no penetration in 30 minutes, C = penetration within 30 minutes, D = penetrated by preliminary flame test)
Spread of flame (second letter)	A, B, C or D
	(A = no spread of flame, B = spread of flame less than 533 mm, C = spread of flame more than 533 mm, D = specimens that continue to burn after the test flame was removed or which had a spread of flame more than 381 mm in the preliminary test.)

A suffix 'X' is added where the material develops a hole or suffers mechanical failure.

Thus a flat roof material classified as EXT. F. AA suffers no fire penetration nor spread of flame during the standard one hour fire test.

Cold-deck, warm-deck and inverted roofs

COLD-DECK ROOFS

In cold-deck roof construction, the weatherproof layer is applied directly onto the roof decking, usually particleboard or plywood, and this is directly supported by the roof structure, frequently timber joists (Fig. 6.1). Thermal insulation is laid over the gypsum plasterboard ceiling, leaving cold void spaces between the structural timbers or steel. In this form of roof construction, there is a significant risk of condensation forming on the underside of the decking, and this may cause deterioration of the structure. Precautions must be taken to ensure adequate ventilation of the cold voids and also the underside of the deck must not cool below the dew point when the external temperature is − 5°C. Any vapour check under the insulation layer is vulnerable to leakage around electrical service cables. In remedial work on cold-deck roofs, if adequate ventilation cannot be achieved, then conversion to a warm-deck or inverted roof system may be advantageous. Cold-deck roof construction is not recommended for new building work in the Code of Practice (BS 8217: 2005).

WARM-DECK ROOFS

In warm-deck roof construction the thermal insulation is laid between the roof deck and the weatherproof covering (Fig. 6.1). This ensures that the roof deck and its supporting structure are insulated from extremes of temperature, thus limiting excessive thermal movement which may cause damage. As the insulating material is directly under the waterproof layer, it must be sufficiently strong to support any foot traffic associated with maintenance of the roof. The waterproof and insulation layers will require mechanical fixing or ballasting to prevent detachment in strong winds. Surface condensation on the underside of a roof deck within warm-deck roof construction would normally indicate insufficient thermal insulation. Warm-deck construction is the preferred method for lightweight roofs.

INVERTED ROOFS

In inverted roof construction, both the structural deck and the weatherproof membrane are protected by externally applied insulation (Fig. 6.1). This ensures that the complete roof system is insulated from extremes of hot and cold, also from damage by solar radiation and maintenance traffic. The insulation layer is usually ballasted with either gravel or fully protected with paving slabs. Disadvantages of inverted roof construction are the greater dead-weight, and the difficulty in locating leaks under the insulation layer. Inverted roof construction is the preferred method for concrete and other heavyweight roof systems.

Built-up roofing

Built-up roofing consists of two or more layers of bitumen sheets bonded together with self-adhesive or hot bitumen. Bitumen is the residual material produced after the removal by distillation of all volatile products from crude oil. The properties of bitumen are modified by controlled oxidation, which produces a more rubbery material suitable for roofing work. In the manufacture of bitumen sheets, the continuous base layer of organic fibres, glass fibre or bitumen-saturated polyester is passed through molten oxidised bitumen containing inert filler; the material is then rolled to the required thickness. The bitumen sheet is coated with sand to prevent adhesion within the roll or with mineral chippings to produce the required finish. A range of thicknesses is available and for ease

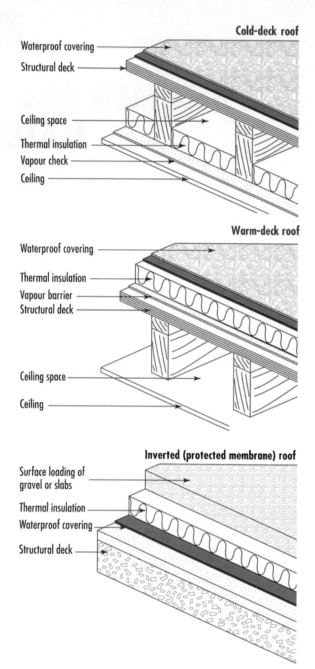

Fig. 6.1 Cold-deck, warm-deck and inverted roofs

of recognition the classes relating to the base fibre layer are colour coded along one edge of the rolls.

TYPES OF BITUMEN ROOFING SHEETS

Reinforced bitumen sheets for roofing are classified by BS 747: 2000 according to their base fibre and primary function. Classes are subdivided into types according to their surface finish and their intended use within roofing. Class 1 bitumen sheets, manufactured from organic fibres incorporating a layer of jute hessian, are only recommended for use under tiles and slates; they may be aluminium faced on the underside to provide a heat-reflecting surface. However, where the risk of condensation is high, sarking sheets, which are waterproof but vapour permeable, may be appropriate.

Only Class 3 and Class 5 felts are tear-resistant and durable and therefore appropriate for use in built-up roofing systems (Table 6.1). Class 3 sheets have a glass-fibre base. The Type 3B sheets have a fine surface finish and are suitable for the lower layers of a built-up roofing system, while Type 3E has a granular finish for the exposed layer. Type 3G sheets are perforated for the first layer where partial bonding and/or venting is required. Class 5 sheets have a high performance polyester base. Type 5U sheets are suitable as the underlayer for a built-up roofing system, while Type 5B sheets form the lower layers and the exposed layer, if additional surface treatment is to be applied. Type 5E sheets are suitable for the exposed finish layer including flashings and vertical surfaces. Class 4 (Type 4A) reinforced bitumen sheets incorporating jute fibre are sheathing felts used as underlay for mastic asphalt roofing and flooring on a timber base. However, on solid substrates such as concrete, glass fibre tissue, which does not compress under load, is the more appropriate separating layer.

Built-up roofing systems using the Class 3 glass-fibre based products are standard, but the Class 5 polyester-based products have greater strength and durability at a higher initial cost. A typical three-layer system is illustrated in Figure 6.2. All oxidised bitumen sheets, when exposed to ultraviolet light and ozone, gradually age harden and become less resistant to fatigue failure.

ROOFING SYSTEMS

Built-up roofing sheets may be applied to roofs constructed from precast or *in-situ* reinforced concrete, plywood (exterior grade), timber (19 mm tongued and grooved, with preservative treatment

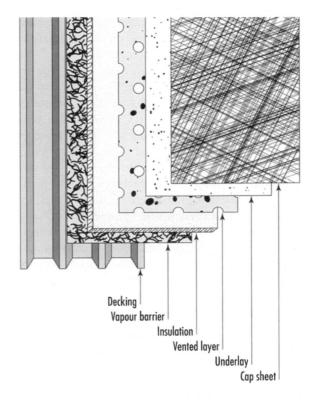

Decking
Vapour barrier
Insulation
Vented layer
Underlay
Cap sheet

Fig. 6.2 Typical three-layer built-up roofing system

Table 6.1 Reinforced bitumen sheets for roofing to BS 747: 2000

Class	Base Material	Type	Use	Colour Code
Class 1	organic fibres and jute hessian base	Type 1F	underslating sheet	white
Class 3	glass-fibre base	Type 3B	fine granule surface	red
		Type 3E	mineral surface	red
		Type 3G	venting base layer	none
Class 4	flax or jute fibre base	Type 4A: black	underlay to mastic asphalt	none
		Type 4A: brown	underlay to mastic asphalt	none
Class 5	polyester base	Type 5B	fine granule surface	blue
		Type 5E	mineral surface	blue
		Type 5U	fine granule underlayer	blue

(Class 2 sheets with asbestos-fibre bases are no longer manufactured.)

and conditioning), particleboard (except in conditions of high humidity where only the cement-bonded product is appropriate), oriented strand board, wood wool slabs or profile metal decking (galvanised steel or aluminium). Certain proprietary composite decking systems, such as units comprising plywood, rigid urethane foam and aluminium foil, are also appropriate except in areas of high humidity. In warm roof applications a vapour check is applied over the decking. Appropriate vapour check materials include lapped and bonded bitumen membranes, lapped polythene sheets or 12 mm mastic asphalt on glass-fibre tissue depending upon the structural material.

Insulation materials include cork, rigid mineral (MW) or glass wool, perlite (EPB), cellular foamed glass (CG), rigid polyurethane foam (RUP) extruded (XPS) or expanded moulded (EPS) polystyrene, phenolic foam (PF), polyisocyanurate (PIR), bitumen impregnated fibreboard and various proprietary composite systems. Most manufacturers are now able to supply CFC-free insulation products. The heat-sensitive expanded plastics are frequently supplied pre-bonded to a cork, perlite or fibreboard layer to receive the *pour and roll* hot bitumen or *torch-on* sheet systems. Where suitable falls are not incorporated into the roof structure, the insulation can be supplied ready cut-to-falls. A minimum in service fall of 1 in 80 is required to prevent *ponding*; this requires a design fall of 1 in 40 to allow for settlement.

On sloping roofs the first layer of built-up roofing sheet is applied down the slope, but on flat roofs (less than 10°) the direction of the first layer need not relate to the falls. The first layer is either partly- or fully-bonded depending upon the substrate. Perforated sheet is frequently laid loose as the first layer, and becomes spot-bonded as the hot bitumen for the second layer is applied. On timber the first layer is nailed. Partial bonding permits some thermal movement between the felt system and the decking, and also allows for the escape of any water vapour trapped in the decking material. The use of proprietary breather vents on large roofs allows the escape of this entrapped air from the roof structure by migration under the partially bonded layer. Side laps of 50 mm and end laps of typically 100 mm should all be staggered between layers. On sloping roofs the first layer should be nailed at the top of each sheet at 50 mm centres and higher melting point 115/15 bitumen should be used for bonding the subsequent layers to prevent slippage. Protection from the effects of ultraviolet light is afforded either by the factory-applied mineral surface finish to the *cap* sheet, the application of reflective paint, or on flat roofs typically by a 12 mm layer of reflective white spar stone chippings.

POLYMER-MODIFIED BITUMEN SHEETS

High-performance bitumen sheets based on polyester bases for toughness and polymer-modified bitumen coatings for increased flexibility, strength and fatigue resistance, offer considerably enhanced durability over the standard oxidised bitumen sheets. The two types are based on styrene butadiene styrene (SBS) and atactic polypropylene (APP) modified bitumen.

Elastomeric SBS high-performance sheets

Styrene-butadiene-styrene polymer-modified bitumen sheets have greater elasticity than standard oxidised bitumen sheets. They are laid either by the traditional pour-and-roll technique, which is used for standard bitumen sheets, or by torching-on (Fig. 6.3). In the pour-and-roll process, bonding bitumen is heated to between 200 and 250°C and poured in front of the sheet as it is unrolled, giving continuous adhesion between the layers. In the torching-on process as the sheet is unrolled, the backing is heated to the molten state with propane burners. Alternatively, cold adhesive or mechanical fastening may be used.

Plastomeric APP high-performance sheets

Atactic polypropylene polymer-modified bitumen contains typically 25% atactic polymer in bitumen with some inert filler. The product is more durable than oxidised bitumen and has enhanced high-temperature resistance and low-temperature flexibility. Sheets are manufactured with a polyester and/or glass-fibre core. Some additionally have a glass-fibre reinforced weathering surface. The APP polymer-modified bitumen sheets are bonded with cold adhesive or by torching the heat-sensitive backing, as the temperature of hot-poured bitumen is too low to form a satisfactory bond.

METAL-FACED ROOFING SHEETS

Copper-faced high-performance SBS sheets give visual quality and enhanced durability compared with standard mineral surfaced built-up sheet roofing systems. The small squared pattern of indentations

Hot bitumen bonding roofing sheet

Torching-on high-performance roofing sheet

Fig. 6.3 Laying built-up roofing sheet with poured hot bitumen and by torching-on

(Fig. 6.4) allows for thermal movement between the bonded sheet and metal finish. The material weathers similarly to traditional copper roofing systems producing a green patina. The metal foil thickness is typically 0.08 mm.

INVERTED ROOFS

In the inverted roof the built-up sheet waterproofing membrane is laid directly onto the roof deck. Non-absorbent insulation, such as extruded polystyrene, is laid onto the membrane, which is then covered with a filter sheet to prevent the ingress of excessive organic material. River-washed ballast or pavings on supports protect the system from mechanical and wind damage.

Inverted roofs have the advantage that the waterproof membrane is protected from thermal stress by the insulating layer. This in turn is protected from damage by the paving or ballast finish. High-performance built-up sheet systems are suitable for inverted roofs.

Mastic asphalt

TYPES OF MASTIC ASPHALT

Mastic asphalt is a blended bitumen-based product. It is manufactured either from the bitumen produced by the distillation of crude oil, or from lake asphalt, a naturally occurring blend of asphalt containing 36% by weight of finely divided clay, mainly imported from Trinidad. The bitumen is blended with limestone powder and fine limestone aggregate to produce the standard roofing types specified in BS 6925: 1988 (Table 6.2).

The effect of the finely divided clay particles within lake asphalt type BS 988T confers better laying characteristics and enhanced thermal properties; these are advantageous when the material is to be exposed to wide temperature changes, particularly in warm roof construction systems.

Mastic asphalt is usually delivered as blocks for melting on site prior to laying, although hot molten asphalt is occasionally supplied for larger contracts. Laid mastic asphalt is brittle when cold but softens in hot sunny weather. The hardness is increased by the re-melting process, and also by the addition of further limestone aggregate. Polymer-modified mastic asphalts, usually containing styrene butadiene styrene block copolymers, are more durable and have enhanced flexibility and extensibility at low temperatures, allowing for greater building movements and better resistance to thermal shock. Where mastic asphalt roofs are subjected to foot or vehicle access, then paving-grade mastic asphalt (BS 1447: 1988) should be applied as a wearing layer over the standard roofing grade material. Two key grades S and H are

Table 6.2 Mastic asphalt grades to BS 6925: 1988

Type	Composition
BS 988B	100% bitumen
BS 988T25	75% bitumen, 25% lake asphalt
BS 988T50	50% bitumen, 50% lake asphalt
Specified by manufacturers	polymer-modified grades

Fig. 6.4 Copper-faced bitumen roofing sheet

available; the softer grade is suitable for footways and rooftop car parks, the harder grade for heavily stressed areas. For standard flooring, type F1076 mastic asphalt is required; for coloured flooring type F1451, and for tanking and damp-proof courses type T1097 is necessary. The flooring types of mastic asphalt are available in four grades (hard, light, medium and heavy duty) according to the required wearing properties.

ROOFING SYSTEMS

Mastic asphalt may be laid over a wide variety of flat or pitched roof decking systems, either as warm or cold roof constructions, although the latter is generally not recommended due to condensation risks (BS 8218: 1998). As mastic asphalt is a brittle material it requires continuous firm support. Appropriate decks are concrete (*in-situ* or precast), plywood (exterior grade), timber boarding (19 mm minimum thickness) particleboard, wood wool slabs (50 mm minimum thickness) and profiled metal decking. A typical concrete warm roof system is illustrated in Figure 6.5.

In dense concrete construction, a sand and cement screed is laid to falls over the *in-situ* slab. The falls should be designed such that even with inevitable variations on site, they are never less than 1 in 80, as this is essential to ensure the immediate removal of the surface water and to prevent ponding. Plywood, profiled metal and other decking systems would be similarly laid to falls. A layer of bitumen-bonded Type 3B glass-fibre roofing sheet is applied over the structure to act as a vapour check.

Insulation

Thermal insulation, to provide the necessary roof U-value, is bonded with hot bitumen. A wide variety of insulation boards or blocks including compressed cork, high-density mineral wool, fibreboard, perlite, cellular foamed glass, high-density extruded polystyrene and polyisocyanurate are suitable, although where insufficient rigidity is afforded by the insulation material, or if it would be affected by heat during application of the hot asphalt, it must be overlaid with firm heat-resistant boards to prevent damage during maintenance or construction. Where falls are not provided by the structure, the insulation may be set appropriately, providing that the thinnest section gives the required thermal properties. A separating layer of loose-laid geotextile material or Type 4A sheathing felt (BS 747: 2000) is then applied to allow differential thermal movement between the decking system and the mastic asphalt waterproof finish.

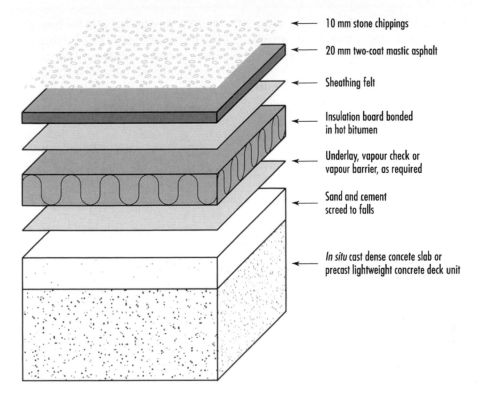

←— 10 mm stone chippings

←— 20 mm two-coat mastic asphalt

←— Sheathing felt

←— Insulation board bonded in hot bitumen

←— Underlay, vapour check or vapour barrier, as required

←— Sand and cement screed to falls

←— *In situ* cast dense concete slab or precast lightweight concrete deck unit

Fig. 6.5 Typical mastic asphalt roofing system

Mastic asphalt application

Mastic asphalt is laid to 20 mm in two layers on roofs up to 30° and in three layers to 20 mm on slopes greater than 30°. Upstands of 150 mm are required to masonry, rooflights, pipes, etc. where they penetrate the roof membrane. Where adhesion on vertical surfaces is insufficient, expanded metal lathing should be used to support the mastic asphalt. An apron flashing should protect the top of the upstand. A layer of sand is rubbed into the top of the final layer whilst it is hot to break up the skin of bitumen-rich material which forms at the worked surface.

Surface protection

Mastic asphalt gradually hardens over a period of a few years and should be protected from softening under bright sunlight by the application of surface protection. Reflective paint coatings – for example, titanium oxide in polyurethane resin or aluminium pigmented bitumen – are effective until they become dirty, but for vertical surfaces are the only appropriate measure. Reflective coatings are available in a range of colours giving differing levels of solar reflectivity. For horizontal surfaces and pitches up to 10°, a layer of 10–14 mm white stone chippings will give better protection not only from sunlight but also from ultraviolet light, which gradually degrades bitumen products. Additionally a layer of stone will act to reduce the risk of thermal shock during very cold periods. Where traffic is anticipated, the mastic asphalt should be protected with glass-fibre reinforced cement (GRC) tiles or concrete pavings.

INVERTED ROOFS

Mastic asphalt forms a suitable waterproof membrane for externally insulated or inverted roofs. The application of the insulating layer over the mastic asphalt has the advantage that it protects the waterproof layer from thermal shock, impact damage and degradation by ultraviolet light. The insulation, usually extruded polystyrene boards, is held down either by gravel or precast concrete paving slabs.

Single-ply roofing systems

Single-ply roofing systems, consist of a continuous membrane usually between 1 and 3 mm thick, covering any form of flat or pitched roof (Fig. 6.6). As water-proofing is reliant on the single membrane, a high quality of workmanship is required and this is normally provided by the specialist installer. In refurbishment work where the substrate may be rough, a polyester fleece may be used to prevent mechanical damage to the membrane from below. Life expectancies are typically quoted as 25 years. The wide range of membrane materials used may generally be categorised into thermoplastic, elastomeric and modified bitumen products. In many cases the single-layer membrane is itself a laminate, incorporating either glass fibre or polyester to improve strength and fatigue resistance or dimensional stability respectively. Both thermoplastic and elastomeric products are resistant to ageing under the severe conditions of exposure on roofs. Fixings offered by the proprietary systems include fully bonded, partially bonded, mechanically fixed and loose laid with either ballast or concrete slabs. Joints are lapped and either heat or solvent welded, usually with THF (tetrahydrofuran). A final seal of the plastic in solvent may be applied to the

joint edge after the lap joint has been checked for leaks. Most manufacturers provide a range of purpose-made accessories such as preformed corners, rainwater outlet sleeves and fixings for lightning protection.

THERMOPLASTIC SYSTEMS

Thermoplastic systems, made from non-cross-linked plastics, can be joined by solvent or heat welding. They generally exhibit good weathering properties and chemical resistance. The dominant thermoplastic systems are based on plasticised PVC (polyvinyl chloride) which is normally available in a range of colours. Certain PVC products contain up to 35% by weight of plasticisers, which can migrate to adjacent materials, leaving the membrane less flexible and causing incompatibility with extruded polystyrene insulation or bitumen products. Products manufactured from VET (vinyl ethylene terpolymer, a blend of 35% ethyl vinyl acetate and 65% PVC with only 4% plasticiser added as lubricant) are more compatible with bitumen and

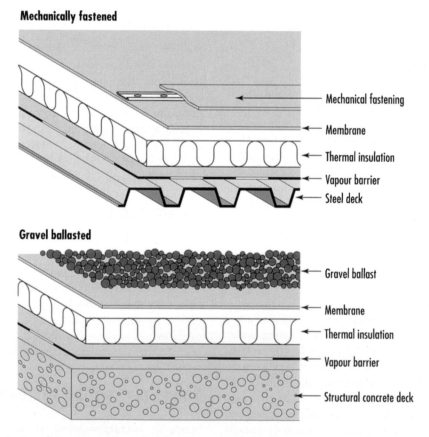

Mechanically fastened

- Mechanical fastening
- Membrane
- Thermal insulation
- Vapour barrier
- Steel deck

Gravel ballasted

- Gravel ballast
- Membrane
- Thermal insulation
- Vapour barrier
- Structural concrete deck

Fig. 6.6 Typical single-ply roofing system, mechanically fastened and gravel ballasted

polystyrene-based insulation products. Other products include CPE (chlorinated polyethylene) which has enhanced chemical-resistant properties, CSPE (chlorosulphonated polyethylene) which is highly weather-resistant and PIB (polyisobutylene) a relatively soft material which is easily joined. FPO (flexible polyolefine alloy) membranes based on an alloy of EPR (ethylene propylene rubber) and polypropylene are halogen-free and combine the temperature, chemical resistance and weldability of PVC and the flexibility of the elastomeric single-ply systems.

ELASTOMERIC SYSTEMS

Elastomeric systems are dominated by EPDM (ethylene propylene diene monomer) which is a cross-linked or cured polymer. It is characterised by high elongation and good weathering resistance to ultraviolet light and ozone. The standard material is black or grey in colour but white is also available. Most products are seamed with adhesives or applied tapes as EPDM cannot be softened by solvents or heat; however, EPDM laminated with thermoplastic faces can be heat welded on site. Products may be mechanically fixed, ballasted or adhered to the substrate.

MODIFIED-BITUMEN SYSTEMS

Most modified-bitumen systems are based on SBS (styrene butadiene styrene), APP (atactic polypropylene) or rubber-modified bitumen. Some products are now combining the durability of APP modified bitumens with the enhanced flexibility of SBS modified bitumen. Systems usually incorporate polyester or glass-fibre reinforcement for increased dimensional stability. Modified-bitumen roofing membranes are usually thicker (e.g. 5 mm) than polymer-based single-ply systems.

Liquid coatings

A range of bitumen-based and polymer-based materials is used in the production of liquid-roof waterproofing membranes. Whilst some products are installed in new work, many products are used for remedial action on failed existing flat roofs as an economic alternative to re-roofing (Fig. 6.7). They may be appropriate when the exact location of water ingress cannot be located or when re-roofing is not practicable due to the disruption that it would cause.

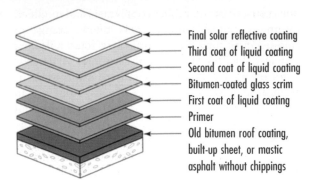

Final solar reflective coating
Third coat of liquid coating
Second coat of liquid coating
Bitumen-coated glass scrim
First coat of liquid coating
Primer
Old bitumen roof coating, built-up sheet, or mastic asphalt without chippings

Fig. 6.7 Typical refurbishment of a built-up sheet or asphalt roof with a liquid coating system

It is essential that the nature of the existing roof is correctly determined so that an appropriate material can be applied; also, failures in the substrate must be identified and rectified. The surface of the existing material must be free of loose material and dust to ensure good adhesion with the liquid coating which can be applied by brush, roller or airless spray. While achieving a uniform thickness is difficult, the systems have the advantage of being seamless. Solar reflective white and a range of colours are available. Without further protection, roofs should only be subjected to light maintenance pedestrian traffic. Installation should normally be carried out by specialist roofing contractors.

BITUMEN-BASED SYSTEMS

Most bitumen systems require a primer to seal the existing roof membrane and provide a base for the liquid coating. Two or three coats of bitumen solution or emulsion will normally be required for the waterproofing layer, and a solar reflective finish should be applied after the membrane has fully dried. A layer of glass-fibre reinforcement is usually incorporated during application of the waterproofing membrane to give dimensional stability. Two-component systems mixed during the spraying process cure more rapidly, allowing a seamless 4 mm elastomeric coat to be built up in one layer on either flat or pitched roofs. Where the material is ultraviolet light-resistant, a solar reflective layer may not be necessary.

POLYMER-BASED SYSTEMS

The range of polymers used for liquid roof finishes is extensive, including acrylic resins, polyurethanes, polyesters, silicones, rubber copolymers and modified bitumens. Some manufacturers offer a range of colours incorporating the necessary solar reflecting properties. Glass-fibre or polyester mat is used as reinforcement within the membrane layer, which is applied in a minimum of two coats. Additives to improve fire resistance such as antimony trioxide and bromine compounds may be incorporated into the formulations and good fire ratings in respect of flame penetration, and surface spread can be achieved with some products. Solvent-based products have the advantage of rapid drying times, whilst solvent-free products have 'greener' credentials and some are odour-free, reducing the disruption caused during the refurbishment of occupied buildings.

Green roofs

Green roofs are flat or low pitched roofs which are landscaped over the waterproofing layer. The landscaping may include some hard surfaces and have access for leisure and recreational functions as well as the necessary routine maintenance. Green roofs offer not only increased life expectancy for the waterproofing layer by protecting it from physical damage, ultraviolet light and temperature extremes, but also increased usable space. Environmental advantages include reduced and delayed rainwater run-off, also considerable environmental noise, air quality and wildlife habitat benefits.

Green roofs (Fig. 6.8) may be waterproofed using modified-bitumen high performance built-up systems, single-ply membrane systems or mastic asphalt. Under planting, T-grade mastic asphalt should be laid to 30 mm in three layers rather than the usual two layers to 20 mm thickness. Green roofs are divided between the *intensive* and the *extensive* systems.

EXTENSIVE GREEN ROOFS

Extensive green roofs are designed to be lighter in weight, relatively cheap, not open to recreational use and to require the minimum of maintenance. Their prime purpose is either ecological or for the environmental masking of buildings. Planting should be of drought-tolerant, wind- and frost-resistant species such as sedums, herbs and grasses. Instant cover can be created by the installation of pre-cultivated vegetation blankets where the immediate visual effect is required. Alternatively, a mixture of seeds, plant cuttings, mulch and fertiliser is sprayed onto the growing medium, and this will mature into the finished green roof over a period between one and two years. The complete system with planting, soil, filter sheet, drainage, moisture retention layer and root barrier will add between 60 and 200 kg/m^2 loading to the roof structure, which must be capable of this additional imposed load. Limited maintenance is required to remove unwanted weeds, fill bare patches and apply organic fertilisers in the spring and to remove dead

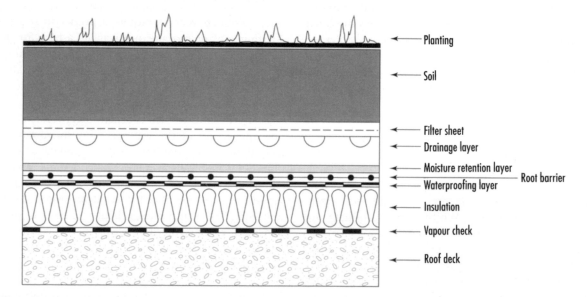

Fig. 6.8 Typical green roof system

Fig. 6.9 Extensive green roof – Westonbirt Arboretum, Gloucestershire. Photograph: Arthur Lyons

plants and weeds in the autumn. Figure 6.9 illustrates a typical low maintenance green roof at the Westonbirt Arboretum, Gloucestershire; a similar effect is achieved at the Earth Centre, Doncaster which is planted with sedum (Fig.15.1 [Chapter 15]).

INTENSIVE GREEN ROOFS

Intensive green roofs are generally designed to accept recreational activity and to include the widest range of vegetation from grass to shrubs and semi-mature trees. Depths of soil are typically between 200 and 300 mm, which together with the necessary minimum 50 mm of water reservoir and drainage systems generate an additional imposed load of typically 400 kg/m^2 on the existing or proposed structural system. Intensive green roofs may incorporate both soft and hard landscaping and slopes up to 20° are practicable. To conform to Health and Safety requirements edge protection (e.g. hand ail) or a fall arrest system (e.g. harness attachment points) must be incorporated into the design.

A typical intensive green roof system requires the following construction:

- soil, compost and planting – hard and soft landscaped areas;
- filter fleece to prevent soil blocking the drainage system;
- moisture retention material and drainage system;
- protection mat (to prevent damage to root barrier and waterproofing layers);
- polyethylene foil (isolating layer);
- root barrier;
- waterproof layer;
- insulation;
- vapour barrier.

Where trees are required, soil depths may need to be increased to typically 750 mm, with the associated increase in overall weight. Care must be taken to ensure that the roof membrane is not damaged by gardening implements.

References

FURTHER READING

British Flat Roofing Council. 1993: *Flat roofing: Design and good practice.* London: CIRIA/BFRC.
British Flat Roofing Council. 1994: *The assessment of lifespan characteristics and as-built performance of flat roofing systems.* Nottingham: BFRC.
Coates, D.T. 1993: *Roofs and roofing: Design and specification handbook.* Caithness: Whittles.
Guertin, M. 2003: *Roofing with asphalt shingles.* USA: Taunton Press.

McDonough, W. 2004: *Green roofs, ecological design and construction.* USA: Schiffer Publishing.

Metal Cladding and Roofing Manufacturers Association. 1991–4: *Design guides* – Technical Papers 1–8, Birkenhead: MCRMA.

National Federation of Roofing Contractors. 1989: *Liquid roof coatings.* 2nd ed. London: National Federation of Roofing Contractors.

Ruberoid. 2003: *Flat roofing – A guide to good practice.* London: Ruberoid.

Ruberoid and Permanite 2004: *The Waterproofers Handbook.* IKO Group.

STANDARDS

BS 476 Fire tests on building materials and structures:
  Part 3: 2004 Classification and method of test for external fire exposure to roofs.
BS 594 Hot-rolled asphalt for roads and other paved areas:
  Part 1: 2005 Specification for constituent materials and asphalt mixtures.
  Part 2: 2003 Specification for transport, laying and compaction of hot rolled asphalt.
BS 743: 1970 Materials for damp-proof courses.
BS 747: 2000 Reinforced bitumen sheets for roofing.
BS 1446: 1973 Mastic asphalt (natural rock asphalt fine aggregate) for roads and footways.
BS 1447: 1988 Mastic asphalt (limestone fine aggregate) for roads, footways and pavings in building.
BS 1521: 1972 Waterproof building papers.
BS 3690 Bitumens for building and civil engineering.
  Part 3: 1990 Specification for mixtures of bitumen with pitch, tar and Trinidad lake asphalt.
BS 4016: 1997 Specification for flexible building membranes (breather type).
BS 4841 Rigid polyurethane (PUR) and polyisocyanurate (PIR) foam for building applications:
  Part 3: 1994 Specification for two types of laminated board (roofboards) with auto-adhesively bonded reinforcing facings for use as roofboard thermal insulation for built-up roofs.
BS 5250: 2002 Code of practice for control of condensation in buildings.
BS 6229: 2003 Code of practice for flat roofs with continuously supported coverings.
BS 6398: 1983 Specification for bitumen damp-proof courses for masonry.
BS 6925: 1988 Mastic asphalt for building and engineering (limestone aggregate).
BS 8204 Screeds, bases and *in-situ* floorings:

  Part 5: 2004 Code of practice for mastic asphalt underlays and wearing surfaces.
BS 8217: 2005 Reinforced bitumen membranes for roofing – code of practice.
BS 8218: 1998 Mastic asphalt roofing – code of practice.
BS EN 495: 2001 Flexible sheets for waterproofing – plastic and rubber sheets for roof waterproofing.
BS EN 544: 2005 Bitumen shingles with mineral and/or synthetic reinforcements.
BS EN 1107 Flexible sheets for waterproofing:
  Part 1: 2000 Bitumen sheets for roof waterproofing.
  Part 2: 2001 Plastic and rubber sheets for roof waterproofing.
BS EN 1108: 2000 Flexible sheets for waterproofing – bitumen sheets for roof waterproofing – form stability.
BS EN 1109: 2000 Flexible sheets for waterproofing – bitumen sheets for roof waterproofing – flexibility.
BS EN 1110: 2000 Flexible sheets for waterproofing – bitumen sheets for roof waterproofing – flow resistance.
pr EN 1297–1: 1994 Flexible sheets for roofing – deterioration of resistance to UV and water ageing.
BS EN 1848 Flexible sheets for waterproofing. Determination of length, width and straightness:
  Part 1: 2000 Bitumen sheets for roof waterproofing.
  Part 2: 2001 Plastic and rubber sheets for roof waterproofing.
BS EN 1849 Flexible sheets for waterproofing. Determination of thickness and mass:
  Part 1: 2000 Bitumen sheets for roof waterproofing.
  Part 2: 2001 Plastic and rubber sheets for roof waterproofing.
BS EN 1850 Flexible sheets for waterproofing. Determination of visible defects:
  Part 1: 2000 Bitumen sheets for roof waterproofing.
  Part 2: 2001 Plastic and rubber sheets for roof waterproofing.
BS EN 1931: 2000 Flexible rubber sheets for waterproofing – vapour transmission.
BS EN 12039: 2000 Flexible rubber sheets for waterproofing – adhesion of granules.
BS EN 12310 Flexible rubber sheets for waterproofing. Determination of resistance to tearing:
  Part 1: 2000 Bitumen sheets for roof waterproofing.
  Part 2: 2000 Plastic and rubber sheets for roof waterproofing.
BS EN 12591: 2000 Bitumen and bituminous binders – specifications for paving grade bitumens.
BS EN 12697: 2005 Bituminous mixtures – test methods for hot mix asphalt.

BS EN 13055 Lightweight aggregates:
 Part 2: 2004 Lightweight aggregates for bituminous mixtures.
BS EN 13416: 2001 Flexible rubber sheets for waterproofing – sampling.
BS EN 13501 Parts 1, 2, 3 and 5 Fire classification of construction products and building elements.
BS EN 13583: 2001 Flexible rubber sheets for waterproofing – hail resistance.
BS EN 13924: 2006 Bitumen and bituminous binders – specification for hard paving grade bitumens.
pr EN 13948: 2001 Flexible sheets for waterproofing – bitumen, plastic and rubber sheets for roof waterproofing – determination of resistance to root penetration.
CP 153 Windows and rooflights:
 Part 2: 1970 Durability and maintenance.
DD ENV 1187: 2002 Test methods for external fire exposure to roofs.

BUILDING RESEARCH ESTABLISHMENT PUBLICATIONS

BRE Digests

BRE Digest 180: 1986 Condensation in roofs.
BRE Digest 311: 1986 Wind scour of gravel ballast on roofs.
BRE Digest 312: 1986 Flat roof design: the technical options.
BRE Digest 324: 1987 Flat roof design: thermal insulation.
BRE Digest 419: 1996 Flat roof design: bituminous waterproof membranes.
BRE Digest 493: 2005 Safety considerations in designing roofs.

GBG Good building guide

GBG 36: 1999 Building a new felted flat roof.
GBG 43: 2000 Insulated profiled metal roofs.
GBG 51: 2002 Ventilated and unventilated cold pitched roofs.

BRE Information papers

BRE IP 15/82 Inspection and maintenance of flat and low pitched timber roofs.
BRE IP 19/82 Considerations in the design of timber flat roofs.
BRE IP 13/87 Ventilating cold deck flat roofs.
BRE IP 2/89 Thermal performance of lightweight inverted warm deck flat roofs.
BRE IP 8/91 Mastic asphalt for flat roofs: testing for quality assurance.
BRE IP 7/95 Bituminous roofing membranes: performancc in use.
BRE IP 7/04 Designing roofs with safety in mind.

BRE Report

BR 302: 1996 Roofs and roofing.

ADVISORY ORGANISATIONS

European Liquid Roofing Association, Fields House, Gower Road, Haywards Heath, West Sussex RH16 4PL (01444 417458).
Flat Roofing Alliance, Fields House, Gower Road, Haywards Heath, West Sussex RH16 4PL (01444 440027).
Institute of Asphalt Technology, Paper Mews Place, 290 High Street, Dorking, Surrey RH4 1QT (01306 742792).
Mastic Asphalt Council, PO Box 77, Hastings, Kent TN35 4WL (01424 814400).
Metal Cladding and Roofing Manufacturers Association, 18 Mere Farm Road, Prenton, Wirral, Chesire CH43 9TT (0151 652 3846).
National Federation of Roofing Contractors Ltd., 24 Weymouth Street, London W1G 7LX (020 7436 0387).
Roofing Industry Alliance, Fields House, Gower Road, Haywards Heath, West Sussex RH16 4PL (01444 440027).
Single-Ply Roofing Association, The Building Centre, 26 Store Street, London WC1E 7BT (0115 914 4445).

7

GLASS

—

Introduction

The term *glass* refers to materials, usually blends of metallic oxides, predominantly silica, which do not crystallise when cooled from the liquid to the solid state. It is the non-crystalline or amorphous structure of glass (Fig. 7.1) that gives rise to its transparency.

Glass made from sand, lime and soda ash has been known in Egypt for 5000 years, although it probably originated in Assyria and Phoenicia. The earliest man-made glass was used to glaze stone beads, and later to make glass beads (*circa* 2500 BC), but it was not until about 1500 BC that it was used to make hollow vessels.

For many centuries glass was worked by drawing the molten material from a furnace. The glass was then rolled out or pressed into appropriate moulds and finally fashioned by cutting and grinding. Around 300 BC the technique of glass blowing evolved in Assyria, and the Romans developed this further by blowing glass into moulds. Medieval glass produced in the Rhineland contained potash from the burning of wood rather than soda ash. Together with an increase in lime content this gave rise to a less durable product which has caused the subsequent deterioration of some church glass from that period.

The various colours within glass derived from the addition of metallic compounds to the melt. Blue was obtained by the addition of cobalt, whilst copper produced blue or red and iron or chromium produced green. In the fifteenth century white opaque glass was produced by the addition of tin or arsenic, and by the seventeenth century ruby-red glass was made by the addition of gold chloride. Clear glass could only be obtained by using antimony or manganese as a

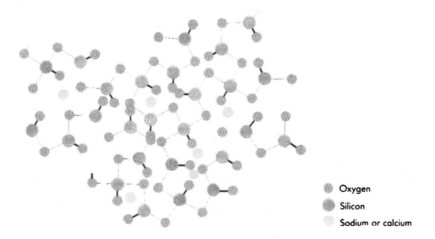

Oxygen
Silicon
Sodium or calcium

Fig. 7.1 Structure of glass (after Button, D. and Pye, B. (eds) 1993. *Glass in building.* Butterworth Architecture)

decolouriser to remove the green colouration caused by iron impurities within the sand.

In the late twentieth century, with the advent of fully glazed facades as illustrated by the Faculty of Law building at the University of Cambridge (Fig. 7.2), the construction industry has become a major consumer of new glass, and a proactive force in the development of new products. The wide range of glass materials used within the building industry is classified in the standard BS 952–1: 1995.

Within the UK, thousands of tonnes of glass are recycled each year, but this is mainly domestic waste, which cannot be used for the production of window glass as this requires pure raw materials. Even architectural waste glass is of variable composition with contamination from wire, sealants and special glasses making it not useable as cullet in the manufacturing process without careful sorting. However, excess recycled glass not required for remaking bottles has been used for making decorative paving surfaces.

Fig. 7.2 Glazed facade – Faculty of Law, University of Cambridge. Architects: Foster and Partners. Photograph: Arthur Lyons

Manufacture

COMPOSITION

Soda lime silicate glass

Modern glass is manufactured from sand (silica), soda ash (sodium carbonate) and limestone (calcium carbonate), with small additions of salt-cake (calcium sulfate) and dolomite (magnesian limestone). This gives a final composition of typically 70–4% silica, 12–16% sodium oxide, 5–12% calcium oxide, 2–5% magnesium oxide with small quantities of aluminium, iron and potassium oxides. The addition of 25% broken glass or *cullet* to the furnace mix accelerates the melting process and recycles the production waste. Most raw materials are available within the UK, although some dolomite is imported. The production process is relatively energy intensive at 15 000 kWh/m^3 (c.f. concrete – 625 kWh/m^3), but the environmental pay-back arises from its appropriate use in energy-conscious design. Soda lime silicate glass may be chemically strengthened by an ion exchange process which replaces small surface ions by larger ones, thus putting the surfaces and edges into compression.

Alkaline earth silicate, borosilicate and ceramic glass

Other products used within construction include the alkaline earth silicate and borosilicate glasses; these have significantly different chemical compositions giving rise to their particular physical properties. The composition of alkaline earth silicate glass is typically 55–70% silica, 5–14% potassium oxide, 3–12% calcium oxide, 0–15% aluminium oxide, with quantities of zirconium, strontium and barium oxides. Borosilicate glass is typically 70–87% silica, 0–8% sodium oxide, 0–8% potassium oxide, 7–15% boron oxide, 0–8% aluminium oxide, with small quantities of other oxides. A particular characteristic of borosilicate glass is that it has a coefficient of expansion one-third that of standard soda lime silicate glass, making it significantly more resistant to thermal shock in case of fire.

Glass ceramics are characterised by a zero coefficient of expansion making them highly resistant to thermal shock. The composition of glass ceramics is complex comprising typically 50–80% silica, 15–27% alumina, with small quantities of sodium oxide, potassium oxide, barium oxide, calcium oxide, magnesium oxide, titanium oxide, zirconium oxide, zinc oxide, lithium oxide and other minor constituents. The glass is initially produced by a standard float or rolling technique, but subsequent heat treatment converts part of the normal glassy phase into a fine-grained crystalline form, giving rise to the particular physical properties.

FORMING PROCESSES

Early methods

Early crown glass was formed by spinning a 4 kg cylindrical gob of molten glass on the end of a blow pipe. The solid glass was blown, flattened out and then transferred to a solid iron rod or *punty*. After reheating it was spun until it opened out into a 1.5 m diameter disc. The process involved considerable wastage including the bullion in the centre, which nowadays is the prized piece. An alternative process involved the blowing of a glass cylinder which was then split open and flattened out in a kiln. This process was used for the manufacture of the glass for the Crystal Palace in 1851.

Subsequently in a major development, a circular metal bait was lowered into a pot of molten glass and withdrawn slowly, dragging up a cylindrical ribbon of glass 13 m high, the diameter of the cylinder being maintained with compressed air. The completed cylinder was then detached, opened up and flattened out to produce flat window glass.

It was only by the early twentieth century with the development of the Fourcault process in Belgium and the Colburn process in America that it became possible to produce flat glass directly. A straight bait was drawn vertically out of the molten glass to produce a ribbon of glass, which was then drawn directly up a tower, or in the Colburn process turned horizontally, through a series of rollers; finally, appropriate lengths were cut off. However, such drawn sheet glass suffered from manufacturing distortions. This problem was

overcome by the production of plate glass, which involved horizontal casting and rolling, followed by grinding to remove distortions and polishing to give a clear, transparent but expensive product. The process was ultimately fully automated into a production line in which the glass was simultaneously ground down on both faces. The plate glass manufacturing process is now virtually obsolete having been replaced by the *float process*, which was invented in 1952 by Pilkington and developed into commercial production by 1959. Drawn glass is only manufactured for conservation work, where gaseous and solid inclusions in the glass are required to emulate the historic material.

Float glass

A furnace produces a continuous supply of molten glass at approximately 1100°C, which flows across the surface of a large shallow bath of molten tin contained within an atmosphere of hydrogen and nitrogen (this prevents oxidation of the surface of the molten metal) (Fig. 7.3). The glass ribbon moves across the molten metal, initially at a sufficiently high temperature for the irregularities on both surfaces to become evened out leaving a flat and parallel ribbon of glass. The temperature of the glass is gradually reduced as it moves forward until, at the end of the molten tin, it is sufficiently solid at 600°C not to be distorted when supported on rollers. Thickness is controlled by the speed at which the glass is drawn from the bath. Any residual stresses are removed as the glass passes through the 200 m annealing lehr or furnace, leaving a fire-polished material. The glass is washed and substandard material discarded for recycling. The computer-controlled cutting, firstly across the ribbon, then the removal of the edges, is followed

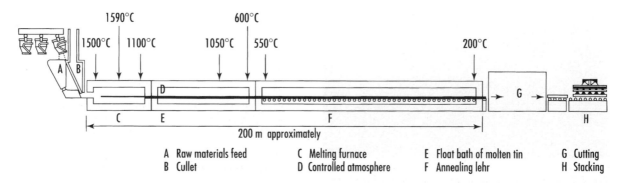

Fig. 7.3 Float glass process

by stacking, warehousing and dispatch. A typical float glass plant will manufacture 5000 tonnes of glass per week, operating continuously for several years.

Float glass for the construction industry is made within the thickness range 2–25 mm, although 0.5 mm is available for the electronics industry. Many surface modified glasses are produced, either by incorporating metal ions into the glass within the float process, or under vacuum by magnetically enhanced cathodic sputtering or by spraying the surface with metal oxides or silicon. Body-tinted glass, which is of uniform colour, is manufactured by blending additional metal oxides into the standard melt. The use of all-electric melting processes offers higher quality control and less environmental pollution than previously produced from earlier oil or gas-fired furnaces.

Non-sheet products

GLASS FIBRES

Continuous filament

Continuous glass fibres are manufactured by constantly feeding molten glass from a furnace into a forehearth fitted with 1600 accurately manufactured holes through which the glass is drawn at several thousand metres per second. The fibres (as small as 9 microns in diameter) pass over a size applicator and are gathered together as a bundle prior to being wound up on a collet. The material may then be used as rovings, chopped strand or woven strand mats for the production of glass-fibre reinforced materials such as GRP (glass-fibre reinforced polyester), GRC (glass-fibre reinforced cement), or GRG (glass-fibre reinforced gypsum) – see Chapter 11.

Glass wool

Glass wool is made by the Crown process, which is described in Chapter 13.

CAST GLASS

Glass may be cast and pressed into shape for glass blocks and extruded sections.

Profiled sections

Profile trough sections in clear or coloured 6 or 7 mm cast glass are manufactured in sizes ranging from 232 to 498 mm wide, 41 and 60 mm deep and up to 7 m

long, with or without stainless steel longitudinal wires (Fig. 7.4) (BS EN 572–7: 2004). The system can be used horizontally or vertically, as single or double glazing, and as a roofing system spanning up to 3 m. A large radius curve is possible as well as the normal straight butt jointed system, and the joints are sealed with one-part translucent silicone. The standard double-glazed system has a U-value of 2.8 W/m^2 K, but this can be enhanced to 1.8 W/m^2 K by the use of low-emissivity coated glass. Amber or blue tinted versions are available for solar control or aesthetic reasons. The double-glazed system produces a sound reduction within the 100–3200 Hz range of typically 40 dB.

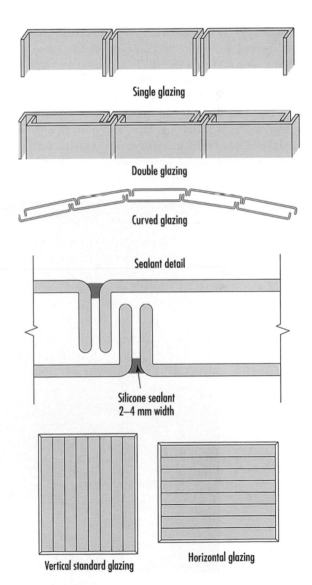

Fig. 7.4 Profiled glass sections

The incorporation of 16-mm-thick translucent aerogel insulated polycarbonate panels into profiled trough sections significantly increases the thermal insulating properties of the units. Airborne sound insulation is enhanced, particularly at frequencies below 500 Hz. Light transmission through the translucent aerogel panels is approximately 50%, but UV transmission is zero. (Aerogels are described in Chapter 13 Insulating Materials.)

Glass blocks

Glass blocks for non-load-bearing walls and partitions are manufactured by casting two half blocks at 1050°C, joining them together at 800°C, followed by annealing at 560°C. The void space is partially evacuated, giving a U-value of 2.5 W/m² K. The standard blocks (Fig. 7.4) are 115,190, 240 and 300 mm square with thicknesses of 80 and 100 mm, although rectangular and circular blocks are also available. Walls may be curved as illustrated in Figure 7.5. The variety of patterns offering differing degrees of privacy, include clear, frosted, Flemish, reeded, and crystal designs with colours from blue, green and grey to pink and gold. Blocks with solar reflective glass or incorporating white glass fibres offer additional solar control; colour may be added either to the edge coating or the glass itself. Special blocks are also available to form corners and ends, also for ventilation.

For exterior and fire-retarding applications natural or coloured mortar (2 parts Portland cement, 1 part lime, 8 parts sand) is used for the jointing. Walls may be straight or curved; in the latter case, the minimum radius varies according to the block size and manufacturer's specification. Vandal- and bullet-proof blocks are available for situations requiring higher security. For interior use, blocks may be laid with sealant rather than mortar. Glass blocks are now available in preformed panels for speedier installation.

Glass blocks jointed with mortar give a fire resistance of 30 or 60 minutes with respect to stability and integrity but not insulation, according to the BS 476: Part 22: 1987 or BS EN 1364–1: 1999. Fire stop blocks (F-30) manufactured from 26 mm rather than 8 mm glass offer the increased fire protection of two hours' integrity and 30 minutes' thermal insulation. Sound reduction over the 100–3150 Hz range is typically 40–2 dB for standard blocks, but up to 45–9 dB for the

Fig. 7.5 Glass Blocks. Photograph: Courtesy of Glass Block Technology-(www.glassblocks.co.uk)

F-30 blocks. Visible light transmission ranges downwards from 85 to 77% depending upon the pattern and block size, but this is reduced to 60% for coloured glass.

Glass pavers

Glass pavers are cast either as single layer blocks or shells, or as hollow blocks where insulation is required. The standard sizes are 120, 150 and 190 mm square, with a depth of 55, 60 or 80 mm for single layer shells and 80, 100 and 160 mm for double layer insulating blocks. Single layer square blocks range in size from 90 to 300 mm with thicknesses typically between 18 and 38 mm. Surfaces may be clear or textured for non-slip action. On-site installation requires reinforced concrete construction allowing a minimum 30 mm joint between adjacent blocks and appropriate expansion joints around panels. Precast panels offer a high standard of quality control and rapid installation. They may also be used to create architectural features such as naturally illuminated domes or archways.

CELLULAR OR FOAMED GLASS

The manufacture of cellular or foamed glass as an insulation material is described in Chapter 13.

Sheet products

STANDARD FLOAT GLASS SIZES

The standard thicknesses for float glass are 3, 4, 5, 6, 8, 10, 12, 15, 19 and 25 mm to maximum sheet sizes of 3×12 m. Thicker grades are available to smaller sheet sizes. (The U-value for standard 6 mm float glass is 5.7 W/m^2 K.)

TRADITIONAL BLOWN AND DRAWN GLASSES

Traditional blown and drawn glasses are available commercially both clear and to a wide range of colours. Drawn glass up to 1600×1200 mm is optically clear but varies in thickness from 3 to 5 mm. It is suitable for conservation work where old glass requires replacement. Blown glass contains variable quantities of air bubbles and also has significant variations in thickness giving it an antique appearance. Where laminated glass is required, due to variations in thickness, these traditional glasses can only be bonded to float glass with resin.

CURVED SHEET GLASS

BS 952–2: 1980 refers to standard glass curvatures. Curved glass can be manufactured by heating annealed glass to approximately 600°C, when it softens and sag-bends to the shape of the supporting mould. Sheets up to $3 \text{ m} \times 4 \text{ m}$ can be bent with curvature in either one or two directions. Tight curvatures can be produced for architectural feature glass. Patterned, textured, and pyrolytic-coated solar control glasses can all be curved by this technique. Bent glass can subsequently be sandblasted, toughened, or laminated, even incorporating coloured interlayers if required.

SELF-CLEANING GLASS

Self-cleaning glass has an invisible hard coating 15 microns thick, which incorporates two special features. The surface incorporating titanium dioxide is photocatalytic, absorbing ultraviolet light, which with oxygen from the air then breaks down or loosens any organic dirt on the surface. Additionally, the surface is hydrophilic, causing rainwater to spread evenly over the surface, rather than running down in droplets, thus uniformly washing the surface and preventing any unsightly streaks or spots appearing when the surface dries. Self-cleaning glass has a slightly greater mirror effect than ordinary float glass, with a faint blue tint. It is available as normal annealed glass also in toughened or laminated form. The surface coating, which reduces the transmittance of the glass by about 5%, is tough but as with any glass can be damaged by scratching. A blue solar control version, suitable for conservatory roofs, reduces the solar heat gain by approximately 60% depending upon the thickness of glass used and its combination in a double-glazing system.

CLEAR WHITE GLASS

Standard float glass is slightly green due to the effect of iron oxide impurities within the key raw material sand. However, clear white glass can be produced, at greater cost, by using purified ingredients. The light transmittance of clear white glass is approximately 2% greater than standard glass for 4 and 6 mm glazing. Unlike standard glass which appears green at exposed polished edges, clear white glass is virtually colourless.

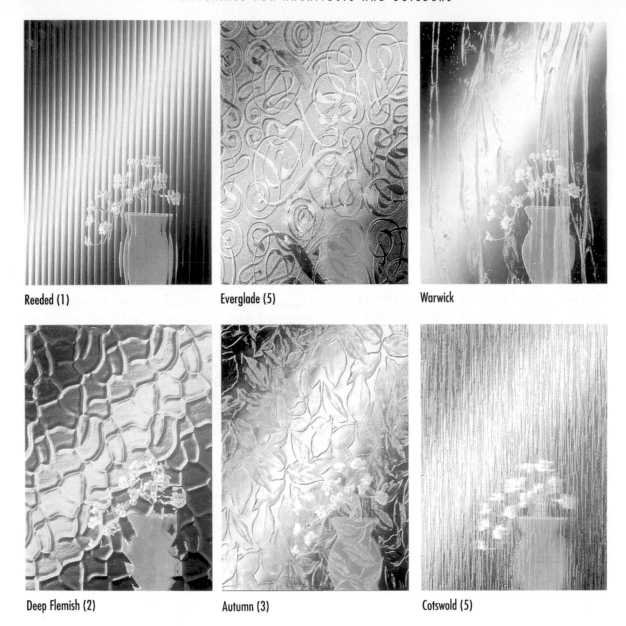

Reeded (1)   Everglade (5)   Warwick

Deep Flemish (2)   Autumn (3)   Cotswold (5)

Fig. 7.6 Typical embossed patterned glasses and relevant obscuration factors

EMBOSSED PATTERN GLASS

A wide range of 3, 4, 6, 8 and 10 mm patterned glass is commercially available, offering a range of obscuration factors depending upon the depth and design of the pattern as illustrated in Figure 7.6. The classification of obscuration factors differs between manufacturers with 1 (lowest) to 5 (highest) for Pilkington and 1 (highest) to 10 (lowest) for Saint Gobain. The degree of privacy afforded by the various glasses is not only dependent upon the pattern but also upon the relative lighting levels on either side and the proximity of any object to the glass. The maximum stock sheet size is 3300 × 1880 mm.

Patterned glass is manufactured to BS EN 572–5: 2004, from a ribbon of molten cast glass which is passed through a pair of rollers, one of which is embossed. Certain strong patterns, such as *Cotswold* or *reeded*, require client choice of orientation, whereas the more flowing designs may need appropriate

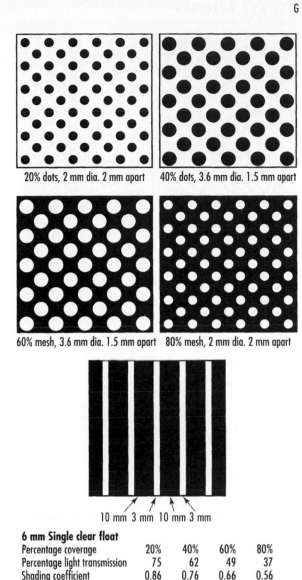

20% dots, 2 mm dia. 2 mm apart 40% dots, 3.6 mm dia. 1.5 mm apart

60% mesh, 3.6 mm dia. 1.5 mm apart 80% mesh, 2 mm dia. 2 mm apart

10 mm 3 mm 10 mm 3 mm

6 mm Single clear float

Percentage coverage	20%	40%	60%	80%
Percentage light transmission	75	62	49	37
Shading coefficient	0.86	0.76	0.66	0.56

Fig. 7.7 Screen printed glass

matching. Patterned glasses may be toughened, laminated or incorporated into double-glazing units for thermal, acoustic or safety considerations; a limited range is available in bronze tinted glass. Wired patterned glass (BS EN 572–6: 2004) is available with a 12.5 or 25 mm square mesh.

SCREEN PRINTED GLASS

White or coloured ceramic frit is screen printed onto clear or tinted float glass, which is then toughened and heat soaked, causing the ceramic enamel to fuse permanently into the glass surface. Standard patterns (Fig. 7.7) or individual designs may be created, giving the required level of solar transmission and privacy. Screen printed glass, which is colourfast and abrasion-resistant, is usually installed with the printed side as the inner face of conventional glazing.

DECORATIVE ETCHED AND SANDBLASTED GLASS

Acid etched glass, 4 mm and 6 mm in thickness, is available to a maximum sheet size of 2140 × 1320 mm with a small range of patterns (Fig. 7.8). These glasses have a low obscuration factor and should not be used in areas of high humidity, as condensation or water causes temporary loss of the pattern. Etched glasses need to be handled carefully on site, as oil, grease and finger marks are difficult to remove completely. Etched glasses may be toughened or laminated; when laminated, the etched side should be outermost to retain the pattern effect and when incorporated into double glazing, the etched glass forms the inner leaf with the etched face towards the air gap. As with embossed glass, pattern matching and orientation is important. Similar visual effects can be achieved by sandblasting techniques, although the surface finish is less smooth. Patterns may be clear on a frosted background or the reverse, depending upon the aesthetic effect and level of privacy required. Additionally, glass with both a textured and etched finish is available.

DECORATIVE COLOURED GLASS

Traditional coloured glass windows constructed with lead cames, soldered at the intersections and wired to saddle bars, are manufactured from uniform pot, surface flashed or painted glasses. For new work, additional support is afforded by the use of lead cames with a steel core, and non-corroding saddle bars of bronze or stainless steel should be used. A three-dimensional effect is achieved by fixing with ultraviolet sensitive adhesive, coloured bevelled glass to clear or coloured sheet glass; the thin edges (1.5–2.5 mm) being covered with adhesive lead strip. Such effects can also be simulated by the use of coloured polyester or vinyl film and lead strip acrylic bonded to a single sheet of clear glass. The base glass may be toughened or laminated as appropriate, and the decorative coloured glass laminate can be incorporated into standard double-glazing units. The effect of coloured glass can also be achieved by using a coloured polyvinyl butyral

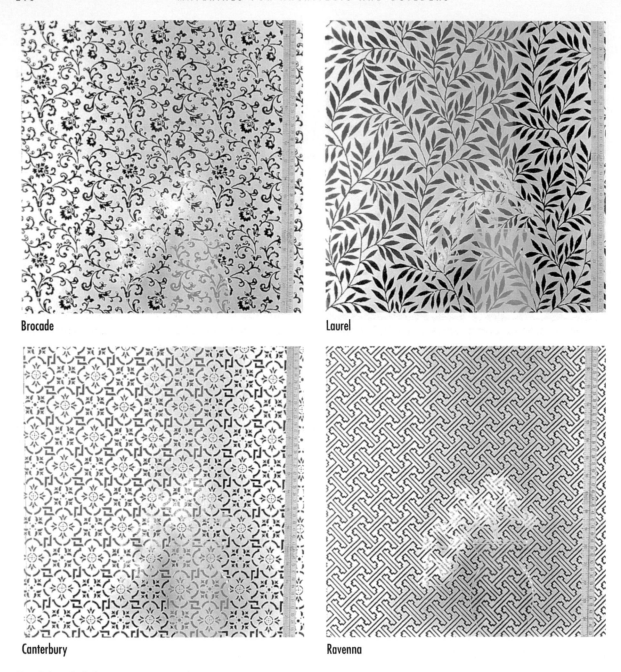

Brocade

Laurel

Canterbury

Ravenna

Fig. 7.8 Etched glass

interlayer within laminated glass. Various colour effects and tones can be achieved by combining up to four different coloured interlayers and, if required, a variety of patterns including spots, stripes, squares and dots. These laminated coloured and patterned glasses have the same impact resistance and acoustic insulation as standard clear laminated glass of the same dimensions.

GEORGIAN WIRED GLASS

Georgian wired glass (BS EN 572–3: 2004) is produced by rolling a sandwich of a 13 mm electrically welded steel wire mesh between two ribbons of molten glass. This produces the standard cast 7 mm sheet, suitable when obscuration is required. For visual clarity the cast product is subsequently ground with sand and water then polished with jeweller's rouge to 6 mm sheet (Fig. 7.9). Both the cast and polished grades have a light transmission of 80%. Wired glass is not stronger than the equivalent thickness of annealed glass; however, when cracked, the pieces remain held together.

On exposure to fire, the wire mesh dissipates some heat, but ultimately Georgian glass will crack, particularly if sprayed with water when hot. However, the wire mesh holds the glass in position, thus retaining its integrity and preventing the passage of smoke and flame. Accidental damage may cause the breakage of the glass, but again it is retained in position by the mesh, at least until the wires are affected by corrosion.

Georgian wired glass is available in sheet sizes up to 1985 × 3500 mm (cast) and 1985 × 3300 mm (polished). It is easily cut and can be laminated to other glasses but cannot be toughened. Standard Georgian glass is not considered to be a *safety glass* to BS 6206: 1981, which defines three classes with decreasing impact resistance down from Class A to Class C. However, certain laminates or products with increased wire thickness do achieve the impact resistance standards for safety glass to BS 6206: 1981 and should be marked accordingly. They may therefore be used in locations requiring safety glass according to Part N of the Building Regulations and BS 6262–4: 2005.

TOUGHENED GLASS

Toughened glass (Fig. 7.10) is up to four or five times stronger than standard annealed glass of the same thickness. It is produced by subjecting preheated annealed glass at 650°C to rapid surface cooling by the application of jets of air. This causes the outer faces to be set in compression with balancing tension forces within the centre of the glass. As cracking within glass commences with tensile failure at the surface, much greater force can be withstood before this critical point is reached.

Toughened glass cannot be cut or worked, and therefore all necessary cutting, drilling of holes and grinding or polishing of edges must be completed in advance of the toughening process. In the *roller hearth* horizontal toughening process, some bow roller wave distortions and end edge sag may develop, but these will be within narrow tolerances; however,

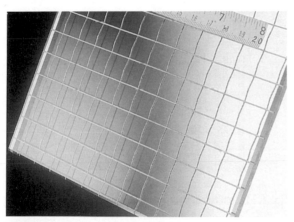

Polished

Cast

Fig. 7.9 Georgian wired glass

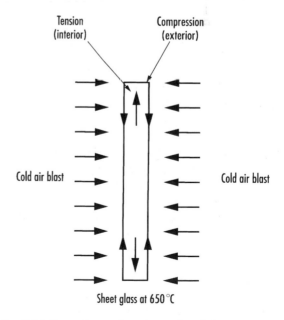

Fig. 7.10 Process for manufacturing toughened glass

they may be observed in the more reflective glasses when viewed from the outside of a building. In the vertical toughening process the sheet is held by tongs which leave slight distortions where they have gripped the glass.

Toughened glass will withstand considerable extremes of temperature and sudden shock temperatures. If broken, it shatters into small granules which are not likely to cause the serious injuries associated with the accidental breakage of annealed glass. To be classified as a *safety glass*, toughened glass must be tested and marked according to required standard BS 6206: 1981. When toughened glass is specified for roof glazing, balustrades and spandrel panels, it is subjected to a heat soaking at 290°C, a process which is destructive to any sub-standard units. This removes the low risk of spontaneous breakage of toughened glass on site, caused by the presence of nickel sulfide inclusions within the material. All standard float, coated, rough-cast and some patterned glasses can be toughened.

HEAT-STRENGTHENED GLASS

Heat-strengthened glass is manufactured by a similar process to toughening, but with a slower rate of cooling which produces only half the strength of toughened glass. On severe impact, heat-strengthened glass breaks into large pieces like annealed glass, and therefore alone is not a safety glass. It does not require heat soaking to prevent the spontaneous breakage which occurs occasionally with toughened glass. Heat-strengthened glass is frequently used in laminated glass where the residual strength after fracture gives some integrity to the laminate. Typical applications include locations where resistance to wind pressure is necessary, such as the upper storeys and corners of high buildings, and also in spandrels where there is an anticipated higher risk of thermal cracking. Modern applications of laminated heat-strengthened glass include the use of 12-mm-thick 1.5 × 3 m roofing panels developed by Arup in Flintholm, Copenhagen. The residual strength of the laminate prevents the glazing falling out of the frame if broken, reducing the hazard to users of the transport interchange building.

LAMINATED GLASS

Laminated glass (Fig. 7.11) is produced by bonding two or more layers of glass together with a plastic interlayer of polyvinyl butyral (PVB) sheet or a polymethyl methacrylate low-viscosity resin. The

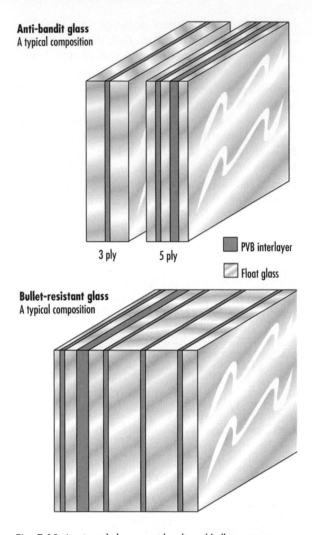

Fig. 7.11 Laminated glass – anti-bandit and bullet-resistant

low-viscosity resin is more versatile as it allows for the manufacture of curved laminates or the incorporation of patterned glasses. The lamination process greatly increases the impact resistance over annealed glass of the same thickness. Furthermore on impact, whilst the glass laminations crack, they do so without splintering or disintegration, being held together by the interlayer. Therefore, laminated glass can be defined as a *safety glass* providing it achieves the appropriate class standard to BS 6206: 1981.

The all-glass facade of the Prada Store in Tokyo by Herzog & de Meuron (Fig. 7.12) is manufactured from rhomboidal panes of PVB-bonded laminated glass; some units are flat, but others are curved with either the convex or concave surface to the outside giving rise to varied colours by reflection. The

Fig. 7.12 Curved rhomboidal glass panes – Prada Store, Tokyo. Architects: Herzog and de Meuron. Photograph: Courtesy of Sam Glynn

The incorporation of specialist film interlayers within laminated glass offers further diversity. Interlayers may have variable optical and thermal transmission properties, by incorporating photochromic, thermochromic, thermotropic or electrochromic materials. Alternatively the interlayer may diffract the incident light through specified angles, as within prisms and holograms. Thus, within a deep room, natural light can be refracted up to a white ceiling for dispersion further back within the space.

PLASTIC FILM LAMINATES

A range of transparent and translucent plastic films are readily applied internally or externally to modify the properties of glass. These include patterned films to create privacy, manifestation films to prevent people accidentally walking into clear glass screens or doors, and reflective films to reduce solar gain and glare. Safety films, as applied to overhead glazing on the Eurostar Waterloo International Terminal (Fig. 5.1 Chapter 5), remove the risk of injury from falling glass when nickel sulfide inclusions cause the spontaneous failure of toughened glass. Similarly, security films ensure that glass damaged by accidental impact or vandalism remains in place.

FIRE-RESISTANT GLASS

The ability of a particular glass to conform to the criteria of integrity and insulation within a fire is a measure of its fire resistance (Table 7.1). However, to achieve a specified performance in fire it is necessary to ensure that the appropriate framing, fixings and glass have all been used, as fire resistance is ultimately dependent upon the whole glazing system and not the glass alone.

The European specification (BS EN 13501–2: 2003) for the fire-resistance of a material or an assembly is classified by its performance against the criteria; integrity (E), insulation (I), radiation (W) and also, not normally relevant to glass, load-bearing capacity (R). The standard time periods are 15, 20, 30, 45, 60, 90, 120, 180 and 240 minutes.

Typical classifications:

E30	integrity only 30 minutes
EW30	integrity and radiation protection for 30 minutes
EI30	integrity and insulation for 30 minutes
E60EI30	integrity 60 minutes and insulation 30

perimeters of the double-glazed units are necessarily flat to ensure correct positioning and sealing.

The impact resistance of laminated glass may be increased by the use of thicker interlayers, increasing the numbers of laminates or by the inclusion of polycarbonate sheet. Typically, anti-bandit glass (BS 5544: 1978) has two or three glass laminates while, depending on the anticipated calibre and muzzle velocity, bullet-resistant glass (BS EN 1063: 2000) has four or more glass laminates. To prevent spalling, the rear face of bullet-resistant glass may be sealed with a scratch-resistant polyester film and for fire protection Georgian wired glass may be incorporated. Laminated glasses made from annealed glass can be cut and worked after manufacture.

Specialist properties for X-ray or ultraviolet light control can be incorporated into laminated glasses by appropriate modifications to the standard product. The latter reduces transmissions in the 280–380 nm wavelength ranges, which cause fading to paintings, fabrics and displayed goods.

Table 7.1   Typical fire resistance properties of glass

Fire resistance of non-insulating glass - integrity only

Type	Thickness (mm)	E Integrity (min)
Georgian wired glass – cast	7	30–60
Georgian wired glass – polished	6	30–90
Toughened standard glass	6–12	30–90
Toughened borosilicate glass	6–12	30–120
Glass blocks	80	30–60

Fire resistance of insulating glass - integrity and insulation

Type	Thickness (mm)	EI Integrity and Insulation (min)
Laminated intumescent glass	11	15
	15	30
	19	45
	23	60
	23 × 2 (double glazing unit)	120
Laminated gel interlayer glass	22–74	≤90

Fire resistance data are significantly dependent upon glass thickness, glazing size, aspect ratio and the metal, timber or butt-jointed glazing system.

Non-insulating glass

Non-insulating glass products will prevent the passage of flame, hot gases and smoke, but will allow heat transmission by radiation and conduction, thus ultimately further fire spread may occur through the ignition of secondary fires. Intense radiation through glass areas may render adjacent escape routes impassable.

Georgian wired glass offers up to a 120 minutes' fire resistance rating with respect to integrity, depending upon the panel size and fixings. If the glass cracks within a fire, its integrity is retained as the wire mesh prevents loss of the fractured pieces. Georgian wired glass is cheaper than insulated fire-resistant glasses and may be cut to size on site.

Toughened calcium-silica based glasses can achieve 90 minutes' fire resistance with respect to integrity. The glazing remains intact and transparent, but will break up into harmless granules on strong impact if necessary for escape. Toughened glass cannot be cut or worked after manufacture.

Boro-silicate glass, with a low coefficient of expansion, is more resistant to thermal shock than standard annealed glass and does not crack on exposure to fire. It can be thermally strengthened to increase its impact resistance. Certain ultra-heat-resistant ceramic glasses have a zero coefficient of thermal expansion. As a result, they can resist temperatures up to 1000°C, and the thermal shock of a cold water spray when heated by fire.

Insulating glass

Insulating glasses are manufactured from float glass laminated with either intumescent or gel materials. Intumescent laminated glass has clear interlayers, which on exposure to fire expand to a white opaque material, inhibiting the passage of conductive and radiant heat (Fig. 7.13). The glass layers adjacent to the fire crack but retain integrity owing to their adhesion with the interlayers. The fire resistance, ranging between 30 and 120 minutes for insulation and integrity, depends on the number of laminations, usually between 3 and 5. To avoid the green tint associated with thick laminated glass, a reduced iron-content glass may be used to maintain optimum light transmission. For exterior use the external grade has an additional glass laminate with a protective ultraviolet filter interlayer. Laminates may be manufactured with tinted glass or combined with other patterned or solar control glasses. Insulating glass is supplied cut to size and should not be worked on site.

Double-glazed units with two leaves of intumescent laminated glass give insulation and integrity ratings of 120 minutes; alternatively, units may be formed with one intumescent laminate in conjunction with specialist solar control glasses. The fire-resistant laminated glasses conform to the requirements of BS 6206: 1981, in respect of Class A impact resistance. Whilst most fire-resistant materials are relatively thick, one 7 mm product, with a single clear intumescent layer, giving

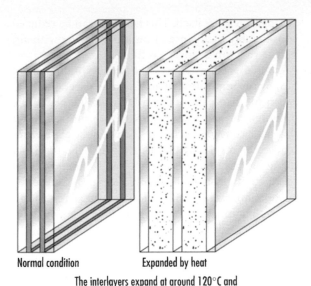

Normal condition | Expanded by heat

The interlayers expand at around 120°C and transform into a rigid and opaque shield

Fig. 7.13 Fire-resistant glass with intumescent material laminates

both fire and impact resistance, achieves a rating of E30/EW30/EI20, i.e. 30 minutes' integrity, 30 minutes' reduced heat radiation and 20 minutes' insulation.

Gel insulated glasses are manufactured from laminated toughened glass with the gel layer sandwiched between two or more glass layers. In the event of fire, the gel interlayer, which is composed of a polymer-containing aqueous inorganic salt solution, absorbs heat by the evaporation of water and produces an insulating crust. The process is repeated layer by layer. Depending upon the thickness of the gel layer, fire resistance times of 30, 60 or 90 minutes are achieved.

Partially insulating glass

Partially insulating glass consisting of a 10 mm triple laminate of float glass with one intumescent interlayer and one polyvinyl butyral layer offers a modest increase in fire resistance over non-insulating glass. Surface treatments can also increase the heat reflectance of both glass faces.

ENERGY CONSERVATION

The Building Regulations (Approved Document Part L1A: Conservation of fuel and power in new dwellings [2006 edition]) require that all new dwellings have predicted carbon dioxide emissions – Dwelling Emission Rate (DER) – not greater than

the Target Emission Rate (TER) determined for the equivalent notional building of the same size and shape. The TER, expressed as the quantity of CO_2 in kg emitted per m^2 of floor area per year, is calculated for dwellings of less than or equal to 450 m^2 total floor area by the Standard Assessment Procedure (SAP 2005 edition). For all dwellings above 450 m^2 floor area, the Simplified Building Energy Model (SBEM) calculation is required. For reasons of energy conservation and climate change, the target emissions are reduced to 80% of those from equivalent buildings compliant to the previous 2002 regulations.

The following reference figures are area-weighted average limits for guidance. However, for most buildings enhanced specifications are likely to be required in order to achieve the required carbon dioxide Target Emission Rate.

Wall	0.35 W/m² K
Floor	0.25 W/m² K
Roof	0.25 W/m² K
Windows/roof window/door	2.2 W/m² K

U-values better than the following are considered to be positive design features:

Wall	0.28 W/m² K
Floor	0.20 W/m² K
Roof	0.15 W/m² K
Window/door	1.8 W/m² K

The Building Regulations also require provision to prevent excessive summer solar gains causing high internal temperatures. This can be achieved by the appropriate use of window size and orientation, solar control with shading or specialist glazing systems, ventilation and high thermal capacity. The Building Regulations set limits on services performance and envelope airtightness (including sample pressure testing) and in addition they require each new building to be supplied with guidance on its energy efficient operation.

The TER method gives a good degree of flexibility in relation to the design by allowing trade-offs between different energy-saving factors. The benefits of useful solar radiation gains may also be taken into account.

The Building Regulations (Approved Document Part L2A: Conservation of fuel and power in new buildings other than dwellings [2006 edition]) requires the predicted carbon dioxide emissions – Building Emission Rate (BER) to be no greater than the Target Emission

Rate (TER) calculated by the Simplified Building Energy Model (SBEM) or other approved software tools. The area-weighted average limits for elements are set at the same standards as for new dwellings. However, the appropriate Building Emission Rate, rather than individual element U-values, is required for compliance.

Building Regulations Approved Documents Part L1B and L2B refer to work on existing dwellings and other buildings respectively. Guidance is quoted on the reasonable provision and standards for new extensions and replacement of existing thermal elements, but again, full compliance in all aspects of the regulations is required.

Standards for new elements in existing buildings:

Wall	0.30 W/m² K
Floor	0.22 W/m² K
Pitched roof – insulation at ceiling level	0.16 W/m² K
Pitched roof – insulation at rafter level	0.20 W/m² K
Flat roof – integral insulation	0.20 W/m² K
Windows/roof window/ roof light	1.8 W/m² K or Band D or centre pane 1.2 W/m² K

Standards for replacement elements in existing buildings:

Wall	0.35 W/m² K
Floor	0.25 W/m² K
Pitched roof – insulation at ceiling level	0.16 W/m² K
Pitched roof – insulation at rafter level	0.20 W/m² K
Flat roof – integral insulation	0.25 W/m² K
Windows/roof window/roof light	2.0 W/m² K (dwellings) 2.2 W/m² K (non dwellings) or Band E or centre pane 1.2 W/m² K

Double and triple glazing

Whenever the internal surface of exterior glazing is at a lower temperature than the mean room surface temperature and the internal air temperature, heat is lost by a combination of radiation exchange at the glass surfaces, air conduction and air convection currents inside and out, and also by conduction through the glass itself. This heat loss can be reduced considerably by the use of multiple glazing with air, partial vacuum or inert gas fill (Fig. 7.14).

Double glazing reduces the direct conduction of heat by the imposition of an insulating layer of air between the two panes of poorly insulating glass. The optimum air gap is approximately 16 mm, as above this value convection currents between the glass panes reduce the insulating effect of the air. The use (as a filling agent) of argon, which has a lower thermal conductivity than air, further reduces heat transfer by conduction. The use of krypton or even xenon within a 16 mm double-glazing gap in conjunction with low-emissivity glass can achieve a U-value of 0.8 W/m² K. Similar reductions in conducted heat can be achieved by the incorporation of an additional air space within triple glazing. Thin low-emissivity films suspended within the cavity can further reduce the U-values of double-glazing units to as low as 0.6 W/m² K. Typical U-values are shown in Table 7.2.

Window energy ratings

A European system of energy rating bands for complete window units based on the range A (best) to G (poorest) gives guidance to specifiers on energy efficiency. The ratings take into account a combination of the three key factors which affect performance; namely, U-value, solar gain and heat loss by air infiltration. The

Table 7.2 Typical U-values for single and multiple-glazing systems

Glass system	U-value (W/m² K)
Single clear glass	5.4
Double clear glass	2.8
Triple clear glass	1.9
Double clear glass with hard low emissivity coating (e.g. *K Glass*)	1.9
Double clear glass with soft low emissivity coating (e.g. *Kappafloat*)	1.8
Double clear glass with low emissivity coating and argon fill	1.5
Double solar glass with low emissivity coating and argon fill	1.2
Triple clear glass with two low emissivity coatings and two argon fills	0.8
Double clear glass with hard low emissivity coating mounted in timber or PVC-U frame (dependent on size of glazing unit)	2.4

The data relate to 6 mm glass and 12 mm spacing.

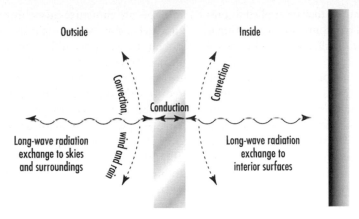

Mechanism for heat loss, through single glass

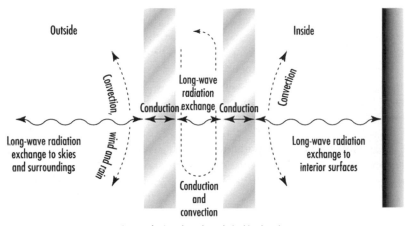

Mechanism for heat loss, through double-glazed units

Fig. 7.14 Mechanism of heat loss through single and double glazing (after Button, D. and Pye, B. (eds) 1993. *Glass in building*. Butterworth Architecture)

solar heat gain and U-values relate to the whole unit not just the glazed areas. The leakage rate is taken for average conditions. The rating bands are colour coded from green (A) through yellow (D) to red (G) for ease of recognition and they compare the overall energy performance of the windows measured as the total annual energy flow (kWh/m²/yr). Some windows in the A band may give an overall positive energy contribution to the buildling, whilst G band windows contribute an energy loss in excess of 70 kWh/m²/yr.

An appropriately insulated and sealed PVC-U framed system with triple glazing, using 4 mm clear white glass for the outer and inner panes, 4 mm hard coat low-emissivity glass as the centre pane, and two 16 mm cavities filled with argon gas can achieve the highest 'A' rating. Typically, a timber-frame double-glazed window with 16 mm argon fill, corrugated metal strip

spacer and soft coat low-emissivity glass would achieve a C rating, whilst a thermal-break aluminium frame double-glazed window with 16 mm argon fill, silicone rubber spacer and soft coat low-emissivity glass would achieve a D rating.

Low-emissivity glass

Low-emissivity glasses are manufactured from float glass by the application of a transparent low-emissivity coating on one surface. The coating may be applied either on-line, within the annealing lehr at 650°C, as a pyrolytic hard coat as in *K-Glass* or *Eko-plus* or off-line after glass manufacture by magnetic sputtering under vacuum which produces a softer coat as in *Optitherm* or *Cool-lite*. Only on-line manufactured low-emissivity glasses may normally be toughened after coating, but off-line low-emissivity coatings may be applied to previously toughened glass. The on-line surface coating is more durable and is not normally damaged by careful handling.

Low-emissivity glass functions by reflecting back into the building the longer wavelength heat energy associated with the building's occupants, heating systems and internal wall surfaces, whilst allowing in the transmission of the shorter wavelength solar energy (Fig. 7.15). The incoming solar energy is absorbed by the internal walls and re-radiated as longer wavelength energy, which is then trapped by the low-emissivity coating on the glass.

Low-emissivity coatings can reduce by three-quarters the radiant component of the thermal transfer between the adjacent surfaces within double glazing. The reduction in emissivity of standard uncoated glass from 0.84 to below 0.16 gives a decrease in U-value from 2.8 W/m^2 K for standard double glazing to 1.8 W/m^2 K with low-emissivity glass. Frequently, low-emissivity glass is protected in use within sealed double-glazed units. The outer leaf in the double-glazing system may be clear or any other specialist glass for security or solar control. Pyrolytic low-emissivity coatings are suitable for incorporation into secondary glazing for existing windows. The emissivities of low-E coatings range from 0.15 to 0.20 for hard coats and from 0.05 to 0.1 for soft coatings.

With the changes in the requirements for energy conservation, low-emissivity glass double glazing will become the standard for all new building works, as in the Swiss Re building in central London (Fig. 7.16).

Double-glazing units

Hermetically sealed double-glazing units are usually manufactured with aluminium spacers which incorporate moisture-adsorbing molecular sieve or silica gel and are sealed typically with polyisobutylene, polyurethane, polysulfide or epoxysulfide and a protective cap (Fig. 7.17). Frequently a second seal of two-part silicone is used to prevent leakage. To reduce cold bridging and condensation, thin stainless steel spacers offer greater thermal efficiency. Further, particularly with small glazing units, the overall energy efficiency of the unit can be considerably affected by the thermal conductivity of the frame. UK Building Regulations require that the overall U-values for the whole window including frames and glazing be taken into consideration. Timber frames offer good insulation, plastics less so due to their higher conductivity and the incorporation of steel reinforcement, whilst aluminium frames require the inclusion of a thermal break to reduce the risk of surface condensation and significant heat loss (Table 7.3).

SOLAR CONTROL GLASSES

Solar glasses offer a modified passage of light and heat energy compared to clear glass of the same thickness. A descriptive code indicates the relative quantities of light and heat transmitted for a particular glass (e.g. 50/62 for 6 mm bronze body-tinted glass) and this can be related to the equivalent data for clear float glass (87/83 for 6 mm clear). Additionally, as a guide to solar heat radiation control, the shading coefficient relates the solar radiant heat transmitted by a particular glass to that for 3–4 mm clear glass (Table 7.4). The two key methods of control are increased solar absorption as in body-tinted glasses or increased solar reflection, although certain solar glasses combine both methods of control. Additionally, double-glazed units can incorporate adjustable blinds or louvres. Solar control reflective film is available in a range of grades, thickness and colours for installation onto existing glazing.

Body-tinted glass

Body-tinted glasses have a uniform through colour of grey, bronze, green, blue, pink or amber. They reduce both the transmitted heat and light compared to equivalent clear float glass, as illustrated in Figure 7.18. They function by absorbing some of the incident solar radiation, causing the glass to warm. The glass then dissipates this absorbed heat both towards the inside and outside of the building, but owing to the greater movement of air externally, a greater proportion

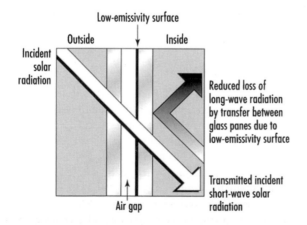

Fig. 7.15 Mechanism of heat loss control with low-emissivity glass

Fig. 7.16 Glazing and detail – Swiss Re Building, London. Architects: Foster and Partners. Photographs: Arthur Lyons

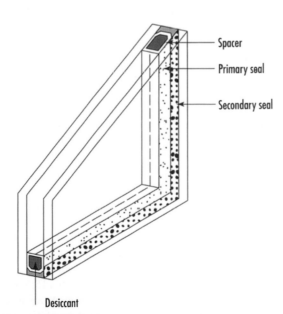

Spacer

Primary seal

Secondary seal

Desiccant

Fig. 7.17 Typical double-glazing unit

is expelled. Thus in double-glazed units, the body-tinted glass must form the outer pane. Additional environmental control can be achieved by the use of low-emissivity glass for the inner pane, when a proportion of the heat re-radiated inwards from the body-tinted glass is reflected back out again by the low-emissivity coating.

Body-tinted glasses may be toughened or laminated for use in hazardous areas. The degree of solar control offered by body-tinted glass is categorised as low to medium, being dependent upon both the thickness and colour of the product.

Reflective-coated glass

Reflective-coated glasses offer medium to high solar gain control by the action of a range of reflective coatings. Coatings may be applied on-line during the float process or subsequently as sputtered surface treatments to clear or body-tinted glass.

Table 7.3   Typical overall U-values (W/m² K) for windows, doors and roof windows

	Type of frame							
	Wood		Metal		Thermal break		PVC-U	
Glass separation (mm)	6	12	6	12	6	12	6	12
Windows								
Double glazed	3.3	3.0	4.2	3.8	3.6	3.3	3.3	3.0
Double glazed, low E glass	2.9	2.4	3.7	3.2	3.1	2.6	2.9	2.4
Double glazed, argon fill	3.1	2.9	4.0	3.7	3.4	3.2	3.1	2.9
Double glazed, low E, argon fill	2.6	2.2	3.4	2.9	2.8	2.4	2.6	2.2
Triple glazed	2.6	2.4	3.4	3.2	2.9	2.6	2.6	2.4
Double glazed doors								
Half glazed	3.1	3.0	3.6	3.4	3.3	3.2	3.1	3.0
Fully glazed	3.3	3.0	4.2	3.8	3.6	3.3	3.3	3.0
Roof windows less than 70° from horizontal								
Double glazed	3.6	3.4	4.6	4.4	4.0	3.8	3.6	3.4
	Wood		Metal		Thermal break		PVC-U	
Other windows/doors								
Single glazed windows/doors	4.7		5.8		4.3		4.7	
Solid timber panel door	3.0							
Half glazed/half timber panel door	3.7							

Available colours in reflection are blue, green, grey, silver, gold and bronze, although colours by transmission may be different (e.g. *Reflectafloat* is silver in reflection but bronze in transmission). By combining the range of coatings with different body-tinted glasses a wide range of solar control properties is achieved.

Where reflective coatings are to be applied after the float manufacturing process, all other working such as toughening and bending must be completed prior to the surface treatment. While many mineral coatings are durable in normal use as single glazing, they are damaged by abrasives, and may also exhibit minor imperfections, although these are considered acceptable if not observed at a distance closer than 3 m and are less than 2 mm in diameter.

ACOUSTIC CONTROL

The level of sound reduction by glazing is influenced by the mass of the glass and the extent of air leakage around the opening lights. Sound insulation for single glazing follows the mass law – doubling the glass thickness reduces sound transmission by approximately 4 dB. Toughened, patterned and wired glass of the same thickness respond as for plain glass, but laminated glasses based on a 1 mm layer of polymethyl methacrylate (PMMA) or a thick polyvinyl butyral (PVB) interlayer have enhanced sound insulation properties. The plastic interlayers, because they are soft material, change the frequency response of the composite sheet in comparison with the same weight of ordinary glass, and also absorb some of the sound energy.

Double glazing for sound insulation should be constructed with the component glasses differing in thickness by at least 30% to reduce sympathetic resonances; typically 6 mm and 10 mm would be effective. Where the passage of speech noise (630–2000 Hz) is to be reduced, the procedure of filling the double-glazing units with sulphur hexafluoride has been used, but this is detrimental to attenuation of traffic and other low frequency noises within the range 200–50 Hz; also it is now considered environmentally inappropriate. A typical high performance double-glazing unit giving a U-value of 1.3 W/m² K and a sound reduction of 35 dB, would be constructed from 6.4 mm laminated glass with a 15.5 mm argon filled cavity and a 4 mm inner pane of low-emissivity glass. For enhanced sound insulation an

Table 7.4 Characteristics of typical heat-absorbing and heat-reflecting solar glasses

Heat-absorbing glass (6 mm – body tinted)

Colour	Code	Light transmittance	Light reflectance	Heat transmittance	Shading coefficient
Green	72/62	0.72	0.06	0.62	0.72
Blue	54/62	0.54	0.05	0.62	0.72
Bronze	50/62	0.50	0.05	0.62	0.72
Grey	42/60	0.42	0.05	0.60	0.69

Heat-reflecting glass (6 mm – pyrolytic coating)

Colour	Coating face	Code	External colour in reflectance	Light transmittance	Light reflectance	Heat transmittance	Shading coefficient
Clear	out	43/55	bright silver	0.43	0.45	0.55	0.63
Clear	in	43/58	silver	0.43	0.40	0.58	0.66
Blue/green	out	33/40	bright silver	0.33	0.45	0.40	0.46
Blue/green	in	33/46	blue/green	0.33	0.30	0.46	0.53
Grey	out	20/41	bright silver	0.20	0.45	0.41	0.47
Grey	in	20/49	grey	0.20	0.13	0.49	0.56
Bronze	out	25/41	bright silver	0.25	0.45	0.41	0.48
Bronze	in	25/49	bronze	0.25	0.18	0.49	0.56
Silver		33/53	silver (bronze in transmission)	0.33	0.43	0.53	0.61

Heat-reflecting glass (6 mm – sputter coating)

Colour	Code	Light transmittance	Light reflectance	Heat transmittance	Shading coefficient
Silver	10/23	0.10	0.38	0.23	0.26
Silver	20/34	0.20	0.23	0.34	0.39
Silver	30/42	0.30	0.16	0.42	0.48
Bronze	12/32	0.12	0.11	0.32	0.37
Bronze	10/24	0.10	0.19	0.24	0.27
Bronze	26/40	0.26	0.17	0.40	0.46
Blue	13/32	0.13	0.12	0.32	0.37
Blue	20/33	0.20	0.20	0.33	0.38
Blue	30/39	0.30	0.16	0.39	0.45
Blue	40/50	0.40	0.10	0.50	0.57
Green	8/25	0.08	0.27	0.25	0.29
Green	17/32	0.17	0.17	0.32	0.37
Grey	10/32	0.10	0.09	0.32	0.37

Standard clear float glass (6 mm)

Colour	Light transmittance	Light reflectance	Heat transmittance	Shading coefficient
Clear float glass	0.87	0.08	0.83	0.95

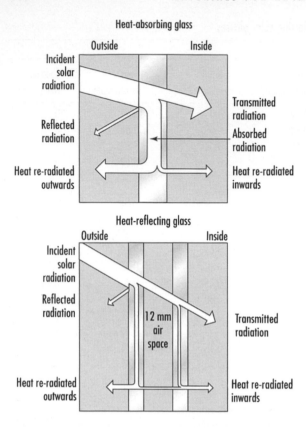

Fig. 7.18 Mechanism of solar gain control with heat-absorbing and heat-reflecting glasses

air gap of at least 100 mm is required, with the economical optimum being 200 mm. The reveals should be lined with sound-absorbing material such as fibreboard, to reduce reverberation within the air space. All air gaps must be fully sealed with opening lights closed by multipoint locking systems and compressible seals. An air gap corresponding to only 1% of the window area can reduce the efficiency of sound insulation by 10 dB.

COLOURED ENAMELLED GLASS

Opaque or translucent enamelled glass spandrel panels for curtain walling can be manufactured to match or contrast with the range of vision area solar control glasses. Manufactured from toughened heat-soaked glass for impact and thermal shock resistance, they may be single or double glazed, with integral glass fibre or polyurethane foam insulation and an internal finish. Panels are colourfast and scratch-resistant, and may be manufactured from plain, screen-printed or decorative glass.

SPECIALIST GLASSES

One-way observation glass

Where unobserved surveillance is required, one-way observation mirror glass can be installed. In order to maintain privacy, the observer must be at an illumination level no greater than one seventh that of the observed area and wear dark clothing. From the observed area the one-way observation glass has the appearance of a normal mirror. One-way observation glass is available annealed, toughened or laminated.

Mirror glass

Standard mirror glass is manufactured by the chemical deposition onto float glass of a thin film of silver, from aqueous silver and copper salt solutions. The film is then protected with two coats of paint or a plastic layer. A recent development is the production of mirror glass by chemical vapour deposition within the float glass process. Mirror glass is produced by the on-line application of a three-layer coating of silicon-silica-silicon, which acts by optical interference to give the mirror effect. Mirror glass manufactured by this process is less prone to deterioration and may be more easily toughened, laminated or bent than traditional mirror glass.

Anti-reflection glass

Treatment of standard float glass can reduce surface reflection from 0.09 to 0.025, thus increasing the transmittance. The coating is applied equally to both faces. Although used mostly for the protection of displayed art works, this material can be used for interior display windows and dividing screens, and also to reduce multiple reflections from the surfaces of double-glazed units, where both sheets of glass must be anti-reflective.

Alarm glass

Glass containing either a ceramic loop or a series of straight wires can be incorporated into an intruder alarm system, which is activated when the glass is broken. Usually straight-wired alarm glass is incorporated into a laminated system, while a ceramic loop circuit would be fixed to the inner face of the toughened outer pane within a double-glazing system.

Electrically-heatable laminated safety glass

Electrically-heatable laminated safety glass incorporates fine electrically conducting wires which may be switched on when there is the risk of condensation. Typical applications are in areas of high humidity, such as swimming pools, kitchens and glass roofs, particularly when there are significant differences between the internal and external temperatures. Power consumption ranges from 100 W/m² for homes to 500 W/m² for industrial applications depending upon the internal environment and external ambient conditions. Electrical connections are made at the perimeter of the glass.

Dichroic glass

Dichroic glass has a series of coatings which create optical interference effects. These cause the incident light to be split into the spectral colours, which depending upon the angle of incidence of the light are either reflected or transmitted. This effect can be used to create interesting colour patterns, which vary with both the movement of the sun and the observer.

Sphere glass

This material consists of an array of glass hemispheres, typically 4 mm to 8 mm in diameter fixed to standard toughened glass, and is suitable for use in feature walls, partitions and ceilings. The hemispheres and the base material each may be clear glass, standard or custom coloured.

Electromagnetic radiation-shielding glass

Electromagnetic radiation-shielding glass can be used to protect building zones containing magnetically-stored data from accidental or deliberate corruption by external electric fields. For maximum security, the conducting laminates within the composite glass should be in full peripheral electrical contact with the metal window frames and the surrounding wall surface screening.

X-ray protection glass

X-ray protection glass contains 70% lead oxide, which produces significant shielding against ionising radiation. The glass is amber in colour due to the high lead content. A 6 mm sheet of this lead glass has the equivalent shielding effect of a 2 mm lead sheet against X and γ radiation.

Sound-generating glass

Terfenol-D is a magnetostrictive material which when stimulated by a magnetic field expands and contracts rapidly producing a large physical force. If a device containing Terfenol-D is attached to the smooth surface of glass and an audio input is fed into the system, then the whole sheet of glass will vibrate, acting as a loudspeaker. Thus shop windows can be turned into loudspeakers, producing across their surface a uniform sound, which can be automatically controlled to just greater than the monitored street noise level, thus avoiding sound pollution. Two devices, appropriately positioned, will generate stereo sound. Magnetostrictive devices will operate similarly on any flat rigid surfaces such as tabletops, work-surfaces and rigid partitions. Terfenol-D is named from the metallic elements iron, terbium and dysprosium from which it is manufactured.

Manifestation of glass

Where there is a risk that glazing might be unseen, and thus cause a hazard to the users of a building, particularly large areas at entrances and in circulation spaces, the presence of the glass should be made clear with a solid or broken line, decorative feature or company logo at a height between 600 mm and 1500 mm above floor level. In such circumstances the risk of impact injury must be reduced by ensuring that the glass is either robust, protected, in small panes, or breaks safely. Critical locations with a risk of human impact (BS 6262–4: 2005) are clear glazed panels from floor level to 800 mm, also floor level to 1500 mm for glazed doors and glazed side panels within 300 mm of doors.

VARIABLE TRANSMISSION (SMART) GLASSES

Variable transmission or *smart* glasses change their optical and thermal characteristics under the influence of light (photochromic), heat (thermochromic) or electric potential (electrochromic). These glasses offer the potential of highly responsive dynamic climate control to building facades. These smart materials including thermotropic products are also available as plastic laminates for incorporation into laminated glass systems.

Photochromic glass

Photochromic glasses incorporate silver halide crystals, which are sensitive to ultraviolet or short-wave visible light. The depth of colour is related directly to the intensity of the incident radiation and is fully reversible. For use in buildings, these materials have the disadvantage that they respond automatically to changes in solar radiation, rather than to the internal environment within the building.

Thermochromic glass

Thermochromic glasses change in transmittance in response to changes in temperature. Like photochromic glass, these materials have the disadvantage of responding to local conditions, rather than to the requirements of the building's internal environment.

Electrochromic glass

Electrochromic glasses change their transmittance in response to electrical switching and are therefore likely to become the basis of *smart* windows. Electrochromic multi-layer thin-film systems become coloured in response to an applied low voltage, and are then cleared by reversal of the electric potential. The depth of colouration is dependent upon the magnitude of the applied d.c. voltage. Optically stable materials, which exhibit electrochromism, are the oxides of tungsten, nickel and vanadium. Electrochromic thin-film systems may be laminated to any flat sheet glass.

Electro-optic laminates

Electrically operated vision-control glass *(Priva-lite)* consists of a laminated system of glass and polyvinyl butyral layers containing a polymer dispersed liquid crystal layer, which can be electrically switched from transparent to white/translucent for privacy.

Intelligent glass

Conventional glass coatings reduce both light and heat transmission. However, a coating based on tungsten-modified vanadium dioxide allows visible light through at all times, but reflects infrared radiation at temperatures above 29°C. Thus, at this temperature, further heat penetration through the glass is blocked. Therefore the intelligent glass, which has a slight yellow/green colour, admits useful solar gain in cooler conditions but cuts out excessive infrared solar gain under hot conditions.

Intelligent glass facades

An intelligent glass building facade changes its physical properties in response to sensors detecting the external light and weather conditions, thus reducing the energy consumption necessary to maintain the appropriate internal environment. Therefore intelligent facades have ecological significance in reducing global greenhouse emissions and also in reducing operational building costs to clients and users.

Truly intelligent facades capitalise on the incident solar energy striking the facade of a building, adapt the skin functionality to the appropriate thermal control and solar protection, and in addition may generate electricity through photovoltaic cell systems. Solar control may be provided by switched electrochromic glass or by using laminated prismatic or holographic films which deflect the solar radiation according to its angle of incidence. Additionally, intelligent facades respond to air flows or ground heat sources to ensure appropriate and responsive ventilation. This function is usually achieved by the use of a double-skin facade, which acts as a ventilation cavity. During the heating season the double skin can pre-warm the incoming fresh air, and when cooling is required it can remove, by convection, built-up excess heat from the double-glazing unit. Furthermore, excess heat energy can be stored for redistribution when required.

Smart shading systems respond to reduce excess incident solar radiation. Electrically controlled louvres or blinds located between two glass panes open and close according to either a solar detector or to a range of weather sensing devices. Such mechanical systems, including the prototype iris diaphragm shading devices used by Jean Nouvel in the Institute de Monde Arabe in Paris, demand considerable maintenance for continued operation.

Glass supporting systems

The fixing of glazing, and particularly solar control glasses, should be sufficiently flexible to allow for tolerances and thermal movements. A minimum edge clearance of at least 3 mm is required for single glazing and 5 mm for double-glazing units. Edge cover should be sufficient to cope with the design wind loading, with a minimum normally equal to the glass or unit thickness to ensure a neat sight line. Glass thickness

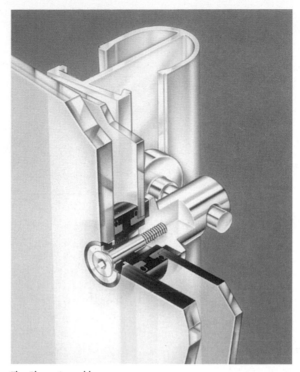

The *Planar* Assembly

Fig. 7.19 Typical facade glazing system

should be checked for suitability against predicted wind speeds, modified appropriately by consideration to the effects of local topography, building height and size of the glazing component.

The Pilkington *Planar System* (Fig. 7.19) offers the designer a flush and uninterrupted facade of glass. The only fixings to be seen on the external facade are the countersunk bolt heads. The system (which can be used for single, double or triple glazing, vertically or sloping) is designed such that each glazing unit is separately supported by the mullion system, so there is no restriction on the height of the building. Thermal and wind movement is taken up by the fixing plate, which is sufficiently flexible to allow some rotation of the glass. In the double-glazed system the units are principally supported by the outer pane. Glass to glass butt joints are sealed with silicone.

The recently developed *Planar Triple Glazing System* (Fig. 7.20), with one solar control, two low-emissivity glazings and two 16 mm airspaces, has a standard U-value of 0.8 W/m^2 K, but with two argon fills and high performance coated glass a U-value of 0.5 W/m^2 K can be achieved.

The *Financial Times* building in Docklands, London (Fig. 7.21) is designed with a long clear facade sandwiched between two aluminium-clad solid ends. The glazed section, 96 × 16 m, consists of a single-glazed suspended toughened-glass wall, bolted by circular plate assemblies to external aerofoil forms and intermediate cantilevered arms. This creates a wall of uninterrupted glass which is striking by day and transparent at night when the illuminated printing presses can be seen clearly.

STRUCTURAL GLASS

Glass columns are frequently used as fins to restrain excessive deflection caused by wind and other lateral loads to glass facades. The fixings between the facade glazing and fin units are usually stainless steel clamps bolted through preformed holes in the toughened or laminated glass, although silicone adhesives can also be used. Typically storey-height fins are 200–300 mm wide in 12–15 mm toughened glass, fixed into aluminium or stainless steel shoes to the floor and/or glazing head. A soft interlayer between the metal fixing and the glass is incorporated

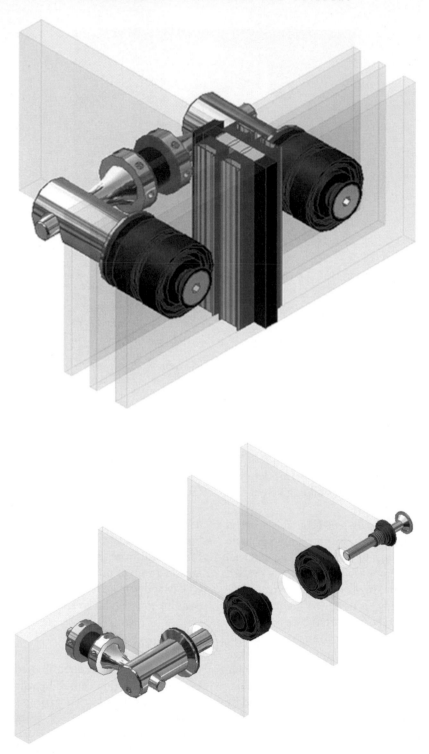

Fig. 7.20 *Planar* triple-glazing system. Diagram: Courtesy of Pilkington plc

Fig. 7.21 Glazing system – Financial Times, Docklands, London. Architects: Nicholas Grimshaw and Partners. Photographs: Courtesy of Nicholas Grimshaw and Partners, Jo Reid & John Peck

Fig. 7.22 Structural glazing detail – Gateshead Millennium Bridge Pavilions. Architects: Wilkinson Eyre Architects. Photograph: Courtesy of Wilkinson Eyre Architects

to prevent stress concentrations on the glass surface and to allow for differential thermal movement between the glass and metal.

Glass is strong in compression and therefore an appropriate material for load-bearing columns and walls, providing that the design ensures sufficient strength, stiffness and stability. Generally, consideration of buckling is the critical factor, although safety factors must be considered in relation to robustness and protection against accidental damage. Column sections need not be rectangular as, for example, a cruciform section manufactured from laminated toughened glass gives both an efficient and elegant solution.

Glass beams are usually manufactured by laminating toughened glass. Typically, a 4 m × 600 mm deep beam manufactured from three 15 mm toughened-glass laminates could carry a load of over 5 tonnes, thus supporting at 2 m centres a 4 m span glass roof. Glass beams can be jointed to glass columns by mortice and tenon jointing fixed with adhesive.

Single-storey all-glass structures, such as the small pavilions at the Gateshead Millennium Bridge (Fig. 7.22), entrance foyers and cantilevered canopies have been constructed using a combination of laminated and toughened-glass walls, columns and beams. Usually metal fixings have been used, but where the purity of an all-glass system is required, high-modulus

Fig. 7.23 Glass staircase – house in Notting Hill, London. Architects: Alan Power Architects. Photograph: Courtesy of Chris Gascoyne/View

structural adhesives such as modified epoxy resins are used for invisible fixing.

Where clear double-glazed structural units are required, standard aluminium spacers can be replaced by glass spacers sealed with clear silicone, although the edges may require etching to conceal the necessary desiccant. It should be remembered that all-glass constructions require careful design consideration in relation to excessive solar gain and other environmental factors.

Other structural applications for glass include stairs, walkways, floors and balustrades. For stairs, laminated glass, either annealed or toughened or with an acrylic interlayer, may be used, with a typical thickness of 30 mm depending upon the span and the support system. Glass is, however, slippery when wet,

even if treated by sandblasting, and should not normally be considered for external locations. An all-glass staircase in Notting Hill, London (Fig. 7.23) uses glass risers and treads joined with structural silicone. The complete stair unit is supported on one side by a masonry wall but on the other by a full height laminated glass wall, to which it is fixed only with metal brackets and structural silicone. The steps incorporate sand-blasted dots to prevent slipping. Freestanding glass balustrades are usually manufactured from heat-soaked toughened glass ranging in thickness from 12–25 mm depending upon their height and anticipated loading.

Recent innovations also include the structural use of glass rods in web compression members, and glass tubes in structural compression elements.

Fig. 7.24 The Great Court at the British Museum, London. General interior view of the Great Court and Reading Room. Architects: Foster and Partners. Photograph: Courtesy of Nigel Young, Foster and Partners

Glazing check list

Perhaps more than any other building component glazing is expected to perform many functions. It is therefore necessary to ensure that all factors are taken into consideration in the specification of glasses. It is evident that many of the environmental control factors are closely inter-related and the specifier must check the consequences of a design decision against all the parameters. Many of these factors are illustrated by the feature roof to the Great Court of the British Museum in London (Fig. 7.24).

- view in and out by day and night;
- visual appearance by day and by night – colour and reflectivity;
- energy-conscious balance between daylight and artificial lighting;
- sky and reflected glare;
- overheating and solar control;
- shading;
- passive solar gain and energy efficiency;
- thermal comfort, U-values and condensation;
- ventilation;
- acoustic control;
- security – impact damage, vandalism and fire spread.

References

FURTHER READING

Amstock, J.S. 1997: *Handbook of glass in construction.* Maidenhead: McGraw.

Behling, S. and Behling, S. (eds). 1999: *Glass, structure and technology in architecture.* München: Prestel.

Button, D. and Pye, B. (ed.) 1993: *Glass in building: A guide to modern architectural glass performance.* Oxford: Butterworth Architecture.

Compagno, A. 1999: *Intelligent glass façades: Material, practice, design.* 4th ed. Boston: Birkhäuser.

Crosbie, M.J. 2005: *Curtain walls: Recent developments by Cesar Pelli.* Basle: Birkhäuser.

DEFRA. 2005: *The Government's Standard Assessment Procedure for energy rating of dwellings SAP 2005 edition.* Watford: BRE.

Doremus, R.H. 1994: *Glass Science.* 2nd ed. New York: Wiley.

Dutton, P. and Rice, H. 1995: *Structural glass.* Oxford: Spon.

Energy Saving Trust. 2006: *Windows for new and existing housing.* Publication CE66, London: Energy Saving Trust.

Glass and Glazing Federation. 2005: *A guide to best practice in the specification and use of fire-resistant glazing systems.* London: Glass and Glazing Federation.

Hyatt, P. and Hyatt, J. 2004: *Designing with glass: Great glass buildings.* Australia: Images Publishing Group.

Institution of Structural Engineers. 1999: *Structural use of glass in buildings.* London: Institution of Structural Engineers.

Juracek, J.A. 2006: *Architectural surfaces: Details for Architects, designers and artists.* London: Thames and Hudson.

Kaltenbach, F. 2004: *Translucent materials: Glass, plastics, metals.* Basle: Birkhäuser.

OPDM. 2006: *Simplified Building Energy Model (SBEM) user manual and calculation tool.* London: Office of the Deputy Prime Minister.

Pilkington. 2000: *The European glass handbook.* St Helens: Pilkington.

Ryan, P., Otlet, M. and Ogden, R.G. 1997: *Steel- supported glazing systems.* SCI publication 193. Ascot: Steel Construction Institute.

Saint Gobain. 2000: *Glass guide.* Goole: Saint Gobain.

Schittich, C. et al. 1999: *Glass construction manual.* Basle: Birkhäuser.

Wigginton, M. 2002: *Glass in architecture.* 2nd ed. London: Phaidon.

Wigginton, M. and Harris, J. 2000: *Intelligent skins.* Oxford: Architectural Press.

STANDARDS

BS 476 Fire tests on building materials and structures.

BS 644: 2003 Timber windows – factory assembled windows of various types.

BS 952 Glass for glazing:
  Part 1: 1995 Classification.
  Part 2: 1980 Terminology for work on glass.

BS 3447: 1962 Glossary of terms used in the glass industry.

BS 4255 Rubber used in preformed gaskets for weather exclusion from buildings:
  Part 1: 1986 Specification for non-cellular gaskets.

BS 4904:1978 Specification for external cladding for building purposes.

BS 5051 Bullet-resisting glazing:
  Part 1: 1988 Bullet resistant glazing for interior use.

BS 5252: 1976 Framework for colour co-ordination for building purposes.

BS 5357:1995 Code of practice for installation of security glazing.

BS 5516: 2004 Patent glazing and sloping glazing for buildings.

BS 5544: 1978 Specification for anti-bandit glazing (glazing resistant to manual attack).

BS 5713: 1979 Specification for hermetically sealed flat double glazing units.

BS 5821 Methods for rating the sound insulation in buildings and of building elements:

Part 3: 1984 Method for rating the airborne sound insulation of facade elements and facades.

BS 6100 Glossary of building and civil engineering terms:

Part 1 Subsec. 1.4.1: 1999 Glazing.

BS 6180: 1999 Code of practice for protective barriers in and about buildings.

BS 6206: 1981 Specification for impact performance requirements for flat safety glass and safety plastics for use in buildings.

BS 6262: 1982 Glazing for buildings:

Part 1: 2005 General methodology for the selection of glazing.

Part 2: 2005 Code of practice for energy, light and sound.

Part 3: 2005 Code of practice for fire, security and wind loading.

Part 4: 2005 Code of practice for safety related to human impact.

Part 6: 2005 Code of practice for special applications.

Part 7: 2005 Code of paractice for the provision of information.

BS 6375 Performance of windows:

Part 1: 2004 Classification of weathertightness.

Part 2: 1987 Specification for operation and strength characteristics.

BS 6399 Part 2: 1997 Loading for buildings – Wind loads.

BS 8000 Workmanship on building sites:

Part 7: 1990 Code of practice for glazing.

BS 8206: 1992 Lighting for buildings – code of practice for daylighting.

BS 8213: 2004 Windows, doors and rooflights – design for safety in use.

BS EN 356: 2000 Glass in building – security glazing – resistance to manual attack.

BS EN 357: 2004 Glass in building – fire resistant glazed elements.

BS EN 410: 1998 Glass in building – determination of luminous and solar characteristics of glazing.

BS EN 572 Glass in building – basic soda lime silicate glass products:

Part 1: 2004 Definition.

Part 2: 2004 Float glass.

Part 3: 2004 Polished wired glass.

Part 4: 2004 Drawn sheet glass.

Part 5: 2004 Patterned glass.

Part 6: 2004 Wired patterned glass.

Part 7: 2004 Wired or unwired channel shaped glass.

Part 8: 2004 Supplied and final cut sizes.

Part 9: 2004 Evaluation of conformity/product standard.

BS EN 673: 1998 Glass in building – determination of thermal transmittance (U-value).

BS EN 1026: 2000 Windows and doors – air permeability – test method.

BS EN 1036: 1999 Glass in building – mirror from silver-coated float glass for internal use.

BS EN 1051–1: 2003 Glass in building – glass blocks and glass pavers.

BS EN 1063: 2000 Glass in building – security glazing – resistance against bullet attack.

BS EN 1096 Glass in building – coated glass:

Part 1: 1999 Definitions and classification.

Part 2: 2001 Class A, B and S coatings.

Part 3: 2001 Class C and D coatings.

Part 4: 2004 Evaluation of conformity.

BS EN 1279 Glass in building – insulating glass units:

Part 1: 2004 Generalities, dimensional tolerances.

Part 2: 2002 Requirements for moisture penetration.

Part 3: 2002 Gas leakage rate.

Part 4: 2002 Methods of test for the physical attributes of edge seals.

Part 5: 2005 Evaluation of conformity.

Part 6: 2002 Factory production control and periodic tests.

BS EN 1364 Fire resistance tests for non-loadbearing elements.

Part 1: 1999 Walls.

Part 2: 1999 Ceilings.

BS EN 1748 Glass in building – special basic products:

Part 1: 2004 Borosilicate glasses.

Part 2: 2004 Glass ceramics.

BS EN 1863 Glass in building – heat strengthened soda lime silicate glass:

Part 1: 2000 Definition and description.

Part 2: 2004 Evaluation of conformity.

BS EN ISO 10077 Thermal performance of windows, doors and shutters:

Part 1: 2000 Calculation of thermal transmittance – simplified method.

Part 2: 2003 Calculation of thermal transmittance – numerical method for frames.

BS EN 12150–1: 2000 Glass in building – thermally toughened soda lime silicate safety glass:

Part 1: 2000 Definition and description.

Part 2: 2004 Evaluation of conformity.

BS EN 12337 Glass in building – chemically strengthened soda lime silicate glass:

Part 1: 2000 Definition and description.

Part 2: 2004 Evaluation of conformity/product standard.

pr EN 12488: 2003 Glass in building – glazing requirments – assembly rules.

BS EN ISO 12543 Glass in building – laminated glass and laminated safety glass:

Part 1: 1998 Definitions and descriptions.

pr Part 2 Laminated safety glass.

Part 3: 1998 Laminated glass.

Part 4: 1998 Test method for durability.

Part 5: 1998 Dimensions and edge finishing.

Part 6: 1998 Appearance.

BS EN ISO 12567: 2005 Thermal performance of windows and doors.

BS EN 12600: 2002 Glass in building – impact test method and classification for glass.

pr EN 12725: 1997 Glass in building – glass block walls.

BS EN 12758: 2002 Glass in building – glazing and airborne sound.

BS EN 12898: 2001 Glass in building – determination of emissivity.

pr EN 13022 Glass in building – structural sealant glazing:

Part 1: 2003 Supported and unsupported monolithic and multiple glazing.

Part 2: 2003 Product standard for ultra-violet resistant sealant and structural sealant.

Part 3: 2003 Assembly rules.

BS EN 13024 Glass in building – thermally toughened borosilicate safety glass:

Part 1: 2002 Specifications.

Part 2: 2004 Evaluation of conformity.

pr EN 13474: 2000 Glass in buildings – design of window panes.

BS EN 13501 Fire classification of construction products and building elements:

Part 1: 2002 Classification using test data from reaction to fire tests.

Part 2: 2003 Classification using test data from fire resistance tests.

BS EN 13541: 2001 Glass in building – security glazing.

BS EN 14178 Glass in building – basic alkaline earth silicate glass products:

Part 1: 2004 Float glass.

Part 2: 2004 Evaluation of conformity.

BS EN 14179 Glass in building – heat-soaked thermally-toughened soda lime silicate safety glass:

Part 1: 2005 Definition and description.

Part 2: 2005 Evaluation of conformity/product standard.

BS EN 14321 Glass in building – thermally toughened alkaline earth silicate safety glass:

Part 1: 2005 Definition and description.

Part 2: 2005 Evaluation of conformity/product standard.

BS EN ISO 14438: 2002 Glass in building – determination of energy balance value.

pr EN ISO 14439: 1995 Glass in building – glazing requirements – glazing blocks.

BS EN 14449: 2005 Glass in building – laminated glass and laminated safety glass – evaluation of conformity.

BS EN 14600: 2005 Doorsets and openable windows with fire resisting and/or smoke control characteristics.

CP 153 Windows and rooflights:

Part 2: 1970 Durability and maintenance.

Part 3: 1972 Sound insulation.

PD 6512 Use of elements of structural fire protection:

Part 3: 1987 Guide to the fire performance of glass.

BUILDING RESEARCH ESTABLISHMENT PUBLICATIONS

BRE Digests

BRE Digest 309: 1986 Estimating daylight in buildings Part 1.

BRE Digest 310: 1986 Estimating daylight in buildings Part 2.

BRE Digest 338: 1988 Insulation against external noise.

BRE Digest 346 The assessment of wind loads.

Part 1: 1992 Background and method.

Part 2: 1989 Classification of structures.

Part 3: 1992 Wind climate in the United Kingdom.

Part 4: 1992 Terrain and building factors and gust peak factors.

Part 5: 1989 Assessment of wind speed over topography.

Part 6: 1989 Loading coefficients for typical buildings.

Part 7: 1989 Wind speeds for serviceability and fatigue assessments.

Part 8: 1990 Internal pressures.

BRE Digest 377: 1992 Selecting windows by performance.

BRE Digest 379: 1993 Double glazing for heat and sound insulation.

BRE Digest 404: 1995 PVC-U windows.

BRE Digest 430: 1998 Plastics external glazing.

BRE Digest 453: 2000 Insulating glazing units.

BRE Digest 457: 2001 The Carbon Performance Rating for offices.

BRE Digest 497: 2005 Factory glazed windows (Parts 1 & 2).

BRE Information paper

BRE IP 12/93 Heat losses through windows.

BRE IP 3/98 Daylight in atrium buildings.

BRE IP 2/02 Control of solar shading.

BRE IP 3/02 Whole life performance of domestic automatic window controls.

BRE IP 11/02 Retrofitting solar shading.

BRE IP 17/03 Impact of horizontal shading devices on peak solar gains through windows.

BRE IP 1/05 Impact standards for glass.

BRE Reports

BR 280: 1995 Double-glazing units: A BRE guide to improved durability.

BR 443: 2006 Conventions for U-value calculations.

ADVISORY ORGANISATIONS

British Glass Manufacturers Confederation, 9 Churchill Way, Chapeltown, Sheffield, South Yorkshire S35 2PY (0114 2901850).

Glass & Glazing Federation, 44–8 Borough High Street, London SE1 1XB (0845 257 7963).

Plastic Window Federation, Federation House, 85–7 Wellington Street, Luton, Bedfordshire LU1 5AF (01582 456147).

Steel Window Association, The Building Centre, 26 Store Street, London WC1E 7BT (020 7637 3571).

8

CERAMIC MATERIALS

Introduction

Ceramic materials, manufactured from fired clay, have been used in construction since at least 4000 BC in Egypt, and represent the earliest manufactured building materials. Whilst the strict definition of ceramics includes glass, stone and cement, this chapter deals only with the traditional ceramics based on clays. The variety of traditional ceramic products used within the building industry arises from the wide range of natural and blended clays used for their production. The roof of the spectacular Sydney Opera House (Fig. 8.1) is surfaced with white ceramic tiles which reflect the changing light associated with the time of day.

CLAY TYPES

Clays are produced by the weathering of igneous rocks, typically granite, which is composed mainly of feldspar, an alumino-silicate mineral. Clays produced within the vicinity of the parent rock are known as *primary clays*. They tend to be purer materials, less plastic and more vulnerable to distortion and cracking on firing. Kaolin ($Al_2O_3.2SiO_2.2H_2O$), which is the purest clay, comes directly from the decomposition of the feldspar in granite. Secondary clays, which have been transported by water, have a higher degree of plasticity, and fire to a buff or brown colour depending upon the nature and content of the incorporated oxides. Generally, secondary clays, laid down by the process of sedimentation, have a narrower size distribution and their particulate structure is more ordered.

The most common clay minerals used in the manufacture of building materials are kaolin, illite (a micaceous clay) and montmorillonite, a more plastic clay of variable composition. Clay crystals are generally hexagonal in form and in pure kaolin the crystals are built up of alternating layers of alumina and silica (Fig. 8.2). However, in illite and montmorillonite clays, the variable composition produced by sedimentation produces more complex crystal structures.

Ball clays are secondary clays containing some organic matter which is burnt off during the firing process; they tend to have a fine grain size which makes them plastic. When fired alone they have a high shrinkage and produce a light grey or buff ceramic, but they are usually blended into other clays such as kaolin to make a workable clay. Terracotta clays contain significant proportions of iron oxide which gives rise to the characteristic red colour on firing. While the major clay materials used in the manufacture of ceramics are kaolin, illite, feldspar and ball clay, other minor constituents as well as chalk and quartz are frequently incorporated to produce required ceramic properties on firing.

WATER IN CLAY

Moist clay contains both chemically- and physically-bonded water. It is the latter which permeates between the clay particles, allowing them to slide over each other during the wet forming processes. As the formed clay slowly dries out before firing, a small proportion of the residual physically bonded water holds the clay in shape. On firing, the last of the physically bonded water is removed as the temperature exceeds 100°C.

Fig. 8.1   Sydney Opera House and ceramic tile roof detail. Architects: Jørn Utzon and Ove Arup. Photographs: Arthur Lyons

MANUFACTURING PROCESSES

Clay products are either formed by wet or dry processes. In the former case the artefacts must be dried slowly prior to firing, allowing for shrinkage without cracking. Where a high level of dimensional accuracy is required, as in wall and floor tiles, a dry process is used in which powdered clay is compressed into the required form.

As the firing temperature is gradually increased, the majority of the chemically-bonded water is removed by 500°C. At 800°C, carbonaceous matter has been burnt off as carbon dioxide, and the sintering process commences, at first producing a highly porous material. As

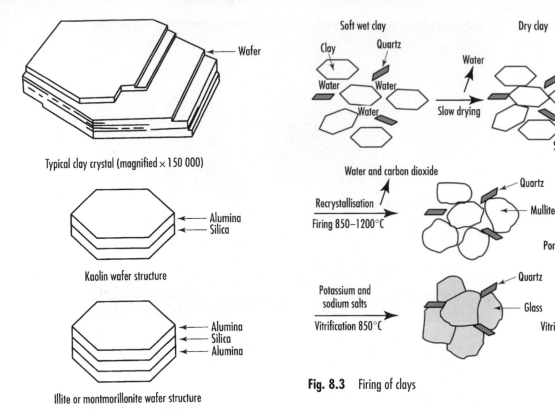

Typical clay crystal (magnified × 150 000)

Kaolin wafer structure

Illite or montmorillonite wafer structure

Fig. 8.2   Structure of clays

Fig. 8.3   Firing of clays

the temperature is further raised towards 1200°C, the alumina and silica components recrystallise to form mullite. With an additional increase in firing temperature, a more glassy ceramic is produced due to further recrystallisation and if the firing temperature reaches 1300°C, any remaining free silica is recrystallised. In the presence of potassium or sodium salts vitrification occurs giving an impervious product (Fig. 8.3).

Ceramic products

FIRECLAY

A range of clays (predominantly blends of alumina and silica) high in silica (40–80%) and low in iron oxide (2–3%) produce fireclay refractory products which will withstand high temperatures without deformation. Dense products have high flame resistance, while the insulating lower-density products are suitable for flue linings. White glazed fireclay is typically used for urinals, floor channels, industrial and laboratory sinks.

BRICKS AND ROOFING TILES

Bricks can be manufactured from a wide range of clays, the principal ones being Keuper marl, Etruria marl, Oxford clay, London clay, Coal Measure shale, Weald and Gault clays with some production from alluvial and fireclay deposits. The composition of the clay varies widely depending upon the type, but typically contains between 40–65% silica, 1–25% alumina and 3–9% iron oxide. The loss on firing may reach 17% in the case of clay containing high levels of organic matter. The production of bricks is described in detail in Chapter 1.

Glazed bricks are manufactured in a wide range of high-gloss, uniform or mottled colours. Colour-fast glazed bricks offer a low maintenance, frost- and vandal-resistant material suitable for light reflecting walls. Standard and purpose-made specials can be manufactured to order. Normal bricklaying techniques are appropriate, but to reduce the visual effect of the mortar joints they may be decreased from the standard 10 to 6 mm. For conservation work, in order to match new to existing it may be necessary to fire the glazed bricks a second time at a reduced temperature to simulate the existing material colour.

Roofing tiles are made from similar clays to bricks, such as Etruria marl, but for both handmade and machine-made tiles, the raw materials have to be screened to a finer grade than for brick manufacture. Traditional red, brown, buff, brindled or 'antique' ceramic roofing tiles are unglazed with a plain or sanded finish. While most interlocking clay tiles can be used to a minimum pitch of 22.5°, one imported product with a double-side interlock and a triple head/tail interlock may be used down to only 10°. This product is available in natural terracotta red or slip-coated brown or grey. Where bright colours are required, high- and low-gloss pantiles are available in a range of strong colours, or to individual specification. For plain tiles, a range of standard fittings is produced for hips, valleys, eaves, ridges, verges, internal and external angles, as shown in Figure 8.4. Tiles are usually shrink-wrapped for protection and ease of handling on site.

Certain floor tiles are also manufactured from Etruria marl. Firing to 1130°C produces sufficient vitrification to limit water absorption to less than 3%, thus giving a highly durable chemical and frost-resistant product. Where high-slip resistance is required, a studded profile or carborundum (silicon carbide) grit may be incorporated into the surface (Fig. 8.5). Figure 8.6 illustrates the appropriate use of ceramic floor tiles.

TERRACOTTA

In order to produce intricately detailed terracotta building components, the clay has to be more finely divided than is necessary for bricks and roof tiles. The presence of iron oxide within the clay causes the buff, brown or red colouration of the fired product. During the latter part of the nineteenth century many civic buildings were constructed with highly decorative terracotta blocks. The material was used because it was cheaper than stone, durable and could be readily moulded. The blocks, which were usually partly hollowed out to facilitate drying and firing, were filled with concrete during construction.

Modern terracotta blocks may still be supplied for new work or refurbishment as plain ashlar, profiled or with sculptural embellishments. Terracotta may be used as the outer skin of cavity wall construction or as 25–40-mm-thick cladding hung with stainless steel mechanical fixings. The production of terracotta blocks requires the manufacture of an oversize model (to allow for shrinkage), from which plaster moulds are made. Prepared clay is then pushed into the plaster mould, dried under controlled conditions and finally fired. Traditional colours together with greens and blues and various textures are produced. For refurbishment work existing terracotta can, subject to natural variations, usually be colour matched. In addition to cladding units, terracotta clay is also used in the manufacture of terracotta floor tiles and an extensive range of decorative ridge tiles and finials (Fig. 8.7). Standard terracotta building blocks made by extrusion are described in Chapter 2.

Terrcotta rainscreen cladding

Rainscreen cladding is the external weathering element to multi-layer rainscreen wall systems. The rainscreen facade is drained and back-ventilated to protect the structural wall from the adverse effects of the sun, wind and rainwater. The design of the joints and the cavity between the facade and the structure result in an equalisation of air pressure between the cavity and the exterior, thus inhibiting the drive of airborne moisture across the cavity. A breather membrane is usually fixed to the structure before the rainscreen system is applied.

Rainscreen systems are appropriate for masonry, concrete, timber-frame and concrete-frame construction. A grid of vertical or horizontal aluminium extrusions is fixed to the facade, creating a minimum air gap of 25 mm. The rainscreen units are then clipped to the support system. The rainscreen cladding units may be manufactured in terracotta or from a wide range of other materials including stone laminate, stainless steel, copper, aluminium or zinc. Rainscreen units are shaped to shed water out of the open drained joints, and individual units may be removed for maintenance or repair. A range of colours and dimensions is available in terracotta units to create the required aesthetic effect. Terracotta rainscreen cladding (Fig. 8.8) is fire-resistant and durable requiring virtually no maintenance except occasional cleaning.

FAIENCE

Faience is glazed terracotta, used either as structural units or in the form of decorative slabs applied as cladding. It was popular in the nineteenth century and was frequently used in conjunction with polychrome brickwork on the facades of buildings such as public houses. Either terracotta may be glazed

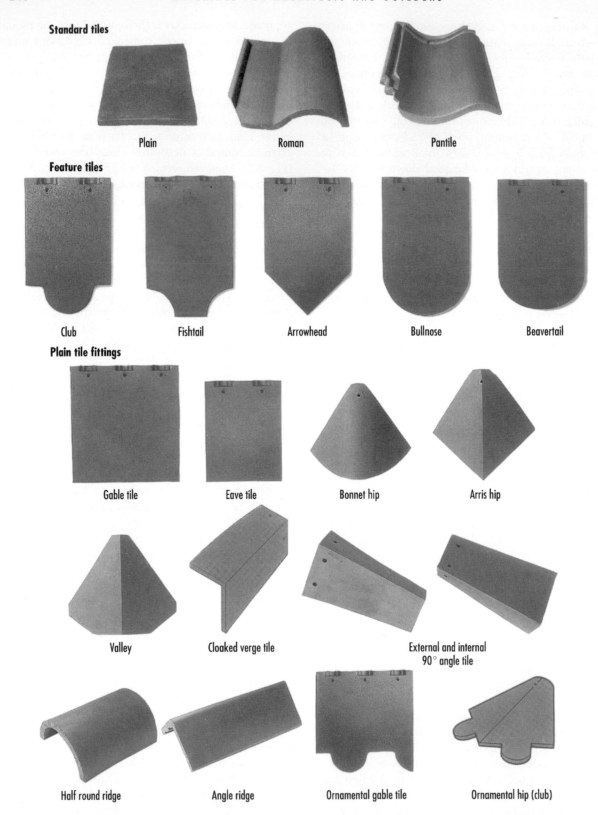

Fig. 8.4 Roof tiles – feature tiles and plain tile fittings

Example layouts

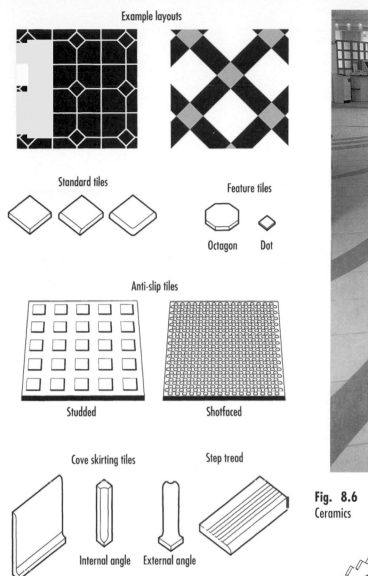

Standard tiles

Feature tiles

Octagon   Dot

Anti-slip tiles

Studded   Shotfaced

Cove skirting tiles   Step tread

Internal angle   External angle

Cove

Fig. 8.5 Floor tiles – textured, smooth and specials

Fig. 8.6 Floor tiles. Photograph: Courtesy of Architectural Ceramics

after an initial firing to the *biscuit* condition or the *slip* glaze may be applied prior to a single firing. The latter has the advantage that it reduces the risk of the glaze crazing although it also restricts the colour range. Faience, with an orange-peel texture, is available with either a matt or gloss finish and in plain or mottled colours. It is a highly durable material being unaffected by weathering, frost or ultraviolet light, but strong impacts can chip the surface causing unsightly damage.

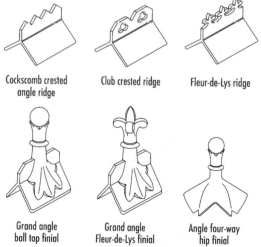

Cockscomb crested angle ridge   Club crested ridge   Fleur-de-Lys ridge

Grand angle ball top finial   Grand angle Fleur-de-Lys finial   Angle four-way hip finial

Fig. 8.7 Terracotta ridge tiles and finials

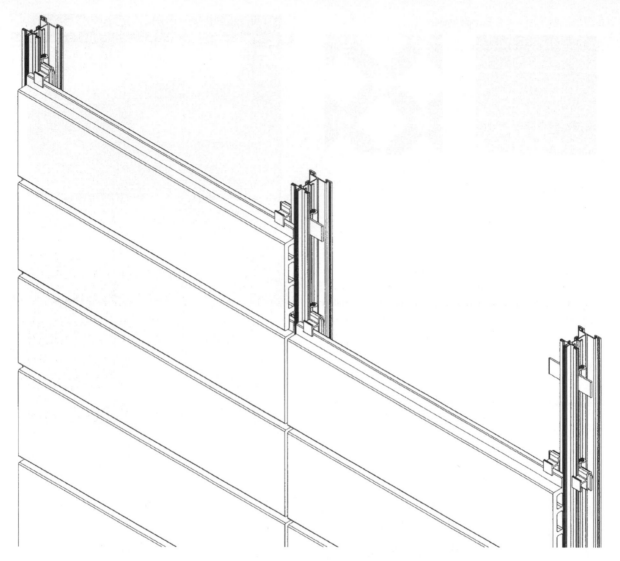

Fig. 8.8 Terracotta rainscreen cladding. Drawing: Courtesy of CGL Comtec

STONEWARE

Stoneware is manufactured from secondary plastic clays, typically fireclays blended with an added flux such as feldspar. On firing to between 1200°C and 1300°C the material vitrifies, producing an impermeable ceramic product with high chemical resistance. The majority of unglazed vitrified clay pipes are stoneware. For most purposes push-fit polypropylene couplings are used which allow flexibility to accommodate ground movement; however, if required, traditional jointed socket/spigot drainage goods are also available in stoneware.

Large stoneware ceramic panels up to 1.2 m square and 8 mm thick are manufactured as cladding units for facades. The units, which are colour-fast, frost- and fire-resistant may be uniform in colour or flecked and glazed or unglazed. Fixing systems are exposed or hidden; the open joint system offers rear ventilation allowing any moisture diffusing from the supporting wall to be dissipated by natural air movement.

Stoneware is also used in the manufacture of some floor tiles. The high firing temperature gives a product of low porosity, typically less than 3%. In one manufacturing process a granular glaze is applied to the tiles within the kiln to produce an impervious vitreous finish.

EARTHENWARE

Earthenware is produced from a mixture of kaolin, ball clay and flint with, in some cases, feldspar as a flux. The material when fired at 1100°C is porous and requires a glaze to prevent water absorption. In the manufacture of traditional glazed drainage goods, the salt glaze is produced by adding damp common salt to the kiln during the firing process. The salt decomposes to form sodium oxide, which then reacts with silica and alumina on the surface of the clay component to produce the salt glaze which is impermeable to moisture.

WALL TILES

Wall tiles (Fig. 8.9) are generally manufactured from earthenware clay to which talc (magnesium silicate) or limestone (calcium carbonate) is added to ensure a white burning clay. To prepare the clay for manufacturing wall tiles by the dry process the components, typically a blend of china clay (kaolin), ball clay and silica sand together with some ground recycled tiles, are mixed with water to form a slip. This is sieved, concentrated to a higher-density slip, then dried to a powder by passage down a heated tower at 500°C. The clay dust, which emerges with a moisture content of approximately 8%, is then pressed into tiles. A glaze is required both to decorate and

Fig. 8.9 Wall tiles. Photograph: Courtesy of Architectural Ceramics

Fig. 8.10 Mosaic. Photograph: Courtesy of Architectural Ceramics

Fig. 8.11   Concert Hall, Tenerife and ceramic mosaic detail. Architect: Santiago Calatrava. Photographs: Arthur Lyons

produce an impermeable product and this may be applied before a single firing process or after the tiles have been fired at 1150°C to the biscuit stage in a tunnel kiln. Either the unfired or biscuit tiles are coated with a slip glaze followed by firing under radiant heat for approximately 16 hours. Damaged tiles are rejected for recycling; the quality-checked tiles are packaged for dispatch. Standard sizes are 108 × 108 mm, 150 – 150 mm, 200 – 150 mm, 200 – 200 mm and 250 × 200 mm.

VITREOUS CHINA

Vitreous china, used for the manufacture of sanitary ware, has a glass-like body which limits water absorption through any cracks or damage in the glaze to 0.5%. It is typically manufactured from a blend of kaolin (25%), ball clay (20%), feldspar (30%) and quartz (25%). For large units such as WCs and wash basins, a controlled drying out period is required before firing to prevent cracking. Glaze containing metallic oxides for colouration is applied before firing to all visually exposed areas of the components.

Vitreous china is also used in the manufacture of some floor tiles due to its impermeable nature. Unglazed floor tiles may be smooth, alternatively studded or ribbed to give additional non-slip properties. Standard sizes are 100 – 100 mm, 150 – 150 mm, 200 – 200 mm and 300 – 300 mm with thickness usually in the range 8 – 13 mm. For lining swimming pools, additional protection against water penetration is given by the application of a glaze.

REPRODUCTION DECORATIVE TILES

Reproduction moulded ceramic wall tiles (encaustic tiles with strong colours burnt into the surface) and geometrical floor tiles can be manufactured to match existing units with respect to form, colour and texture for restoration work. Some manufacturers retain both the necessary practical skills and appropriate detailed drawings to ensure high-quality conservation products, which may be used to replace lost or seriously damaged units. There is also an increasing demand for reproduction decorative tiles in new-build work.

MOSAICS

Mosaics in glazed or unglazed porcelain are hard wearing, frost-proof and resistant to chemicals. Unglazed mosaics may be used for exterior use and other wet areas such as swimming pools, where good slip resistance is important. Mosaics are usually supplied attached to paper sheets for ease of application. Figure 8.10 illustrates a formal mosaic floor, while Figure 8.11 shows the broken tile mosaic finish used by Calatrava on the Tenerife Concert Hall, following the technique developed by Gaudi.

CERAMIC GRANITE

Ceramic granite is a blend of ceramic and reconstituted stone, manufactured from a mixture of feldspar, quartz and clay. The components are crushed, graded, mixed and compressed under very high pressure, followed by firing at 1200°C. The material is produced in 20 and 30 mm slabs, which can be cut and polished to produce a hard shiny finish with the appearance of natural marble or granite, suitable for worktops. Colours range from ochre, off-white and grey to green and blue depending upon the initial starting materials.

References

FURTHER READING

Ashurst, J. and Ashurst, N. 1988: *Brick, terracotta and earth*. Practical Building Conservation 2. Aldershot: Gower Technical Press.

Creative Publishing International. 2003: *The complete guide to ceramic and stone tile*. USA: Creative Publishing International.

Durbin, L. 2004: *Architectural tiles: Conservation and restoration*. Oxford: Elsevier.

Grimshaw, R.W. 1971: *The chemistry and physics of clays and allied ceramic Materials*. 4th ed., London: Ernest Benn.

Hamer, F. and Hamer, J. 1977: *Clays* – Ceramic Skillbooks Series. London: Pitman Publishing.

Hamilton, D. 1978: *The Thames & Hudson manual of architectural ceramics*. London: Thames & Hudson.

Lemmen, H.V. 2002: *Architectural ceramics*. Princes Risborough: Shire Publications.

Ripley, J. 2005: *Ceramics and stone tiling*. Ramsbury: Crowood Press.

Teutonico, J.M. (ed.) 1996: *Architectural ceramics: Their history, manufacture and conservation*. London: James & James Science Publishers.

Wilhide, E. 2003: *Materials; A directory for home design*. London: Quadrille Publishing.

Worrall, W.E. 1986: *Clays and ceramic raw materials.* Kluwer Academic Publishers.

STANDARDS

BS 65: 1991 Specification for vitrified clay pipes, fittings, and ducts, also flexible mechanical joints for use solely with surface water pipes and fittings.

BS 493: 1995 Airbricks and gratings for wall-ventilation.

BS 1125: 1987 Specification for WC flushing cisterns.

BS 1188: 1974 Ceramic washbasins and pedestals.

BS 1196: 1989 Clayware field drain pipes and junctions.

BS 1206: 1974 Fireclay sinks: dimensions and workmanship.

BS 3402: 1969 Quality of vitreous china sanitary appliances.

BS 3921: 1985 Specification for clay bricks.

BS 5385 Wall and floor tiling:

    Part 1: 1995 Code of practice for the design and installation of internal ceramic wall and natural stone wall tiling tiling and mosaics in normal conditions.

    Part 2: 1991 Code of practice for the design and installation of external ceramic wall tiling and mosaics (including terra cotta and faience tiles).

    Part 3: 1989 Code of practice for design and installation of ceramic floor tiles and mosaics.

    Part 4: 1992 Code of practice for tiling and mosaics in specific conditions.

    Part 5: 1994 Code of practice for the design and installation of terrazzo tile and slab, natural stone and composition block flooring.

BS 5503 Specification for vitreous china washdown WC pans with horizontal outlet:

    Part 3: 1990 WC pans with horizontal outlet.

BS 5504 Specification for wall hung WC pan:

    Part 1: 1977 Connecting dimensions.

    Part 2: 1977 Independent water supply. Connecting dimensions.

BS 5506 Specification for wash basins:

    Part 3: 1977 Wash basins (one or three tap holes), materials, quality, design and construction.

BS 5534: 2003 Code of practice for slating and tiling.

BS 6431

    Part 1 Ceramic wall and floor tiles.

BS 8000 Workmanship on building sites:

    Part 11: 1989 Code of practice for wall and floor tiling.

BS EN 295 Vitrified clay pipes and fittings and pipe joints for drains and sewers:

    Part 1: 1991 Requirements.

    Part 2: 1991 Quality control and samples.

    Part 3: 1991 Test methods.

    Part 4: 1995 Requirements for special fittings, adaptors and compatible accessories.

    Part 5: 1994 Requirements for perforated vitrified clay pipes.

    Part 6: 1996 Requirements for vitrified clay manholes.

    Part 7: 1996 Requirements for vitrified clay pipes and joints.

BS EN 538: 1994 Clay roofing tiles for discontinuous laying – Flexural strength test.

BS EN 539 Clay roofing tiles for discontinuous laying – Determination of physical characteristics:

    Part 1: 1994 Impermeability test.

    Part 2: 1998 Test for frost resistance.

BS EN 997: 2003 WC pans and WC suites with integral trap.

BS EN 1304: 1998 Clay roofing tiles for discontinuous laying.

BS EN 1457: 1999 Chimneys clay/ceramic flue liners.

BS EN 1806: 2000 Chimneys – clay/ceramic flue blocks for single wall chimneys.

BS EN ISO 10545 Ceramic tiles:

    Part 1: 1997 Sampling.

    Part 2: 1997 Dimensions and surface quality.

    Part 3: 1997 Water absorption and apparent porosity, relative and bulk density.

    Part 4: 1997 Modulus of rupture and breaking strength.

    Part 6: 1997 Resistance to deep abrasion for unglazed tiles.

    Part 7: 1999 Resistance to surface abrasion for glazed tiles.

    Part 8: 1996 Determination of linear thermal expansion.

    Part 9: 1996 Resistance to thermal shock.

    Part 10: 1997 Determination of moisture expansion.

    Part 11: 1996 Determination of crazing resistance.

    Part 12: 1997 Determination of frost resistance.

    Part 13: 1997 Determination of chemical resistance.

    Part 14: 1997 Determination of resistance to stains.

    Part 16: 2000 Determination of small colour differences.

BS EN 13502: 2002 Requirements and test methods for clay/ceramic flue terminals.

pr EN 14411: 2006 Ceramic tiles – definitions, classification, characteristics and marking.

BS EN 14437: 2004 Determination of the uplift resistance of installed clay or concrete tiles for roofing.

BUILDING RESEARCH ESTABLISHMENT PUBLICATIONS

BRE Digest

BRE Digest 467: 2002 Slate and tile roofs: avoiding damage from aircraft wake vortices.
BRE Digest 499: 2006 Designing roofs for climate change.

BRE Good building guide

GBG 28: 1997 Domestic floors – repairing or replacing floors and flooring.

GBG 64 Part 2: 2005 Tiling and slating pitched roofs: Plain and profiled clay and concrete tiles.

ADVISORY ORGANISATIONS

CERAM Research, Queens Road, Penkhull, Stoke-on-Trent ST4 7QL (01782 76444).
Clay Pipe Development Association Ltd., Copsham House, 53 Broad Street, Chesham, Bucks. HP5 3EA (01494 791456).
Clay Roof Tile Council, Federation House, Station Road, Stoke-on-Trent, Staffordshire ST4 2SA (01782 744631).
Tile Association, Forum Court, 83 Copers Cope Road, Beckenham, Kent BR3 1NR (020 8663 0946).

9

STONE AND CAST STONE

Introduction

The term *stone* refers to natural rocks after their removal from the earth's crust. The significance of stone as a building material is illustrated by widespread prehistoric evidence and its sophisticated use in the early civilisations of the world, including the Egyptians, the Incas of Peru, and the Mayans of Central America.

Geologically, all rocks can be classified into one of three groups: igneous, metamorphic or sedimentary, according to the natural processes by which they were produced within or on the earth's surface.

IGNEOUS ROCKS

Igneous rocks are the oldest, having been formed by the solidification of the molten core of the earth or *magma*. They form about 95% of the earth's crust, which is up to 16 km thick. Depending whether solidification occurred slowly within the earth's crust or rapidly at the surface, the igneous rocks are defined as plutonic or volcanic respectively. In the plutonic rocks, slow cooling from the molten state allowed large crystals to grow which are characteristic of the granites. Volcanic rocks such as pumice and basalt are fine-grained and individual crystals cannot be distinguished by the eye, thus the stones are visually less interesting. Dolerites, formed by an intermediate rate of cooling, exhibit a medium-grained structure.

Apart from crystal size, igneous rocks also vary in composition according to the nature of the original magma, which is essentially a mixture of silicates. A high silica content magma produces acid rocks (e.g. granite) whilst a low silica content forms basic rocks (e.g. basalt and dolerite). Granites are mainly composed of feldspar (white, grey or pink), which determines the overall colour of the stone, but they are modified by the presence of quartz (colourless to grey or purple), mica (silver to brown), or hornblende (dark coloured). The basic rocks such as dolerite and basalt, in addition to feldspar, contain augite (dark green to black) and sometimes olivine (green). Although basalt and dolerite have not been used widely as building stones they are frequently used as aggregates, and cast basalt is now being used as a reconstituted stone.

Granites

Most granites are hard and dense, and thus form highly durable building materials, virtually impermeable to water, resistant to impact damage and stable within industrial environments. The appearance of granite is significantly affected by the surface finish which may be sawn, rough punched, picked, fine tooled, honed or polished. It is, however, the highly polished form of granite which is most effective at displaying the intensity of the colours and reflectivity of the crystals. Additionally, granites may be flamed to a spalled surface, produced by the differential expansion of the various crystalline constituents. Many recent buildings have combined the polished and flamed material to create interesting contrasts in depth of colour and texture. Grey and pink granites are quarried in Scotland, the North of England, Devon and Cornwall, but a wide variety of colours including black, blue, green, red, yellow and brown are imported from other countries (Fig. 9.1) (Table 9.1). Because of the high cost of quarrying and finishing granite, it is frequently used as a cladding material (40 mm externally or 20 mm internally) or alternatively cast directly onto concrete cladding units. Granite is available for flooring and for hard landscaping including pavings, setts and kerbs. Polished granite is also used as a kitchen

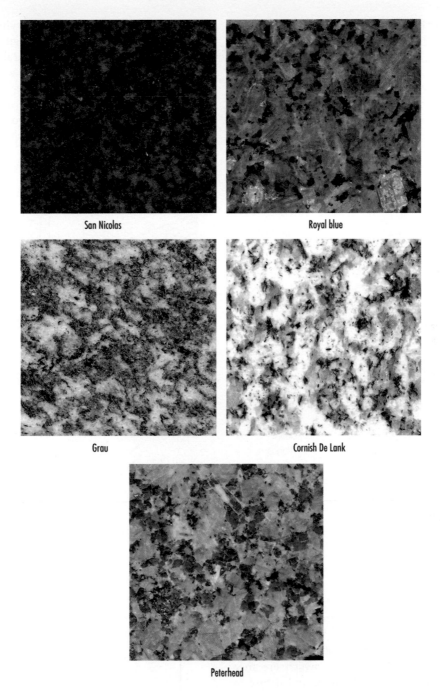

San Nicolas

Royal blue

Grau

Cornish De Lank

Peterhead

Fig. 9.1 Selection of granites

countertop material due to its strength, durability and high-quality finish.

Cast basalt

Basalt is a fine-grained stone nearly as hard as granite. It can be melted at 2400°C and cast into tile units which are deep steel grey in colour. A slightly patterned surface can be created by swirling the molten basalt within the mould. Annealing in a furnace produces a hard virtually maintenance-free shiny textured surface flecked with shades of green, red and bronze. Larger cast units for worktops, in either a honed or polished finish, can be cut to size.

Table 9.1 UK and imported granites

Colour	Name	Country of origin
UK		
Light grey	Merrivale, Devon	England
Silver grey	De Lank &	
	Hantergantick, Cornwall	England
Light and dark pink to brownish red	Shap	England
Pink	Peterhead	Scotland*
Pale to deep red	Ross of Mull	Scotland
Grey	Aberdeen	Scotland*
Black	Hillend	Scotland*
Black	Beltmoss	Scotland*
* available only in limited quantities		
Imported		
Red with black	Balmoral Red	Finland
Red	Bon Accord Red	Sweden
Black	Bon Accord Black	Sweden
Red	Virgo Granite	Sweden
Dark red with blue to purple quartz	Rose Swede	Sweden
Grey	Grey Royal	Norway
Grey	Sardinian Grey	Sardinia
Yellow	Nero Tijuca	Brazil
Beige/brown	Juparana	Brazil
Blue	Blue Pearl	Norway
Green/black	Emerald Pearl	Norway
Pink to red	Torcicoda	Brazil
Beige/brown	Giallio Veneziano	Brazil

SEDIMENTARY ROCKS

Sedimentary rocks are produced by the weathering and erosion of older rocks. In the earliest geological time these would have been the original igneous rocks, but subsequently other sedimentary and metamorphic rocks too will have been reworked. Weathering action by water, ice and wind breaks the rocks down into small fragments which are then carried by rivers and sorted into size and nature by further water action. Most deposits are laid down in the oceans as sedimentary beds of mud or sand, which build up in layers, become compressed and eventually are cemented together by minerals such as calcium carbonate (calcite), quartz (silica), iron oxide or dolomite (magnesium and calcium carbonate) remaining in the groundwater. The natural bedding planes associated with the formation of the deposits may be thick or thin but are potentially weak; this is

used to advantage in the quarrying process. In masonry, to obtain maximum strength and durability, stones should be laid to their natural bed except for cornices, cills and string courses which should be edge-bedded. Stones which are face-bedded will tend to delaminate (Fig. 9.2). When quarried, stones contain *quarry sap* and may be worked and carved more easily than after exposure to the atmosphere.

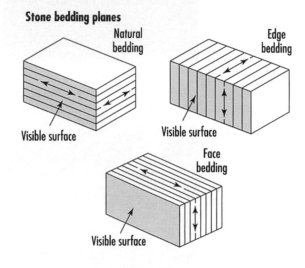

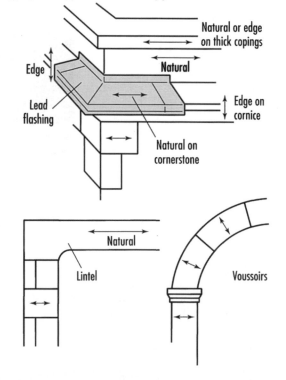

Fig. 9.2 Natural stone bedding planes

Sandstones

Deposits of sand cemented together by calcium carbonate, silica, iron oxide and dolomite produce calcareous, siliceous, ferruginous and dolomitic sandstones respectively. Depending upon the nature of the original sand deposit, the sandstones may be fine or coarse in texture. Sandstones range in colour from white, buff and grey through to brown and shades of red depending upon the natural cement; they are generally frost-resistant. Some common UK sandstones are listed in Table 9.2. Typical finishes are sawn, split faced and clean rubbed, although a range of tooled finishes including broached and droved can also be selected (Fig. 9.3). For cladding, sandstone is normally 75 mm to 100 mm thick and fixed with non-ferrous cramps and corbels. Sandstones are quarried in Scotland, the North of England, Yorkshire and Derbyshire; they include the old and new red sandstones, York Stone and Millstone Grit. Sandstone is imported from Spain and Italy from where *Pietra Serena* is sourced.

Calcareous sandstone

Calcareous sandstones are not durable in acid environments, which may cause the slow dissolution of the natural calcium carbonate cement of the stone. Pure calcite is white, so these sandstones are generally white in colour.

Siliceous sandstone

Siliceous sandstones are predominantly grains of silica (sand) cemented with further natural silica, and are therefore durable even in acid environments. Siliceous sandstones are generally grey in colour.

Ferruginous sandstone

Ferruginous sandstones are bound with oxides of iron which may be brown, ochre or red. They are generally durable.

Dolomitic sandstone

Dolomitic sandstones are bound with a mixture of magnesium and calcium carbonates, and therefore do not weather well in urban environments. They are generally off-white and buff in colour.

Limestones

Limestones consist mainly of calcium carbonate, either crystallised from solution as calcite or formed from accumulations of fossilised shells deposited by various sea organisms. They are generally classified according to their mode of formation. Many colours are available ranging from off-white, buff, cream, grey and blue. Limestones are found in England in a belt from Dorset, the Cotswolds, Oxfordshire and Lincolnshire to Yorkshire. Limestone is also imported from Ireland, France and Portugal to widen the palette of colours. Some common UK limestones are listed in Table 9.3. The standard finishes are fine-rubbed, fine-dragged and split-faced, although tooled finishes are also appropriate. Externally, limestones must not be mixed with or located above sandstones, as this may cause rapid deterioration of the sandstone.

Oölitic limestone

Oöitic limestones are formed by crystallisation of calcium carbonate in concentric layers around small fragments of shell or sand, producing spheroidal grains or oöliths (Fig. 9.4). The oöliths become

Table 9.2 Typical UK sandstones and their characteristics

Name	Colour	Source	Characteristics
Doddington	purple/pink	Northumberland	fine to medium-grained
Darley Dale – Stancliffe	buff	Derbyshire	fine-grained
Birchover gritstone	pink to buff	Derbyshire	medium to coarse-grained
York Stone	buff, fawn, grey, light brown	Yorkshire	fine-grained
Mansfield Stone	buff to white	Nottinghamshire	fine-grained
Hollington	pale pink, dull red pink with darker stripe	Staffordshire	fine to medium-grained
St. Bees	dark red	Cumbria	fine-grained
Blue Pennant	dark grey/blue	Mid-Glamorgan	fine-grained

Broached with draughted margin

Stugged or punched face

Droved

Herringbone

Bats

Fig. 9.3 Typical tooled-stone finishes

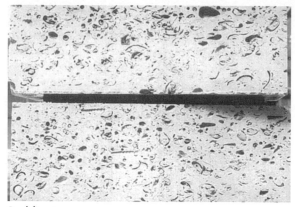

Roach limestone

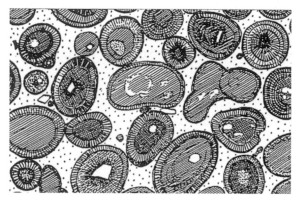

Oölitic limestone (×20) (after Arkwell, W.J. 1946: *Oxford stone*, Faber & Faber)

Fig. 9.4 Roach limestone and oölitic limestone (× 20)

Oölitic limestones are very workable and include Bath Stone and Portland Stone. Clipsham Stone and Ketton Stone have been widely used at Oxford and Cambridge respectively, including the recent Queen's Building of Emmanuel College, Cambridge (Fig. 9.5), which is built of load-bearing Ketton limestone, with appropriately massive columns and flat voussoir arches to the colonnade and window openings. Lime mortar is used to ensure an even spreading of the load between stones. In the case of Foundress Court, Pembroke College, Cambridge (Fig. 9.6), the Bath Stone (Monks Park) is built up three storeys from ground level as a well-detailed cladding, with restraint back to the load-bearing blockwork inner skin. The flexibility of lime mortar is used to reduce the number of visible movement joints.

Organic limestone
Organic limestones are produced in bedded layers from the broken shells and skeletal remains of a wide variety of sea animals and corals. Frequently clay is

cemented together by the further deposition of calcite to produce the rock. Typically the oöliths are up to 1 mm in diameter, giving a granular texture to the stone, which may also incorporate other fossils.

Table 9.3 Typical UK limestones and their characteristics

Name	Colour	Source	Characteristics
Ancaster	cream to buff	Lincolnshire	oölitic limestone — variable shell content; freestone available
Bath Stone	pale brown to light cream	Avon	oölitic limestone
— Westwood Ground			coarse-grained — buff coloured
— Monks Park			fine-grained — buff coloured
Clipsham	buff to cream	Rutland	medium-grained oölitic limestone with shells; some blue stone; best quality stone is durable
Doulting	pale brown	Somerset	coarse textured; fossils uncommon
Hopton Wood	cream or grey	Derbyshire	carboniferous limestone containing many attractive fossils; may be polished
Ketton	pale cream to buff and pink	Lincolnshire	medium-grained oölitic limestone; even-textured; durable stone
Portland Stone	white	Dorset	exposed faces weather white, protected faces turn black.
— Roach			coarse open-textured shelly stone; weathers very well
— Whitbed			fine-grained — some shell fragments; durable stone
— Basebed			fine-grained with few shells; suitable for carving
Purbeck	blue/grey to buff	Dorset	some shells; durable stone.

incorporated into organic limestones and this adversely affects the polish which can otherwise be achieved on the cut stone.

Crystallised limestone
When water containing calcium bicarbonate evaporates, it leaves a deposit of calcium carbonate. In the case of hot springs the material produced is travertine, and in caves stalactites and stalagmites or *onyx-marble* result.

Dolomitic limestone
Dolomitic limestones have had the original calcium carbonate content partially replaced by magnesium carbonate. In general this produces a more durable limestone, although it is not resistant to heavily polluted atmospheres.

METAMORPHIC ROCKS

Metamorphic rocks are formed by the recrystallisation of older rocks, when subjected to intense heat or pressure or both, within the earth's crust. Clay is metamorphosed to slate, limestone to marble and sandstone to quartzite.

Slate

Slate is derived from fine-grained sand-free clay sediments. The characteristic cleavage planes of slate were produced when the clay was metamorphosed and frequently they do not relate to the original bedding planes. Slate can be split into thin sections (typically 4–10 mm for roofing slates) giving a natural riven finish, or it may be sawn, sanded, fine-rubbed, honed, polished, flame-textured or bush hammered. A range of distinctive colours is available: blue/grey, silver grey and green from the Lake District; blue, green, grey and plum red from North Wales; and grey from Cornwall. Slate is also imported from Ireland (grey/green), Canada (blue/grey), France (blue/grey), China (blue/green/grey), Brazil (grey/green/plum) and

Fig. 9.5 Ketton Limestone – The Queen's Building, Emmanuel College, Cambridge. Architects: Hopkins Architects. Photograph: Arthur Lyons

blue/black from Spain, which is the world's largest producer of the material.

Slate is strong, acid- and frost-resistant, lasting up to 400 years as a roofing material. The minimum recommended pitch for slate roofing is 20° under sheltered or moderate exposure and 22.5° under severe exposure, and these situations require the use of the longest slates (460, 560 or 610 mm). Where thick slates (up to 20 mm in thickness) are used for a roof pitch of less than 25°, it should be noted that the slates lie at a significantly lower pitch than the rafters. Fixing nails should be of copper or aluminium. Slate is also used for flooring, cladding, copings, cills and stair treads. When used as a cladding material it should be fixed with non-ferrous fixings or cast directly onto concrete cladding units.

Roofing and external cladding slates satisfy the requirements for the Class A1 characteristic reaction to fire performance, without the need for testing.

Recycled roofing slates, particularly Welsh slate, are generally available in a range of sizes and are appropriate for both conservation work and new build where an immediate weathered appearance is required. Welsh slates have a good reputation for durability, making the recycled product a viable option. Certain regional slates, such as Swithland in Leicestershire are only available as recycled products. This particular type of slate has a single top nail fixing and, unlike most roofing slates which are of a uniform size, is graded from large slates at the eaves to smaller units at the ridge (Fig. 9.7).

Reconstituted slate
Reconstituted slate for roofing is manufactured from slate granules and inert filler, mixed with a thermosetting

Fig. 9.6 Bath Stone – Foundress Court, Pembroke College, Cambridge. Architects: Eric Parry Architects. Photograph: Arthur Lyons

resin and cast into moulds to give a natural riven slate finish. Certain products incorporate glass-fibre reinforcement, and offer a wider range of colours than are available in natural slate. Some interlocking slates may be used down to a pitch of 17.5°, whilst double-lap simulated natural slates can be used down to a pitch of 20° depending upon the degree of exposure. Reconstituted slate is also manufactured in glass-fibre reinforced cement (GRC) as described in Chapter 11.

Marble

Marble is metamorphosed limestone in which the calcium carbonate has been recrystallised into a mosaic of approximately equal-sized calcite crystals. The process, if complete, will remove all traces of fossils, the size of the crystals being largely dependent on the duration of the process. Some limestones which can be polished are sold as marble, but true marble will not contain any fossillised remains. Calcite itself is white, so

a pure marble is white and translucent. The colours and veining characteristics of many marbles are associated with impurities within the original limestone; they range from red, pink, violet, brown, green, beige, cream and white to grey and black. Marble is attacked by acids; therefore honed, rather than highly polished surfaces, are recommended for external applications. Marbles are generally hard and dense, although fissures and veins sometimes require filling with epoxy resins. Most marbles used within Britain are imported from Europe as indicated in Table 9.4; a selection is illustrated in Figure 9.8.

For external cladding above first floor level 40-mm-thick slabs are used, although 20 mm may be appropriate for internal linings and external cladding up to first floor level. Fixing cramps and hooks should be in stainless steel, phosphor bronze or copper. Floor slabs, to a minimum thickness of 30 mm, should be laid on a minimum 25 mm bed. Marble wall and bathroom floor tiles are usually between 7 mm and 10 mm in thickness.

Welsh slate-uniform size

Swithland slate-graded size

Fig. 9.7 Slate roofs

Reconstituted marble

Reconstituted marble is manufactured from marble chippings and resin into tiles and slabs for use as floor and wall finishes. The material has the typical colours of marble but without the veining associated with the natural material.

Quartzite

Quartizite is metamorphosed sandstone. The grains of quartz are recrystallised into a matrix of quartz, producing a durable and very hard wearing stone used mainly as a flooring material. The presence of mica allows the material to be split along smooth

Table 9.4   A selection of imported marbles

Colour	Name	Country of origin
White	White Carrara/Sicilian	Italy
White	White Pentelicon	Greece
Cream	Perlato	Sicily
Cream	Travertine	Italy
Beige	Botticino	Sicily
Pink	Rosa Aurora	Portugal
Red	Red Bilbao	Spain
Brown	Napoleon Brown	France
Green	Verde Alpi	Italy
Black	Belgian Black	Belgium
Black with white veins	Nero Marquina	Algeria

cleavage planes, producing a riven finish. Quartzite is mainly imported from Norway and South Africa and is available in white, grey, grey-green, blue-grey and ochre.

ALABASTER

Alabaster is naturally occurring gypsum or calcium sulfate. Historically it has been used for building as in the Palace of Knossos, Crete, but in the UK its use has been mainly restricted to carved monuments and ornaments. The purest form is white and translucent, but traces of iron oxide impart light brown, orange or red colourations.

Stonework

TRADITIONAL WALLING

Dressed stone may be used as an alternative to brick or block in the external leaf of standard cavity construction. Limestone and sandstone are the most frequently used for walling, but slate is also used where it is available locally. Although random rubble and hand-dressed stone can be supplied by stone suppliers, sawn-bedded (top and bottom) stones are generally the most available. These are normally finished split-faced, pitch-faced, fine-rubbed or sawn. The standard sizes are 100 or 105 mm on bed, with course heights typically 50, 75, 100, 110, 125, 150, 170, 225 and 300 mm (Fig. 9.9). Stones may be to a particular course length, e.g. 300 mm or 450 mm, although they are frequently to random lengths. Quoin blocks, window

and door surrounds, cills and other components are often available as standard. In ashlar masonry, the stones are carefully worked and finely jointed. Stones within horizontal courses are of the same height and are perfectly rectangular in elevation. Joints are generally under 6 mm in width.

The mortar for stone masonry should be weaker than the stone selected. For porous limestones and sandstones, crushed stone aggregate is frequently used as the aggregate in the mortar, typically in a 1 : 3 : 12 mix of Portland cement, lime putty and crushed stone. For ashlar Bath Stone a typical mix would be 1 : 2 : 8 cement, lime and stone dust. Dense sandstones may be bonded with a stronger 1 : 1 : 6 mix, and granite a 1 : 2 or 1 : 3 Portland cement to fine aggregate mix. Jointing should generally be to a similar texture and colour to that of the dressed stone itself, and should be slightly recessed to emphasise the stones rather than the joints. In ashlar work, in which accurately cut squared stones are used, a matching 5 mm flush joint is appropriate.

The David Mellor Cutlery Factory, Hathersage, Derbyshire (Fig. 9.10) illustrates the use of traditionally detailed Derbyshire stone as load-bearing masonry worked in conjunction with precast concrete quoins and padstones. The building takes its form from the base of an old gasholder which provides its foundations.

Gabions

Gabions are wire cages filled with crushed rocks or recycled concrete rubble. They are frequently used in civil engineering applications as retaining walls, and are simply stacked to the required height either vertically or to an appropriate incline. Compressive loads are transmitted through the stones or concrete rubble and any spreading movement is restrained by the tensile forces within the wire cage. Normally the cages are of heavy gauge woven or welded steel mesh, which may be zinc, aluminium/zinc alloy, or PVC-coated, but for use in load-bearing building applications such as walls, stainless steel should be used. Gabions are now being used as significant components in building construction, where the particular rugged aesthetic is required (Fig. 9.11). Gabions may be delivered on site filled or flat packed for filling and fastening, usually with a helical binder in alloy-coated or stainless steel. A range of sizes is available based mainly on a metre module.

STONE CLADDING

For the majority of large commercial buildings, stone is used as a cladding material mechanically fixed to the structural system. The strength of the stone largely determines the appropriate cladding panel thickness. For granites, marbles and slate, 40 mm slabs are usual for external elevations above ground floor level, but for the softer limestones and sandstones a minimum thickness of 75 mm is frequently recommended. However, at heights of less than 3.7 m above ground level, thinner sections are permissible providing that they are of sufficient strength not to suffer distortion and failure. The standard (BS 8298: 1994) gives details of stone thicknesses for external cladding and internal lining.

Fixings (Fig. 9.12) must be manufactured from stainless steel or non-ferrous metal and must be sized to sustain the dead load of the cladding together with applied loads from wind and maintenance equipment. Movement joints are required to accept the differential structural movements of the frame and the thermal and moisture movements of the cladding. Horizontal compression joints of 15 mm minimum should be located at each floor level; vertical movement joints of 10 mm should be at approximately 6 m centres. Polysulfides, polyurethanes and silicones are used as joint sealants, although non-staining silicones should be used on stones which darken by absorption of silicone fluid. Stone cladding systems should ideally be protected from impact damage at ground level by the design detailing.

Concrete-backed stone cladding

An alternative approach to traditional stone cladding is the use of an integral stone veneer on concrete cladding panels. Stone is fixed to the concrete with a series of non-corroding dowels inclined in opposite directions, creating a mechanical fixing, not dependent on the bond between stone and cast concrete. With limestone, a stone veneer of 50 mm is appropriate. The concrete should be cast with appropriate reinforcement and fixings for attachment to the building structure.

Lightweight stone cladding

Thin section stone (approximately 6 mm) can be bonded to lightweight backing materials to reduce the dead weight of stone cladding (Fig. 9.13). The reduction in dead load is significant compared to thick-stone sections which would require traditional stone cladding techniques. One such material, originally used in the aerospace industry, is a sandwich panel

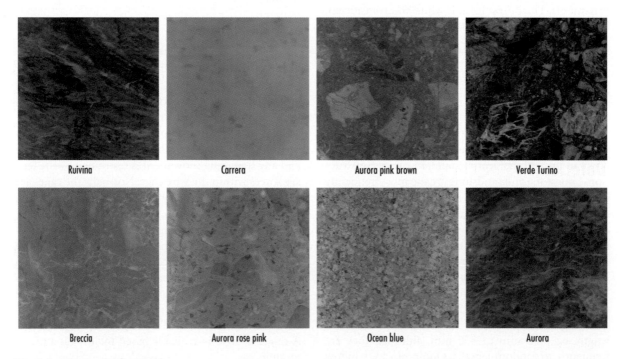

Ruivina   Carrera   Aurora pink brown   Verde Turino

Breccia   Aurora rose pink   Ocean blue   Aurora

Fig. 9.8 Selection of Italian marbles

Roughly squared split-faced random rubble

Polygonal random rubble

Sawn bedded pitched-faced random walling

Sawn-bedded pitched-faced coursed walling

Fig. 9.9 Traditional stone walling

consisting of a core of honeycomb aluminium faced with glass-fibre reinforced epoxy resin skins. The polished stone facing is bonded to one face with epoxy resin to create a lightweight stone-finished panel, which if detailed appropriately, has all the visual qualities associated with solid stone masonry.

Deterioration of stone

The main agencies causing the deterioration of stone are soluble-salt action, atmospheric pollution, frost, the corrosion of metal components and poor design or workmanship.

SOLUBLE-SALT ACTION

If moisture containing soluble salts evaporates from the surface of stonework, then the salts will be left either on the surface as white efflorescence or as crystals within the porous surface layer. If the wetting and drying cycles continue, the crystalline material builds up within the pores to the point at which the pressure produced may exceed the tensile strength of the stone, causing it to crumble. The actual pore size significantly influences the durability of individual stones, but generally the more porous stones, such as limestone and sandstone, are susceptible to soluble-salt action.

ATMOSPHERIC POLLUTION

Stones based on calcium carbonate are particularly vulnerable to attack by acid atmospheric pollutants. Sulfur dioxide in the presence of water and oxygen from the air produces sulfuric acid which attacks calcium carbonate to produce calcium sulfate. Limestones and calcareous sandstones are vulnerable to attack. In the case of limestone, the gypsum (calcium sulfate) produced at the surface is slightly soluble and on exposed surfaces gradually washes away leaving the eroded limestone clean. In unwashed

Fig. 9.10 Load-bearing stone masonry — David Mellor Cutlery Factory, Hathersage, Derbyshire. Architects: Hopkins Architects. Photographs: Arthur Lyons

Fig. 9.11 Rock-filled gabions – London Regatta Centre. Architects: Ian Ritchie Architects. Photograph: Arthur Lyons

areas, the surface becomes blackened with soot producing a hard crust, which eventually blisters exposing powdered limestone. Magnesian limestones react similarly, except that in some cases the recrystallisation of magnesium sulfate under the blackened crust causes a more serious cavernous decay of the stone. Calcareous sandstones, when rain-washed, gradually decay to powder; however, in unwashed areas they produce a hard crust in which the pores are blocked with gypsum. The crust eventually fails due

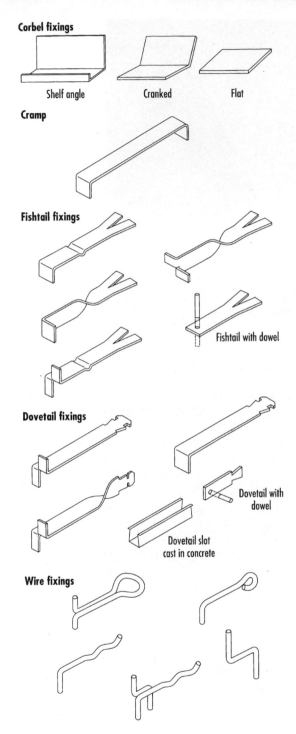

Corbel fixings

Shelf angle   Cranked   Flat

Cramp

Fishtail fixings

Fishtail with dowel

Dovetail fixings

Dovetail with dowel

Dovetail slot cast in concrete

Wire fixings

Fig. 9.12 Typical fixings for stone cladding

to differential thermal expansion. Dolomitic sandstones are less vulnerable to acid attack, unless they contain a significant proportion of vulnerable calcite. Silicious sandstones, which are not attacked directly by atmospheric acids, can be damaged by the calcium sulfate washings from limestone which then cause crystallisation damage to the sandstone surface. Marble, which is essentially calcium carbonate, is also affected by atmospheric acids. Any polished surface is gradually eroded; however, as marble is generally non-porous, crystallisation damage is unusual, and limited to *sugaring* in some cases.

FROST ACTION

Frost damage occurs in the parts of a building which become frozen when very wet, such as copings, cornices, string courses, window hoods and cills. Frost causes the separation of pieces of stone, but it does not produce powder as in crystallisation attack. Generally, limestones and magnesian limestones are more vulnerable to frost damage than sandstones. Marble, slate and granite used in building are normally unaffected by frost due to their low porosities.

CORROSION OF METALS

Rainwater run-off from copper and its alloys can cause green colour staining on limestones. Iron and steel produce rust staining which is difficult to remove from porous stones. Considerable damage is caused by the expansion of iron and steel in stonework caused by corrosion. All new and replacement fixings should normally be manufactured from stainless steel or non-ferrous metals.

FIRE

Fire rarely causes the complete destruction of stonework. In the case of granite, marble and most sandstones, the surfaces may be blackened or spall. Limestones are generally unaffected by fire, although the paler colours may turn permanently pink due to the oxidation of iron oxides within the stone. Reigate stone, a calcareous sandstone, is also resistant to heat, but it is not a durable stone for exterior use.

PLANTS

Generally, large plants including ivy should be removed from old stonework; however, Virginia creeper and similar species are not considered harmful. Lichens may contribute to deterioration of limestones and affected stonework should be treated.

Fig. 9.13 Lightweight veneer stone cladding panels. Photograph from IP 10/01 reproduced by courtesy of BRE

Damp north-facing walls and sloping sandstone surfaces are vulnerable to developing algae and lichen growth.

Maintenance of stonework

CLEANING

External granite, marble and slate claddings require regular washing with a mild detergent solution. In particular, highly polished external marble should be washed at least twice per year to prevent permanent dulling of the surface. Limestone, where it is not self-cleaned by rainwater, should be cleaned with a fine water spray and brushing, removing only deposit and not the gypsum-encrusted surface. However, the washing of limestone may cause a *ginger* staining or efflorescence as the stone dries out, risking the possible corrosion of embedded ferrous cramps, so water quantities should be adequately controlled. Sandstone is usually cleaned mechanically by abrasive blasting or chemical cleaning. Abrasive blasting with sand or grit is satisfactory for hard stones but can seriously damage soft stone and moulded surfaces. Hydrofluoric acid and sodium hydroxide (caustic soda) are used in the chemical cleaning of sandstones, but both are hazardous materials which need handling with extreme care by specialist contractors.

STONE PRESERVATION

Generally, coatings such as silicone water repellents should only be applied to stonework following expert advice and testing. Silicone treatment may in certain cases cause a build-up of salt deposits behind the treated layer, eventually causing failure. Silicone treatment should not be applied to already decayed stone surfaces. Polymeric silanes, such as *Brethane* (alkyl alkoxy-silane) can be used to consolidate decaying stone. The silane is absorbed up to 50 mm into the stone where it polymerises, stabilising the stone but without changing its external appearance. Generally such treatment is appropriate for small artefacts which are in immediate danger of loss if left untreated.

Cast stone

The appearance of natural stones such as Bath, Cotswold, Portland and York can be recreated using a mixture of stone dust and natural aggregates with cement. In certain cases, iron oxide pigments may also be added to match existing stonework as required. Many architectural components such as classical columns, capitals, balustrades and porticos are stock items (Fig. 9.14), but custom-made products may be cast to designers' specifications as illustrated in the facade of the Thames Water building at Reading (Fig. 9.15). High-quality finishes are achieved by the

Baluster

Ball finial

Pier cap

Regency urn

Rail

Pier shaft

Fluted urn

Pier base

Plinth

Fig. 9.14 Typical cast stone units

specialist manufacturers, and cast stone often surpasses natural stone in terms of strength and resistance to moisture penetration. Cast stone may be homogeneous, or for reasons of economy may have the facing material intimately bonded to a backing of concrete, in which case the facing material should be at least 20 mm thick. Untreated and galvanised steel reinforcement should have at least 40 mm cover on exposed faces and corrosion-resistant metals at least 10 mm cover. Most masonry units are designed to be installed with 5 or 6 mm joints and locating holes for dowel joints should be completely filled. Mortars containing lime are recommended rather than standard sand and cement (Table 9.5). Careful workmanship is required to prevent staining of the cast stone surfaces with mortar as it is difficult to remove. Cast stone should weather in a similar manner to the equivalent natural stone.

Table 9.5 Recommended grades of mortar for cast stonework

Exposure	Masonry cement : sand	Plasticised cement : sand	Cement : lime : sand
Severe	1 : 4½	1 : 6	1 : 1 : 6
Moderate	1 : 6	1 : 8	1 : 2 : 9

Fig. 9.15 Reconstructed stone cladding – Thames Water, Reading. Photograph: Courtesy of Trent Concrete Ltd

DRY AND WET CAST STONE

Cast stone is manufactured by either the dry cast or wet cast process. Dry cast stone is formed from zero slump concrete which is densely compacted by vibration. The process is used for the repetitive casting of smaller components, which can be removed from the mould immediately after compaction, allowing many units to be made each day. The wet cast stone system is used for the manufacture of larger units, which remain in the mould for 24 hours, and may incorporate anchor fixings and more complex reinforcement.

References

FURTHER READING

Ashurst, N. and Dimes, F.G. 1994: *Conservation of building and decorative stone*. Oxford: Architectural Press.

Chacon, M.A. 1999: *Architectural stone: fabrication, installation and selection*. New York: John Wiley.

Dernie, D. 2003: *New stone architecture*. London: Laurence King.

Garner, L. 2005: *Dry stone walls*. Princes Risborough: Shire Publications.

Jenkins, J. 2003: *The slate roof bible*. USA: Chelsea Green Publishing.

Kicklighter, C.E. 2003: *Modern masonry, brick, block, stone*. Illinois: Goodheart Willcox.

Mäckler, C. (ed.) 2004: *Material stone: Constructions and technologies for contemporary architecture*. Basel: Birkhäuser.

Pavan, V. 2004: *New stone architecture in Italy*. Basel: Birkhäuser.

Pavan, V. 2005: *New stone architecture in Germany*. Basel: Birkhäuser.

Shadmon, A. 1996: *Stone – An introduction*. London: Intermediate Technology Publications.

Smith, B.J. (ed.) 1996: *Processes of urban stone decay*. Shaftsbury: Donhead Publishing.

Smith. M.R. (ed.) 1999: *Stone in construction*. Engineering Geology Special Publication No.16. Bath: Geological Society.

Studio Marmo 1998: *Natural stone: A guide to selection*. New York: W.W. Norton.

Studio Marmo 2001: *Fine marble in architecture*. New York: W.W. Norton.

STANDARDS

BS 1217: 1997 Specification for cast stone.

BS 5080 Structural fixings in concrete masonry:
  Part 1: 1993 Method of test for tensile loading.
  Part 2: 1986 Method for determination of resistance to loading in shear.

BS 5385 Wall and floor tiling:
  Part 1: 1995 Design and installation of internal ceramic and natural stone wall tiling.
  Part 5: 1994 Code of practice for the design and installation of terrazzo tile and slab, natural stone and composition block flooring.

BS 5534: 2003 Code of practice for slating and tiling.

BS 5628 Code of practice for the structural use of masonry:
  Part 1: 1992 Structural use of unreinfirced masonry.
  Part 2: 2000 Structural use of reinforced and prestressed masonry.
  Part 3: 2001 Materials and components, design and workmanship.

BS 5642 Sills and copings:
  Part 1: 1978 Specification for window sills of precast concrete, cast stone, clayware, slate and natural stone.
  Part 2: 1983 Specification for copings precast concrete, cast stone, clayware, slate and natural stone.

BS 6093: 1993 Code of practice for design of joints and jointing in building construction.

BS 6100 Building and civil engineering terms:
  Part 5.2 1992 Masonry. Stone.

BS 6457: 1984 Specification for reconstituted stone masonry units.

BS 7533 Pavements constructed with clay, natural stone or concrete pavers:
  Part 1: 2001 Guide to the structural design of heavy duty pavements.
  Part 2: 2001 Guide to the structural design of lightly trafficked pavements.
  Part 3: 1997 Code of practice for laying precast concrete paving blocks.
  Part 4: 1998 Construction of pavements of precast concrete flags or natural stone slabs.
  Part 6: 1999 Code of practice for laying natural stone, precast concrete and clay kerb units.
  Part 7: 2002 Construction of pavements of natural stone setts and cobbles.
  Part 8: 2003 Structural design of lightly trafficked pavements of precast concrete and natural stone flags.

Part 10: 2004 Structural design of trafficked pavements constructed of natural stone setts.

BS 8000 Workmanship on building sites:

Part 6: 1990 Code of practice for slating and tiling of roofs and claddings.

Part 11 Code of practice for wall and floor tiling.

Sec. 11.1: 1989 Ceramic tiles, terrazzo tiles and mosaics.

Sec. 11.2: 1990 Natural stone tiles.

BS 8298: 1994 Code of practice for design and installation of natural stone cladding and lining.

BS EN 771 Specification for masonry units:

Part 5: 2003 Manufactured stone masonry units.

Part 6: 2005 Natural stone masonry units.

BS EN 772 Methods of test for masonry units:

Part 4: 1998 Determination of bulk density and porosity.

Part 11: 2000 Determination of water absorption.

Part 13: 2000 Determination of net and gross dry density.

Part 14: 2002 Determination of moisture movement.

Part 20: 2000 Determination of flatness of faces.

BS EN 1341: 2001 Slabs of natural stone for external paving.

BS EN 1342: 2001 Setts of natural stone for external paving.

BS EN 1343: 2001 Kerbs of natural stone for external paving.

BS EN 1469: 2004 Natural stone – finished products, claddings – specifications.

BS EN 1925: 1999 Natural stone test methods – water absorption coefficient.

pr EN 1926: 2006 Natural stone test methods – compressive strength.

BS EN 1936: 1999 Natural stone test methods – real and apparent density.

BS EN 5534: 2003 Code of practice for slating and tiling.

BS EN 12057: 2004 Natural stone products – modular tiles – requirements.

BS EN 12058: 2004 Natural stone products – slabs for floors and stairs – requirements.

BS EN 12326 Slate and stone products for discontinuous roofing and cladding:

Part 1: 2004 Product specification.

Part 2: 2000 Methods of test.

BS EN 12370: 1999 Natural stone test methods – resistance to salt crystallization.

BS EN 12371: 2001 Natural stone test methods – determination of frost resistance.

BS EN 12407: 2000 Natural stone test methods – petrographic examination.

BS EN 12440: 2001 Natural stone – denomination criteria.

BS EN 12670: 2002 Natural stone terminology.

BS EN 13161: 2001 Natural stone test methods – flexural strength.

BS EN 13364: 2002 Natural stone test methods – breaking load at dowel hole.

BS EN 13755: 2002 Natural stone test methods – water absorption.

CP 297: 1972 Precast concrete cladding (non-load-bearing).

BUILDING RESEARCH ESTABLISHMENT PUBLICATIONS

BRE Digests

BRE Digest 370: 1992 Control of lichens, moulds and similar growths.

BRE Digest 420: 1997 Selecting natural building stones.

BRE Digest 448: 2000 Cleaning buildings: legislation and good practice.

BRE Digest 449: 2000 Cleaning exterior masonry (Parts 1 and 2).

BRE Digest 467: 2002 Slate and tile roofs: avoiding damage from aircraft wake vortices.

Good building guide

BRE GBG 64 Part 3: 2005 Tiling and slating pitched roofs: Natural and manmade slates.

BRE Information papers

BRE IP 11/95 Control of biological growths on stone.

BRE IP 6/97 External cladding using thin stone.

BRE IP 7/98 External cladding – how to determine the thickness of natural stone panels.

BRE IP 17/98 Use of lightweight veneer stone claddings.

BRE IP 18/98 Stone cladding panels – *in situ* weathering.

BRE IP 9/99 Cleaning exterior masonry.

BRE IP 10/00 Flooring, paving and setts.

BRE IP 10/01 Lightweight veneer stone cladding panels.

BRE Reports

SO 36: 1989 The building limestones of the British Isles, E. Leary.

BR 62: 1985 The weathering of natural building stones, R.J. Schaffer.

BR 84: 1986 The building sandstones of the British Isles, E. Leary.

BR 134: 1988 The building magnesian limestones of the British Isles, D. Hart.

BR 141: 1989 Durability tests for building stone, K.D. Ross and R.N. Butlin.

BR 195: 1991 The building slates of the British Isles, D. Hart.

ADVISORY ORGANISATIONS

Men of the Stones, Beech Croft, Weston-under-Lizard, Shifnal, Shropshire TF11 8JT (01952 850269).

National Federation of Terrazzo, Marble & Mosaic Specialists, PO Box 2843, London W1A 5PG (0845 609 0050).

Stone Federation Great Britain, Channel Business Centre, Ingles Manor, Castle Hill Avenue, Folkestone, Kent CT20 2RD (01303 856123).

10

PLASTICS

Introduction

The plastics used in the construction industry are generally low-density non-load-bearing materials. Unlike metals, they are not subject to corrosion, but they may be degraded by the action of direct sunlight, with a corresponding reduction in mechanical strength. Many plastics are flammable unless treated; the majority emit noxious fumes in fires. Approximately 20% of plastics production within the UK is used by the building industry. PVC (polyvinyl chloride) which has a high embodied energy content accounts for 40% of this market share, predominantly in pipes, but also in cladding, electrical cable insulation, windows, doors and flooring applications. Foamed plastics for thermal and acoustic insulation are formulated either as open or closed-cell materials, the latter being resistant to the passage of air and water.

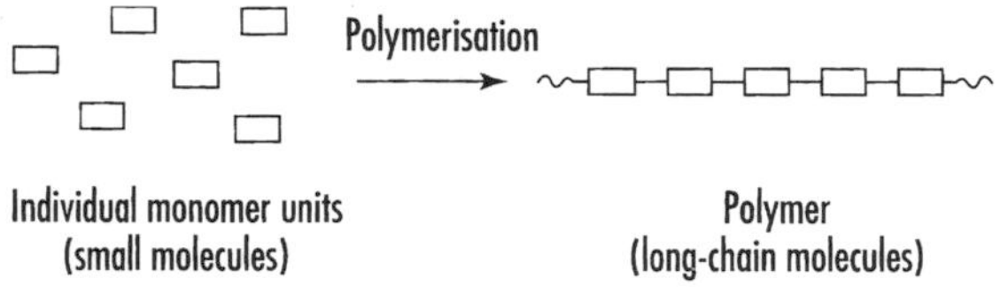

In terms of their chemical composition plastics form a diverse group of materials which have chain-like molecular structures composed of a large number of small repeat units. Whilst some materials such as rubber and cellulose derivatives are based on natural products, the majority of plastics are produced from petrochemical products. The manufacture of polythene, which dates back to 1933, involves the polymerisation of ethylene monomer, a colourless gas, which under high pressure at 200°C is converted into the clear polymer polyethylene or polythene (Fig.10.1).

Polymerisation

In the production of polythene the small molecular units of ethylene are joined end to end by an additional polymerisation process to produce the long-chain macromolecules. A similar process converts vinyl chloride into polyvinyl chloride (PVC) (Fig. 10.2), styrene monomer into polystyrene and tetrafluoroethylene into polytetrafluoroethylene (PTFE).

While the molecular backbones of plastics are predominantly composed of chains of carbon atoms, variations occur, particularly when the polymerisation

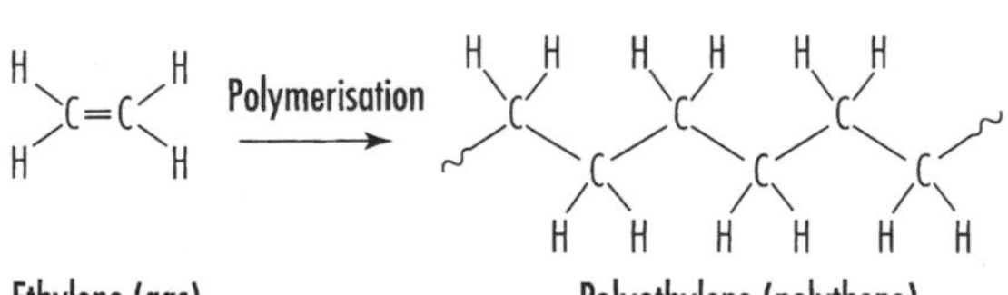

Individual monomer units   Polymer
(small molecules)   (long-chain molecules)

Ethylene (gas)   Polyethylene (polythene)

Fig. 10.1 Polymerisation of ethylene to polyethylene (polythene)

Vinyl chloride   Polyvinyl chloride
(gas)   (PVC)

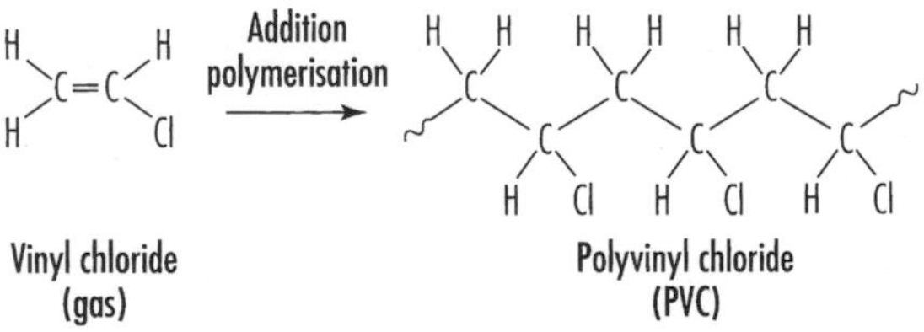

Fig. 10.2 Polymerisation of vinyl chloride to PVC

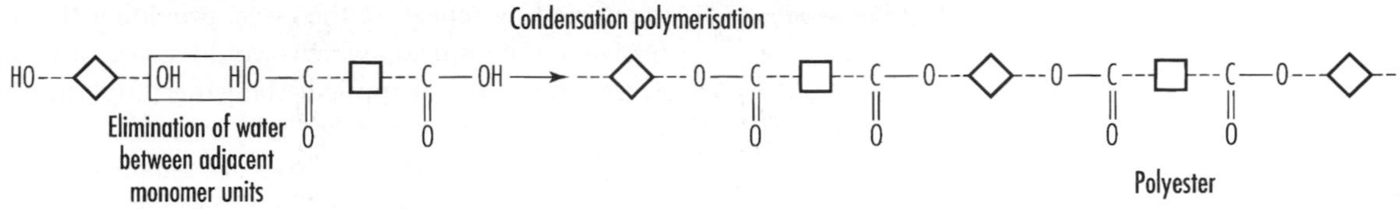

Condensation polymerisation

[figure: condensation polymerisation chemical structure diagram]

Elimination of water
between adjacent
monomer units

Polyester

Fig. 10.3 Condensation polymerisation

process involves the elimination of water between adjacent monomer units. Thus in the case of condensation polymerisation (Fig. 10.3), oxygen or nitrogen atoms are incorporated into the backbone of the macromolecular chains as in the polyesters (resins) and polyamides (nylons).

BRANCHED CHAINS

Depending upon the conditions during the polymerisation process, the polymer chains produced may be linear or branched. In the case of polythene, this affects the closeness of packing of the chains and therefore the bulk density of the material. Thus high-density polythene (HDPE) (s.g. 0.97), which is relatively stiff, has few branched chains compared to low density polythene (LDPE) (s.g. 0.92), which is softer and waxy (Fig. 10.4).

COPOLYMERS

Where two or more different monomers are polymerised together, the product will be a copolymer. The properties of the copolymer will be significantly dependent upon whether the two components have joined together in alternating, random or block sequences (Fig. 10.5).

More complex plastics can be produced for their specific physical properties by combining several components. Thus acrylonitrile butadiene styrene (ABS) is produced by the copolymerisation of the two precursor copolymers: styrene-acrylonitrile and butadiene-styrene rubber.

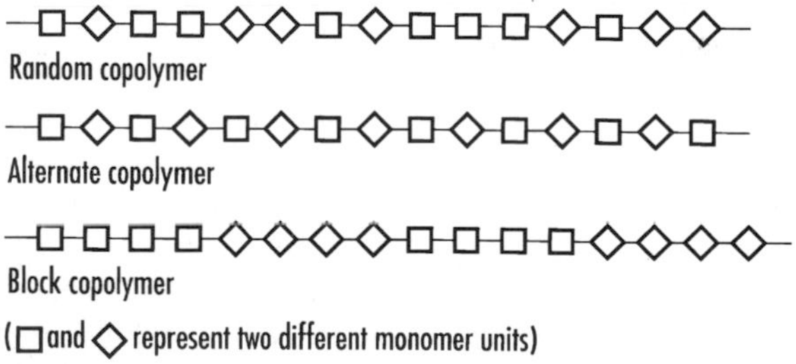

CRYSTALLINITY

In the initial manufactured state, most polymers consist of amorphous randomly-orientated molecular chains. However, if the plastic material is stretched in one direction, such as during the drawing of spun fibres, this causes an alignment of the molecular chains, leading to partial formation of crystalline regions and an associated anisotropy (Fig. 10.6).

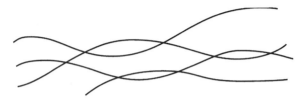

High-density polythene – straight chained

Low-density polythene – branched chained

Fig. 10.4 Straight- and branched-chain polymers

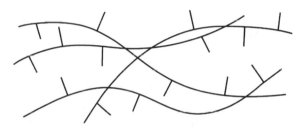

Random copolymer

Alternate copolymer

Block copolymer

(☐ and ◇ represent two different monomer units)

Fig. 10.5 Random, alternate and block copolymers

Crystalline regions may also be produced during the solidification of simple polymers such as polyethylene, but they will be limited in their extents due to the general entanglement of the molecular chains.

GLASS TRANSITION TEMPERATURE

In the molten state, the individual molecular chains of a plastic material move freely relative to each other, allowing the material to be moulded within the various

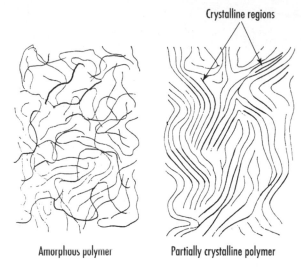

Fig. 10.6 Crystallinity in polymers

forming processes used for the manufacture of components. As the temperature of melted plastic material is lowered, the freedom of movement of the molecular chains is reduced; gradually the plastic becomes more viscous, until eventually it solidifies at its characteristic melting point temperature. However, even when solid, most plastics remain rubbery or flexible, due to rotations within the individual molecular chains. As the temperature is lowered further, the material will eventually become rigid and brittle, as movement can no longer take place within the individual molecular units. The temperature at which a particular plastic changes from flexible to rigid is defined as its characteristic *glass transition temperature*. Depending upon the nature of the particular plastic material this may be above or below normal ambient temperatures. Further, the glass transition temperature for a particular plastic can be significantly changed by, for example, the addition of plasticisers, characterised by the differences in physical properties between PVC-U (unplasticised) and PVC (plasticised polyvinyl chloride).

Polymer types

Polymers are normally categorised in respect of their physical properties as either thermoplastic, thermosetting or elastomeric.

THERMOPLASTICS

Thermoplastics soften upon heating, and reset on cooling. The process is reversible and the material is unaffected by repeating the cycle, providing that excessive temperatures, which would cause polymer degradation, are not applied. Many thermoplastics are soluble in organic solvents, whilst others swell by solvent absorption. Thermoplastics are usually produced initially in the form of small granules for subsequent fabrication into components.

THERMOSETTING PLASTICS

Thermosetting plastics have a three-dimensional cross-linked structure, formed by the linkage of adjacent macromolecular chains (Fig. 10.7). Thermosets are not softened by heating, and will only char and degrade if heated to high temperatures. Thermosets are usually produced from a partially polymerised powder or by mixing two components, such as a resin and a hardener. The resin is essentially the macromolecular component and the hardener cross-links the liquid resin into the thermoset plastic. Curing for epoxy resin adhesives and polyesters as in GRP (glass-fibre reinforced polyester) occurs at room temperature, while for phenolic and formaldehyde-based resins, a raised temperature and pressure are required. Thermosets, because of their three-dimensional structure, are usually solvent-resistant and harder than thermoplastics.

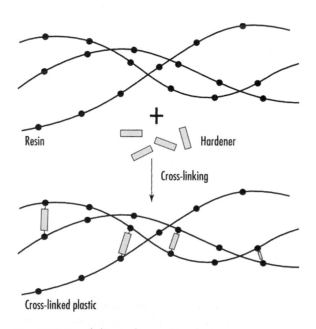

Fig. 10.7 Cross-linking in thermosetting plastics

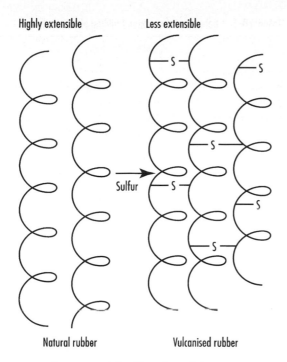

Fig. 10.8 Elastomers and the effect of cross-linking

ELASTOMERS

Elastomers are long-chain polymers in which the naturally helical or zig-zag molecular chains are free to straighten when the material is stretched, and recover when the load is removed. The degree of elasticity depends upon the extensibility of the polymeric chains. Thus natural rubber is highly extensible, but when sulfur is added, the vulcanisation process increasingly restricts movement by locking together adjacent polymer chains (Fig. 10.8). For most uses some cross-linking is required to ensure that an elastomeric material returns to its original form when the applied stress is removed.

Additives

PLASTICISERS

Plasticisers are frequently incorporated into plastics to increase their flexibility. The addition of the plasticiser separates the molecular chains, decreasing their mutual attraction. Thus unplasticised PVC (PVC-U) is suitable for the manufacture of rainwater goods, window units and glazing, whereas plasticised

PVC, is used for flexible single-layer roof membranes, tile and sheet floor coverings and electrical cable insulation. Loss of plasticiser by migration, can cause eventual embrittlement of plasticised PVC components.

FILLERS

Chalk, sand, china clay or carbon black are often added to plastics to reduce costs, improve fire resistance or opacity. Titanium dioxide is added to PVC-U to produce a good shiny surface. Glass fibres are added to polyester resins to give strength to the composite material, glass-fibre reinforced polyester (GRP), as described in Chapter 11.

PIGMENTS AND STABILISERS

Dyes and pigments may be added to the monomer or polymer. Stabilisers are added to absorb ultraviolet light which otherwise would cause degradation. For example, organotin compounds are used in clear PVC sheet to preferentially absorb incident ultraviolet light, in order to prevent degradation by the elimination of hydrogen chloride.

Degradation of plastics

The degradation of plastics is most frequently attributed to the breakdown of the long molecular chains (Fig. 10.9) or, in the case of PVC, the loss of plasticiser. Polymeric molecular chains may be broken by the effect of either heat, ultraviolet light or ozone, or by a combination of any of these factors, thus reducing their average molecular chain length. Discolouration occurs through the production of molecular units with double bonds, usually causing a yellowing of the plastic. Surface crazing and stress cracks may develop where degradation has caused cross-linking, resulting in embrittlement of the surface.

Where plasticiser is lost by migration from PVC, the glass transition temperature is gradually raised, so eventually the material becomes brittle at ambient temperatures. Typically, high-boiling point oils such as dibutyl phthalate and dioctyl phthalate are incorporated into the original PVC, but these gradually evaporate leaving the surface vulnerable to cracking and shrinkage.

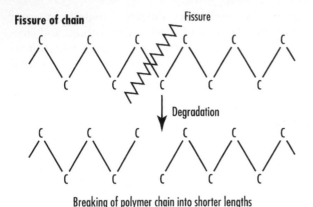

Fissure of chain   Fissure

Degradation

Breaking of polymer chain into shorter lengths

Loss of hydrogen chloride

Loss of hydrogen chloride

Production of double bonds – yellow degraded polymer

Fig. 10.9 Degradation of plastics

Table 10.1	Behaviour of common building plastics in fire
Material	Behaviour in fire
Thermoplastics	
Polythene / Polypropylene	Melts and burns readily
Polyvinyl chloride	Melts, does not burn easily, but emits smoke and hydrogen chloride
PTFE / ETFE	Does not burn, but at high temperatures evolves toxic fumes
Polymethyl methacrylate	Melts and burns rapidly, producing droplets of flaming material
Polystyrene	Melts and burns readily, producing dense black smoke and droplets of flaming material
ABS copolymer	Burns readily
Polyurethane	The foam burns readily producing highly toxic fumes including cyanides and isocyanates
Thermosetting plastics	
Phenol formaldehyde Melamine formaldehyde Urea formaldehyde	Resistant to ignition, but produce noxious fumes including ammonia
Glass-reinforced polyester (GRP)	Burns producing smoke, but flame-retarded grades are available
Elastomers	
Rubber	Burns readily producing black smoke and sulfur dioxide
Neoprene	Better fire resistance than natural rubber

Properties of plastics

FIRE

All plastics are combustible, producing noxious fumes and smoke (Table 10.1). Carbon monoxide is produced by most organic materials, but in addition, plastics containing nitrogen, such as polyurethane foam, generate hydrogen cyanide and PVC produces hydrochloric acid. Some plastics, particularly acrylics and expanded polystyrene have a high surface spread of flame and produce burning droplets; however, others when treated with fire retardant are difficult to ignite and some are self-extinguishing.

STRENGTH

Although plastics have a good tensile strength to weight ratio, they also have a low modulus of elasticity which renders them unsuitable for most load-bearing situations; the only exception being glass-fibre reinforced polyester (GRP) which has been used for some limited load-bearing applications. Generally, thermoplastics soften at moderate temperatures and are subject to creep under ambient conditions.

THERMAL AND MOISTURE MOVEMENT

The thermal expansion of most plastics is high. The expansion of GRP is similar to that of aluminium, but most other plastics have larger coefficients of linear expansion. For this reason, attention must be paid to careful detailing to allow for adequate thermal movement, particularly where weather exclusion is involved. Most plastics are resistant to water absorption, and therefore do not exhibit moisture movement. (Typical coefficients of linear expansion are polythene (HD) 110–130, polypropylene 110, ABS 83–95, PVC 40–80, GRP 20–35 $\times$ 10^{-6} deg C^{-1}.)

Plastics forming processes

Depending upon the nature of the product, plastics may be formed by either continuous or batch processes. With thermoplastics, frequently a two-stage process is most appropriate in which the raw materials, supplied by the primary manufacturer as powder or granules, are formed into an extrusion or sheet which is then reformed into the finished product. However, thermosetting plastics must be produced either from a partially polymerised material or directly from the resin and hardener mix in a single-stage process. Foamed plastics are either blown with internally generated gas, or produced by a vacuum process which reduces reliance on environmentally damaging CFCs and HCFCs.

CONTINUOUS PROCESSES

Extrusion

Plastic granules are fed continuously into the heated barrel of a screw extruder, which forces the molten thermoplastic through an appropriately shaped die to produce rod, tube or the required section (Fig. 10.10). Products include pipes, rainwater goods and fibres.

Film blowing

As a molten thermoplastic tube is produced in the extrusion process, air is blown in to form a continuous cylindrical plastic sheet, which is then rolled flat and trimmed to produce a folded sheet. Adjustment of the applied air pressure controls the sheet thickness.

Calendering

Sheet thermoplastic materials may be produced from plastics granules by compression and fusion between a series of heated rollers. Laminates may be produced by heating together two or more thermoplastic sheets, and during this process, sheet reinforcement material may be incorporated.

BATCH PROCESSES

Injection moulding

Thermoplastic granules are melted in a screw extruder to fill a ram which injects the plastic into an appropriate mould. After cooling, the components are removed from the mould and trimmed as necessary. The process is low cost and rapid. A series of moulds can be attached to the injection moulding machine to ensure continuity of production (Fig. 10.11). Thermosetting polymers can be injection moulded by initial forming at a low temperature followed by heating of the mould to cross-link the liquid plastic.

Compression moulding

In the compression moulding process for thermosetting resins, the appropriate quantity of uncross-linked resin powder is subjected to pressure and heat within the mould. When the polymer has melted and cross-linked, the mould can be opened and the component removed.

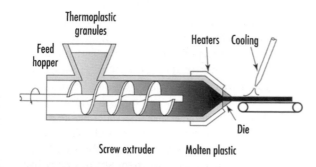

Fig. 10.10 Formation of plastics by extrusion

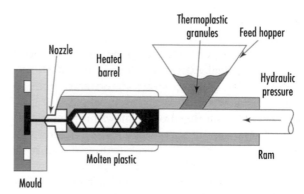

Fig. 10.11 Formation of plastics by injection moulding

Pressing

Pressing is used to form thermoplastic sheet into components. The sheet plastic is initially heated to softening point and then pressed between an appropriately shaped pair of dies.

Vacuum forming and blow moulding

During vacuum forming, thermoplastic sheet is heated over a mould which is then evacuated through a series of fine holes, drawing the soft plastic into the appropriate form. In the similar process of blow moulding, positive air pressure is applied inside a molten polymer tube which is expanded into the shape of the mould.

RAPID PROTOTYPING

New techniques in computer-aided manufacturing enable prototype components to be manufactured to very close tolerances from three-dimensional computer-aided design (CAD) solid modelling images. This has implications not only for the design of building components but also for the manufacture of architectural models.

The systems are based on the successive build-up of very thin layers of solid material to the exact pattern of layered CAD sections. Various lay-up systems have been developed for the deposition of plastic layers. These range from a fine nozzle, to using laser technology to accurately polymerise viscous resin in very thin layers and the use of adhesive-backed paper cut by laser to the required section shapes. Each system produces a highly accurate three-dimensional solid over a period of several hours depending upon the product size. Where any part of the build-up of the solid object needs support during manufacture, the systems automatically produce additional material in a weak form. This can be broken away easily after the whole object is complete, and in the case of laser/resin production finally cured. In all these manufacturing processes the build-up layers are extremely thin, so smooth and accurate surfaces are achieved.

The reverse of this process allows prototype complex shape components or small-scale architectural models to be turned into accurate three-dimensional CAD files, using a delicate probe mechanism which senses all over the object's surfaces. This allows the designer to generate CAD files for highly complex three-dimensional forms which would be virtually impossible to draw directly into a CAD system.

Plastics in construction

The broad range of thermoplastic, thermosetting and elastomeric plastics are collated into families in Figure 10.12. Typical uses in construction are listed in Table 10.2. (Glass-fibre reinforced polyester is described in Chapter 11; foamed plastics as insulation materials in Chapter 13; and plastics used primarily as adhesives in Chapter 14.)

THERMOPLASTICS

Polythene (polyethylene)

Polythene (PE) is one of the cheapest plastics and is available both in the low density (LD) (softening point 90°C) and high density (HD) (softening point 125°C) forms. Polythene is resistant to chemicals, tough at low temperatures, but is rapidly embrittled by ultraviolet light unless carbon black is incorporated. Polythene burns and has a relatively high coefficient of thermal expansion. Low-density polythene is used widely for damp-proof membranes, damp-proof courses and vapour barriers. High-density polythene, which is stiffer than the low-density material, is used for tanking membranes to basements. Polythene is used for the production of cold-water cisterns, but is only suitable for cold-water plumbing applications due to its high thermal expansion; for mains water pressure it requires a significant wall thickness due to its relatively low tensile strength. Cross-linked polyethylene (PEX), manufactured by the action of peroxide catalyst on normal polyethylene, is used for domestic hot water and underfloor heating systems as it can withstand operating temperatures up to 90°C.

Polypropylene

Polypropylene (PP), with a softening point of 150°C, is slightly stiffer than polythene, to which it is closely related chemically. Like polythene, it is resistant to chemicals and susceptible to ultraviolet light, but unlike polythene becomes brittle below 0°C. However, the block copolymer with ethylene does have improved low-temperature impact resistance. Polypropylene is used for pipes, drainage systems, water tanks, DPCs, connecting sleeves for clay pipes and WC cisterns. Polypropylene fibres are used in fibre-reinforced cement to produce an increase in impact resistance over the equivalent unreinforced material. Certain breather membranes used for tile underlay and timber-frame

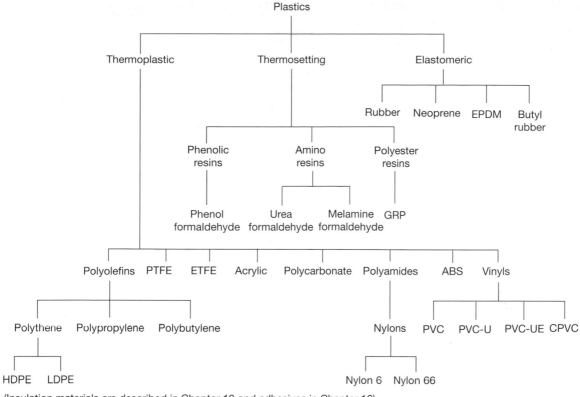

(Insulation materials are described in Chapter 13 and adhesives in Chapter 16)

Fig. 10.12 Plastics used in construction

construction are manufactured from multi-layer systems incorporating polypropylene with polyethylene and glass-fibre reinforcement. Such products are wind and watertight, but vapour-permeable. Many geotextiles for soil stabilisation are manufactured as a mat material from non-woven heat-bonded polypropylene continuous fibres. The material may be reinforced by woven polyester fibres.

Polybutylene

Polybutylene is used for pipework as an alternative to copper. It has the advantage of flexibility and the very smooth internal surface is resistant to the build-up of scale and deposits. It can withstand continuous operating temperatures up to 82°C.

Polyvinyl chloride

Polyvinyl chloride (PVC) is the most widely used plastic material in the construction industry. It is available both in the unplasticised form (PVC-U) and as the plasticised product (PVC). In both forms polyvinyl chloride is combustible giving off noxious hydrogen chloride fumes; however, the unplasticised form tends to burn only with difficulty. PVC begins to soften at 75°C, and therefore cannot be used for hot water systems, although chlorinated PVC (CPVC) can be used at higher temperatures. PVC is soluble in certain organic solvents which, therefore, can be used for the solvent welding of joints, but PVC is unaffected by acids and alkalis.

Plasticised PVC is extensively used in the manufacture of floor coverings, either as individual tile units or as continuously jointed sheet. It is also the most widely used material for single-layer roofing systems due to its durability, colour range and ease of application. It also offers an alternative to bitumen felt for sarking. Plasticised PVC is the standard for electrical cable insulation, and many small building components are made from injection moulded PVC.

PVC-U

PVC-U is widely used for rainwater goods, usually in white, grey, black or brown ,and similarly for

Table 10.2   Typical uses of plastics in construction

Material		Examples of plastics in construction
Thermoplastics		
Polythene	(Low density)	DPC, DPM, vapour checks, roof sarking
	(High density)	Cold-water tanks, cold-water plumbing
Polypropylene		Pipework and fittings, drainage systems, water tanks, WC cisterns, DPCs, fibres in fibre-reinforced concrete
Polybutylene		Hot and cold-water pipework and fittings
Polyvinyl chloride	(PVC-U)	Rainwater goods, drainage systems, pipes and fittings, underground services, window and door frames, conservatories, garage doors, translucent roofing sheets
	(PVC-UE)	Claddings, barge boards, soffits, fascias, window boards
	(PVC)	Tile and sheet floor coverings, single-ply roofing, cable insulation, electrical trunking systems, sarking, tensile membrane structures, glazing to flexible doors, door seals, handrail coatings, vinyl-film finishes to timber products
	(CPVC)	Hot-water systems, window and door frames
ETFE		Inflated systems for translucent wall and roof membranes
PTFE		Sealing tape for plumbing, tensile membrane structures, low-friction movement joints
Polymethyl methacrylate		Baths, shower trays, kitchen sinks, glazing, roof lights, luminaires
Polycarbonate		Vandal-resistant glazing, spa baths, kitchen sinks
Polystyrene		Bath and shower panels, decorative expanded polystyrene tiles
ABS copolymer		Pipes and fittings, rainwater goods, drainage systems, shower trays
Nylons		Electrical conduit and trunking, low-friction components — hinges, brush strips for sealing doors and windows, carpet tiles and carpets shower curtains
Thermosetting plastics		
Phenol formaldehyde		Decorative laminates
Melamine formaldehyde		Laminates for working surfaces and doors, moulded electrical components, WC seats
Urea formaldehyde		Decorative laminates
Glass-reinforced polyester		Cladding and roofing panels, simulated cast-iron rainwater goods,
(GRP)		cold-water tanks, spa baths, garage doors, decorative tiles and panels
Elastomers		
Rubber		Flooring, door seals, anti-vibration bearings
Neoprene		Glazing seals, gaskets
EPDM		Glazing seals, gaskets, single-ply roofing systems
Butyl rubber		Sheet liners to water features and land-fill sites
Nitrile rubber		Tile and sheet flooring

soil and waste pipes. It is also used colour-coded for underground water, gas, electrical and telecommunications systems. PVC-U is used extensively for the manufacture of extruded window frames, door frames and conservatories, usually incorporating sealed double-glazing units. Where insufficient rigidity is achieved with the PVC-U alone, steel inserts within the extruded sections give strength and provide additional protection against forced entry. PVC-U is used in the manufacture of translucent, transparent and coloured profiled sheeting for

domestic structures such as carports and conservatories, where an economical product is required, although eventually the products discolour and craze due the effects of direct ultraviolet light.

PVC-UE

Extruded cellular unplasticised PVC (PVC-UE) is used for cladding, fascias, soffits, window boards, barge boards and other components of uniform section. It is manufactured by the co-extrusion of a high-impact PVC-U surface material over a core of closed cell PVC foam. The foaming agent is usually sodium bicarbonate. The high stiffness to weight ratio arises from the combination of a dense outer skin and a cellular core. Both the cellular core and the wearing surface are stabilised with metallic additives to prevent degradation and discolouration. Contact with bitumen should be avoided. The material will char and melt in fire, but with a limited surface spread of flame. The material is described within the standard BS 7619: 1993.

Tensile membrane structures

PVC-coated polyester is the standard material used for tensile membrane structures and canopies. The durability depends directly upon the degree of translucency; at 15% transmission, 15 years can be reasonably expected. At greater levels of translucency, the expected serviceable lifetime is considerably reduced; however, fluropolymer lacquers to the top surface of the fabric enhance durability. While white fabric is the standard, coloured and patterned membranes are available to client requirements. PVC-coated polyester membranes are a cheaper alternative to PTFE-coated fabrics, but are not non-combustible. PVC-coated polyester is more flexible than PTFE-coated fibreglass and is therefore the preferred material for temporary structures which may be folded for transport and storage. The thermal insulation afforded by single-layer tensile membrane roofs is negligible.

Tensile membrane structures are manufactured from a set of tailored panels stitched or welded together. They are usually tensioned by wires or rods running through edge pockets, or by fixing directly to structural elements. Accurate tensioning is required to generate the correct form and resistance to wind and snow loads. The use of double curvature within the panel elements imparts structural rigidity to the overall membrane structure. Damage by accident or vandalism can usually be repaired on site.

Polytetrafluoroethylene

Polytetrafluoroethylene (PTFE)-coated glass-fibre woven fabrics are used for permanent tensile membrane structures. In a fire, PTFE gives off toxic combustion products, but only at temperatures above which any fabric would have already failed and vented the heat and smoke. With a fire rating of Class 0, PTFE-coated glass-fibre tensile membranes are more expensive than the Class 1 rated PVC-coated polyesters, but

are generally more durable with an anticipated lifespan in excess of 20–25 years. The low friction PTFE surface has good self-cleaning properties.

The Inland Revenue Amenity Building in Nottingham (Fig. 10.13) and the Millennium Dome at Greenwich are roofed with PTFE (*Teflon*) coated glass-fibre tensile membranes. The translucent fabric gives well-lit internal spaces during the daytime, and striking glowing surfaces at night. In the Nottingham building, the membrane roof is suspended from four steel columns, and is linked to the fixed structure below by inflatable elements which absorb any movement.

PTFE tape has a very low coefficient of friction and a high melting point. It is therefore ideal for use as a sealing tape for threaded joints in water and gas pipes. It is also used to form sliding joints in large structures.

Ethylene tetrafluoro ethylene

Ethylene tetrafluoro ethylene copolymer (ETFE) is used as a translucent foil for low-pressure pneumatic metal-framed building envelope cushions. The fluoro-copolymer has the advantages over glass that when used to form two- to five-layer air cushion systems, it offers higher thermal insulation with greater transparency to UV light. ETFE is strong, shatter proof, half the cost, and only one-hundredth the weight of the equivalent glass, thus offering significant economies to the required structural supporting system. ETFE with an anticipated life span of 25 years, can withstand maintenance loads, be easily repaired and is recyclable. It has been used very effectively on the galvanised-tubular steel space-frame envelope for the biomes (domes sheltering plants from around the world) at the Eden Centre, Cornwall (Fig. 10.14). The structure is formed from an icosahedral geodesic outer layer, with a combination of hexagons, pentagons and triangles as the inner layer of the three-dimensional space frame. Only a small pumping system, powered by photovoltaic cells, is required to maintain the air-fill of the ETFE cushions. At the National Space Centre at Leicester, also designed by Nicholas Grimshaw and Partners, the ETFE cushion clad tower houses the main space rocket exhibits (Fig. 10.15).

If automatic smoke venting of an ETFE atrium is required, electrical wiring can be incorporated into the cushion frames, which release the cushions, except at one point, in case of fire. An enclosed atrium is thus turned into a fully open lightwell.

Transmitted light levels through ETFE cushions can be constantly adjusted by the use of partially

Fig. 10.13 Tensile Membrane — Inland Revenue Amenity Building, Nottingham. Architects: Hopkins Architects. Photographs: Courtesy of Martine Hamilton Knight

Fig. 10.14 ETFE – Eden Project, Cornwall. Architects: Nicholas Grimshaw and Partners. Photographs: Arthur Lyons and courtesy of Perry Hooper (interior)

printed internal layers within the cushions, which can be moved closer or further apart by changing the pumped air pressure, thus modifying the shadowing effect. Interesting patterns can be created by the use of coloured cushions, whilst aluminium-coated foils will give a highly reflective effect with reduced sunlight penetration.

Polymethyl methacrylate

Acrylic or polymethyl methacrylate (PMMA) is available in a wide variety of translucent or transparent, clear or brightly coloured sheets. It softens at 90°C, and burns rapidly with falling droplets of burning material. Stress crazing may occur where acrylic has been shaped in manufacture and not fully annealed, but generally the material is resistant to degradation by ultraviolet light. Acrylic is frequently used for decorative signs, roof lights and light fittings. Baths and shower trays are manufactured from acrylic as a lighter alternative to cast iron and ceramics. Although not resistant to abrasion, scratches can usually be polished out with proprietary metal polish.

Fig. 10.15 ETFE – National Space Centre, Leicester. Architects: Nicholas Grimshaw and Partners. Photograph: Arthur Lyons

Polycarbonate

Polycarbonates (PC) are used as vandal-resistant glazing, due to their high-impact resistance, good optical transparency and low ignitability. Polycarbonate blocks offer a lightweight alternative to traditional cast glass blocks. Proprietary extruded cellular systems of double or triple walled polycarbonate offer combined thermal insulation and vandal-resistant properties. The protective outer surface prevents ultraviolet degradation for ten years, and sections may be curved on site within the limits of the manufacturers' specifications.

Acrylonitrile butadiene styrene

Acrylonitrile butadiene styrene (ABS) plastics are a range of complex terpolymers manufactured by combining together the two copolymers, styrene-acrylonitrile and butadiene-styrene. ABS plastics are relatively expensive but tough and retain their strength at low temperatures. They are used to manufacture moulded components, rainwater and drainage goods. A special ABS solvent cement is required for solvent welding.

Nylons

Nylons, usually nylon 66 or nylon 6, are used for the manufacture of small components where low friction is required. Nylons are tough and strong but tend to be embrittled and become powdery on prolonged exposure to sunlight. Carpet tiles in nylon 66 are durable and hard wearing.

Kevlar

Kevlar (polyparabenzamide) fibres are produced by extrusion of a cold solution of the polymer into a cylinder at 200°C, which causes the solvent to evaporate. The resulting fibres are stretched by a drawing process, which aligns the polymer molecules along the fibres to produce a very high modulus material used in ropes and composite plastics.

THERMOSETTING PLASTICS

Phenol formaldehyde

Phenol formaldehyde (PF) was the original, and remains the cheapest thermosetting resin. Currently, its main use is in the production of laminates by the hot pressing of layers of resin-impregnated paper, fabric or glass fibre. The cured resin is brown, but heat-resistant laminates for working surfaces and wallboards are laminated with a decorative printed paper film and coated with a clear melamine formaldehyde finish. Phenol formaldehyde is resistant to ignition, but produces a phenolic smell on burning.

Urea formaldehyde

Urea formaldehyde (UF) is similar to phenol formaldehyde except that because it is clear it can be produced to a range of colours including white. It is used in the manufacture of electrical components and other moulded components such as WC seats. Urea formaldehyde is resistant to ignition, but produces a fishy smell on burning. Urea-formaldehyde foam is no longer used for cavity wall insulation.

Melamine formaldehyde

Melamine formaldehyde (MF) is available clear and in a wide range of colours. When heat cured, it is hard wearing, durable and resistant to heat, and is therefore used as the surface laminate over the cheaper brown phenol formaldehyde layers in the production of working surface and wallboard laminates. Melamine formaldehyde is resistant to ignition, but produces a fishy smell on burning.

ELASTOMERS

Natural rubber

Natural rubber is harvested from the species *Heavea braziliensis*, in Africa, South America and Malaysia. The white latex is predominantly *cis*-polyisoprene, a macromolecule containing some double bonds within the carbon chain. It is these double bonds which permit cross-linking with sulfur when natural rubber is heated under pressure in the vulcanisation process. Natural rubber is usually reinforced with carbon and treated with antioxidants to prevent degradation. It is used for flooring and in antivibration bearings for buildings and large structures.

Neoprene

Unlike natural rubber, Neoprene (polychloroprene) is resistant to chemical attack, and is therefore used for glazing seals and gasket systems. It is available only in black.

EPDM

Unlike neoprene, EPDM (ethylene propylene diene monomer) can be obtained in any colour, and is it characterised by high elongation and good weathering resistance to ultraviolet light and ozone. It is therefore taking over from neoprene as the key material for gaskets and is extensively used in single-ply roofing systems.

Butyl rubber

A copolymer of isobutylene and isoprene, this material has good chemical and weathering resistance. It is used as liners to landfill sites and decorative water features.

COMPOSITE PLASTICS

Composite plastic materials such as glass-fibre reinforced polyester (Chapter 11) have physical properties which differ significantly from the individual component materials. An increasing variety of composite plastics are reaching the construction industry, driven by the demand for product diversity and in some cases recycling.

Wood plastic composites

Wood plastic composites (WPC) encompass a range of materials incorporating polymers such as polyethylene (PE), polypropylene (PP) and polyvinyl chloride (PVC) blended with wood waste from saw mills. Wood chips and saw dust are dried to 2% or 3% moisture and ground down through a hammer mill to wood fibres (<5 mm). Wood flour may be incorporated as a filler giving bulk to the product.

To produce the wood plastic composite, the prepared wood-fibre material and any filler is mixed into the molten polymer, either in a batch or continuous process. The components are then formed by injection moulding for items such as architectural mouldings. Extrusion or pultrusion, a combination of extrusion and pulling, are both used for continuous sections such as window profiles and decking.

Wood plastic composites for outdoor products such as decking, fencing and garden furniture can be manufactured using a proportion of recycled polymer and scrap wood, which potentially will reduce the quantities of these materials in the waste stream.

Wood plastic composites are resistant to rot and insect attack and incorporate pigments to prevent UV fading. Fire resistance is similar to wood of the same density, but can be improved by the incorporation of flame and smoke retardants during the manufacturing process.

Corian

Corian is a composite of natural minerals, pigments and acrylic polymer, which combine together to produce a highly durable and tough material, available in a wide range of colours. The proprietary product is frequently used for kitchen and other countertops as it can be moulded into complex forms and inconspicuously joined into single units with the base plastic composite.

LIGHT-CONDUCTING PLASTICS

A range of light and colour-sensitive materials has been developed by embedding a matrix of light-conducting plastic channels into a substrate of either concrete or acrylic polymer. Each of the light-conducting channels, which operate like fibre-optics, gathers light or shadow from one end and transfers it to the other end creating scintillation or darkening respectively. Overall this creates an optical rippling effect as an object or light passes over the surface. The material, according to the substrate, can be used as floor tiles, walls, partitions, facades or tabletop surfaces. In each case the surface is seen to respond to object movement or changes in light intensity and colour. Units may be individual tiles or larger panels and are available in a range of standard or custom colours.

VARIABLE-COLOUR PLASTIC FILMS

Dichroic plastic films cause the observed colour and opacity to change depending on the viewpoint, also on the direction and intensity of the light source. For example, one dichroic sheet changes between green, gold and orange; another between purple and blue. These films can be used to make the external envelope or internal environment of buildings appear to be active.

Thermochromic pigments in fibre-reinforced plastic sheet cladding change colour with the temperature allowing the building to visually respond to outside temperature. The thermochromic pigment is incorporated into the core fibres and the gel coat to gain maximum effect.

Recycling of plastics

The use of plastics within European countries is approximately 44 million tonnes per year, with products for the construction industry accounting for over a half of the consumption of PVC. Currently most waste disposal is within landfill sites. However, certain thermoplastic products can be recycled into construction products. Expanded polystyrene waste can be recycled by solvent extraction into a material, which has the appearance and many characteristics of wood. PVC bottles can be recycled into plastic pipes by co-extruding new PVC as the inner and outer skins over a recycled PVC core. However, many recycled plastics have a reduced resistance to degradation as stabilisers are lost in reprocessing, and the products would therefore fail to reach the technical standards, which are normally related to the quality achieved by new materials rather than to fitness for purpose.

References

FURTHER READING

APME: 1995. *Plastics: A material of choice in building and construction.* Brussels: Association of Plastics Manufacturers in Europe.

Clough, R. and Martyn, R. 1995: *Environmental impact of building and construction materials: Plastics and elastomers.* London: Construction Industry Research and Information Association.

Cousins, K. 2002. *Polymers in building and construction.* Shrewsbury: RAPRA Technology Ltd.

Johansson, C.M.A. 1991: *Plastics in building.* RAPRA Technology Ltd., Review Report No. 48, **4**(12), Shrewsbury: Rapra Technology Ltd.

Hollaway, L. 1993: *Polymer and polymer composites for civil and structural engineering.* London: Blackie Academic and Professional.

Kaltenbach, F. 2004: *Translucent materials: Glass, plastics, metals.* Basle: Birkhäuser.

Koch, K-M. (ed.) 2004: *Membrane structures.* Munich: Prestel.

Scheuermann, R. and Boxer, K. 1996: *Tensile structures in the urban context.* Oxford: Butterworth-Heinemann.

STANDARDS

BS 476 Fire tests on building materials:
  Part 4: 1970 Non-combustibility test for materials.
  Part 6: 1989 Method of test for fire propagation for products.
  Part 7: 1997 Classification of the surface spread of flame of products.
BS 743: 1970 Materials for damp-proof courses.
BS 1254: 1981 Specification for WC seats (plastics).
BS 2572: 1990 Specification for phenolic laminated sheet and epoxide cotton fabric laminated sheet.
BS 3012: 1970 Low and intermediate density polythene sheet for general purposes.
BS 3284: 1967 Polythene pipes (Type 50) for cold water services.
BS 3505: 1986 Specification for unplasticised polyvinyl chloride (PVC-U) pressure pipes for cold potable water.
BS 3757: 1978 Specification for rigid PVC sheet.
BS 3837: 2004 Expanded polystyrene boards – boards and blocks manufactured from expandable beads.
BS 3953: 1990 Synthetic resin-bonded woven glass fabric laminated sheet.

BS 4023: 1975 Flexible cellular PVC sheeting.

BS 4154 Corrugated plastic translucent sheets made from thermosetting polyester resins (glass fibre reinforced):

Part 1: 1985 Specification for material and performance requirements.

Part 2: 1985 Specification for profiles and dimensions.

BS 4203 Extruded rigid PVC corrugated sheeting:

Part 1: 1980 Specification for performance requirements.

Part 2: 1980 Specification for profiles and dimensions.

BS 4213: 2004 Cisterns for domestic use.

BS 4305 Baths for domestic purposes made from acrylic material:

Part 1: 1989 Specification for finished baths.

BS 4346 Joints and fittings for use with unplasticised PVC pressure pipes:

Part 1: 1969 Injection moulded unplasticised PVC fittings for solvent welding.

Part 2: 1970 Mechanical joint and fittings, principally of unplasticised PVC.

Part 3: 1982 Specification for solvent cement.

BS 4514: 2001 Unplasticised PVC soil and ventilating pipes, fittings and accessories.

BS 4576 Unplasticised polyvinyl chloride (PVC-U) rainwater goods and accessories:

Part 1: 1989 Half-round gutter and pipes of circular cross-section.

BS 4607 Non-metallic conduit fittings for electrical installations:

Parts 1, 3 & 5

BS 4660: 2000 Thermoplastics ancilliary fittings for below ground drainage.

BS 4840 Rigid polyurethane (PUR) foam in slab form.

Parts 1: 1985 & 2: 1994

BS 4841 Rigid urethane foam for building applications:

Part 1: 1993 Laminated board for general purposes.

Part 2: 1975 Laminated board for use as a wall and ceiling insulation.

Part 3: 1994 Specification for two types of laminated board (roofboards).

BS 4901: 1976 Plastics colours for building purposes.

BS 4962: 1989 Specification for plastic pipes and fittings for use as subsoil field drains.

BS 4965: 1999 Specification for decorative laminated plastics sheet veneered boards and panels.

BS 4991: 1974 Specification for polypropylene copolymer pressure pipe.

BS 5241 Rigid polyurethane (PUR) and polyisocyanurate (PIR) foam when dispensed or sprayed on a construction site:

Part 1: 1994 Specification for sprayed foam thermal insulation applied externally.

Part 2: 1991 Specification for dispensed foam for thermal insulation or buoyancy applications.

BS 5254: 1976 Polypropylene waste pipe and fittings.

BS 5255: 1989 Thermoplastics waste pipe and fittings.

BS 5391 Specification for acrylonitrile-butadiene-styrene (ABS) pressure pipe:

Part 1: 1976 Pipe for industrial uses.

BS 5480: 1990 Glass reinforced plastics (GRP) pipes, joints and fittings for use for water supply or sewerage.

BS 5481: 1977 Specification for unplasticised PVC pipes and fittings for gravity sewers.

BS 5608: 1993 Specification for preformed rigid urethane (PUR) and polyisocyanurate (PIR) foams for thermal insulation of pipework and equipment.

BS 5617: 1985 Specification for urea-formaldehyde (UF) foam systems suitable for thermal insulation of cavity walls with masonry or concrete inner and outer leaves.

BS 5618: 1985 Code of practice for thermal insulation of cavity walls by filling with urea-formaldehyde (UF) foam systems.

BS 5955 Plastics pipework (thermoplastics materials):

Part 6: 1980 Installation of unplasticised PVC pipework for gravity drains and sewers.

Part 8: 2001 Specification for the installation of thermoplastics pipes and associated fittings for use in domestic hot and cold water services and heating systems.

BS 6203: 2003 Guide to the fire characteristics and fire performance of expanded polystyrene materials (EPS and XPS) used in building applications.

BS 6206: 1981 Specification for impact performance requirements for flat safety glass and safety plastics for use in building.

BS 6437: 1984 Specification for polyethylene pipes in metric diameters for general purposes.

BS 6515: 1984 Specification for polyethylene damp-proof courses for masonry.

BS 6572: 1985 Specification for blue polyethylene pipes up to nominal size 63 for below ground use for potable water.

BS 6730: 1986 Specification for black polythene pipes up to nominal size 63 for above ground use for cold potable water.

BS 7412: 2002 Plastics windows made from PVC-U extruded hollow profiles.

BS 7414: 1991 White PVC-U extruded hollow profiles with heat welded corner joints for plastics windows: materials type B.

BS 7619: 1993 Specification for extruded cellular un-plasticised PVC (PVC-ME) profiles.

BS 7722: 2002 Surface covered PVC-U profiles for windows and doors.

BS 8203: 2001 Code of practice for installation of re-silient floor coverings.

BS 8204 Screeds, bases and *in situ* floorings:
    Part 6: 2001 Synthetic resin flooring.

BS EN 438 Decorative high-pressure laminates – sheets based on thermosetting resins:
    Part 1: 1991 Specifications.
    Part 2: 1991 Determination of properties.

BS EN 607: 1996 Eaves gutters and fittings made of PVC-U: Definitions, requirements and testing.

BS EN 1115–1: 2001 Plastic piping systems for underground drainage and sewerage – GRP.

BS EN 1013 Light transmitting profiled plastic sheet-ing for single skin roofing:
    Part 1: 1998 General.
    Part 2: 1999 Glass fibre reinforced polyester (GRP).
    Part 3: 1998 Polyvinyl chloride (PVC).
    Part 4: 2000 Polycarbonate (PC).
    Part 5: 2000 Polymethyl methacrylate (PMMA).

BS EN 1329 Plastics piping systems for soil and waste discharge – unplasticised PVC-U:
    Part 1: 2000 Specification for pipes, fittings and the system.

pr EN 1401: 2006 Plastics piping systems for non-pressure underground drainage and sewerage – unplasticised polyvinyl chloride (PVC-U).

BS EN 1451 Plastics piping systems for soil and waste discharge – polypropylene (PP):
    Part 1: 2000 Specification for pipes, fittings and the system.

BS EN 1452: 2000 Plastics piping systems for water supply (PVC-U).

BS EN 1455 Plastics piping systems for soil and waste discharge – ABS:
    Part 1: 2000 Specification for pipes, fittings and the system.

BS EN 1456 Plastic piping systems for buried and above ground drainage – unplasticised polyvinyl chloride (PVC-U):

Part 1: 2001 Specifications for piping components and the system.

BS EN 1519 Plastics piping systems for soil and waste discharge – polyethylene (PE):
    Part 1: 2000 Specification for pipes, fittings and the system.

BS EN 1565–1:2000 Plastics piping systems for soil and waste discharge – styrene copolymer blends.

BS EN 1566 Plastics piping systems for soil and waste discharge – chlorinated polyvinyl chloride (PVC-C):
    Part 1: 2000 Specification for pipes, fittings and the system.

BS EN 1636 Plastic piping systems for non-pressure drainage and sewerage – GRP:
    Part 3: 2001 Fittings.
    Part 5: 2001 Joints.

BS EN 1873: 2005 Prefabricated accessories for roof-ing – individual rooflights of plastics.

BS EN ISO 5999: 2004 Polymeric materials – cellular flexible polyurethane foam for load-bearing applica-tions.

BS EN 12201 Plastics piping systems for water supply – polyethylene (PE):
    Part 1: 2003 General.
    Part 2: 2003 Pipes.
    Part 3: 2003 Fittings.

BS EN 12608: 2003 Unplasticised polyvinyl chloride (PVC-U) profiles for the fabrication of windows and doors.

BS EN 12661-1: 2005 Plastics piping systems for non-pressure underground drainage and sewerage – polyethylene (PE).

DD CEN/TS 12666-2: 2005 Plastics piping systems for non-pressure underground drainage and sewerage – polyethylene - guidance for the assessment of con-formity.

BS EN 13163: 2001 Thermal insulation products for buildings – factory made products of expanded poly-styrene (EPS).

BS EN 13598 Plastics piping systems for non-pressure underground drainage and sewerage – unplasticised polyvinyl chloride (PVC-U), polypropylene (PP) and polyethylene (PE):
    Part 1: 2003 Specification for ancillary fittings.

pr EN 14636-2: 2006 Plastics piping systems for non-pressure drainage and sewerage – polyester resin concrete.

BS EN 14758-1: 2005 Plastics piping systems for non-pressure underground drainage and sewerage – poly-propylene with mineral modifiers.

pr CEN/TR 15438: 2006 Plastics piping systems—guidance for coding of products and their intended uses.

pr BS ISO 15877: 2006 Plastics piping for hot and cold water installations – chlorinated polyvinyl chloride.

BUILDING RESEARCH ESTABLISHMENT PUBLICATIONS

BRE Digests

BRE Digest 161: 1974 Reinforced plastics cladding panels.

BRE Digest 294: 1985 Fire risk from combustible cavity insulation.

BRE Digest 358: 1992 CFCs and buildings.

BRE Digest 382: 1993 New materials in hot climates.

BRE Digest 404: 1995 PVC-U windows.

BRE Digest 405: 2000 Polymer composites in construction.

BRE Digest 430: 1998 Plastics external glazing.

BRE Digest 440: 1999 Weathering of white external PVC-U.

BRE Digest 442: 1999 Architectural use of polymer composites.

BRE Digest 480: 2004 Wood plastic composites and plastic lumber.

BRE Information papers

BRE IP 12/97 Plastics recycling in the construction industry.

BRE IP 7/99 Advanced polymer composites in construction.

BRE IP 8/01 Weathering of plastics pipes and fittings.

BRE IP 12/01 Hot air repair of PVC-U profiles.

BRE IP 2/04 Wood plastic composites: market drivers and opportunities in Europe.

BRE Reports

BR 274: 1994 Fire safety of PTFE-based materials used in building.

BR 405: 2000 Polymer composites in construction.

ADVISORY ORGANISATIONS

British Laminated Fabricators Association, 6 Bath Place, Rivington Street, London EC2A 3JE (020 7457 5025).

British Plastics Federation, 6 Bath Place, Rivington Street, London EC2A 3JE (020 7457 5000).

British Rubber Manufacturers' Association Ltd., 6 Bath Place, Rivington Street, London EC2A 3JE (020 7457 5040).

GLASS-FIBRE REINFORCED PLASTICS, CEMENT AND GYPSUM

Introduction

Composite materials such as the glass-fibre reinforced materials GRP (glass-fibre reinforced polyester), GRC (glass-fibre reinforced cement) and GRG (glass-fibre reinforced gypsum) rely for their utility upon the advantageous combination of the disparate physical properties associated with the individual component materials. This is possible when a strong bond between the glass fibres and the matrix material ensures that the two materials within the composite act in unison. Thus polyester, which alone has a very low modulus of elasticity, when reinforced with glass fibres produces a material which is rigid enough for use as a cladding material. Cement, which alone would be brittle, when reinforced with glass fibres can be manufactured into thin impact-resistant sheets. Similarly glass-fibre reinforcement in gypsum considerably increases its impact and fire resistance.

Glass fibres

The glass fibres for GRP and GRG are manufactured from standard E-glass as shown in Figure 11.1. Molten glass runs from the furnace at 1200°C into a forehearth, and through a spinneret of fine holes from which it is drawn at high speed down to approximately 9 microns in thickness. The glass fibres are coated in size and bundled before winding up on a collet. Subsequently the glass fibre 'cake' is either used as continuous rovings or cut to 20–50 mm loose chopped strand. Glass fibre rovings may be manufactured into woven mats; chopped strand mats are formed with organic binder.

Glass-fibre reinforced plastics

The standard matrix material for glass-fibre reinforced plastics is polyester resin, although other thermosetting resins including phenolic, epoxy and polyurethane may be used. Glass fibres as continuous rovings or chopped strand are used for most purposes; however, the highest-strength products are obtained with woven glass fabrics and unidirectionally aligned fibres. The proportion of glass fibres ranges widely from 20–80% by weight depending upon the strength required. Enhanced performance can be achieved by using the more expensive S-grade high strength and modulus glass fibres used mainly in the aerospace industry. Alternative higher tensile strength fibres include the polyaramids such as *Kevlar* and carbon fibres, but these are considerably more expensive than glass.

FABRICATION PROCESS

A major investment in the manufacture of GRP cladding panels lies within the production of the high-quality moulds. These are usually made from timber, but steel or GRP itself may also be used. Moulds are reused, sometimes with minor variations (e.g. the insertion of a window void within a wall unit), as many times as possible to minimise production costs. The number of different mould designs for any one building is therefore kept to a minimum, and this may be reflected in the repetitiveness of the design.

In the fabrication process, the mould is coated with a release agent to prevent bonding and associated damage to the finished panel surface. A gel coat,

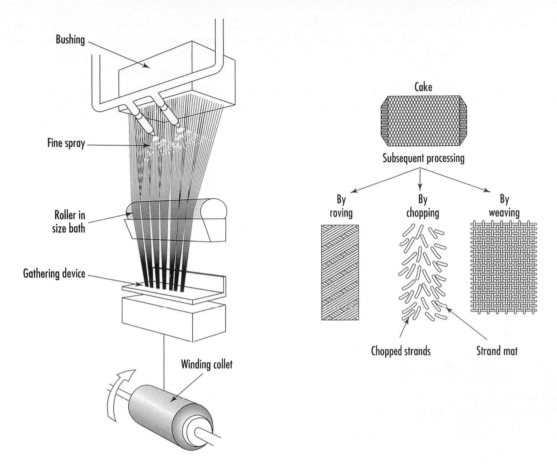

Fig. 11.1 Glass fibre production – rovings, chopped strand and mat

which ultimately will be the weathering surface, is applied to a finished thickness of 0.25–0.4 mm. Early examples of GRP without sufficient gel coat have weathered to a rough surface with consequent exposure of the glass fibres; however, modern gel coats when applied to the correct thickness are durable. The subsequent fabrication involves the *laying-up* of layers of glass fibres and polyester resin to the required thickness, usually with either sprayed rovings or chopped strand mat. Reinforcement and fixings, normally in aluminium due to similarities in coefficients of expansion, may be incorporated and areas requiring additional strength can be thickened as appropriate by the laying-up process. Plastic foam insulation may be encapsulated to give the required thermal properties. Curing may take up to two weeks, after which the unit is stripped from the mould, trimmed around the edges and fitted out.

PHYSICAL PROPERTIES AND DESIGN CONSIDERATIONS

The choice of GRP, for example as a cladding panel, imparts its own aesthetic on a building design. The high strength to weight ratio of GRP allows for the use of large panel units, but cost constraints in the mould-making reduce the number of panel variations to a minimum. Curved edges to panels and openings are preferred to reduce stress-raising points at very sharp corners. The high thermal expansion coefficient of GRP demands careful detailing of movement joints, and their appropriate sealing where necessary with components that retain their flexibility. In some cases the high expansion can be resolved by the use of profiled forms, which also impart strength. Colour fading and yellowing of GRP panels have been problems; however, recent products with ultraviolet light protection are more colour-fast.

Slightly textured finishes are generally more durable than smooth, for exposure to full direct sunlight. GRP can be manufactured with fire-resistant additives; the phenolic resins have the advantage of lower flammability and smoke emissions. Long-term creep precludes the use of GRP as a significant load-bearing material, although single-storey structures, two-storey mobile units and structural overhead walkways are frequently constructed from the material. GRP is vandal-resistant and can be laminated sufficiently to be bullet-resistant. Where both surfaces are to be exposed, the material can be pressed between the two halves of a die.

USES OF GLASS-FIBRE REINFORCED PLASTICS

The lightweight properties of GRP make it eminently suitable for the manufacture of large cladding panels and custom-moulded structures as illustrated in the belfry and spire of St James, Piccadilly, London (Fig. 11.2). Finishes may be self-coloured or incorporate a natural stone aggregate finish. In addition, GRP is frequently used for the production of architectural features such as barge boards, dormer windows, classical columns and entrance canopies (Fig. 11.3). GRP may be pigmented to simulate various timbers, slate, Portland or Cotswold stone and lead or copper. It

Fig. 11.2 GRP replacement spire – Church of St James, London. Photograph: Courtesy of Smith of Derby Ltd

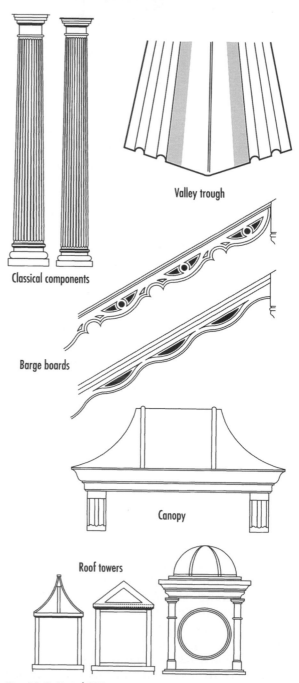

Valley trough

Classical components

Barge boards

Canopy

Roof towers

Fig. 11.3 Typical GRP components

is also used to produce a wide range of small building components including baths, valley troughs, flat roof edge trim and water drainage systems. In addition, a wide range of composite cladding panels are manufactured from glass-fibre reinforced resins incorporating stone granules within the core of the material. These products, which are impact and fire-resistant, are available with either a granular stone, painted or gel-coat finish.

CARBON-FIBRE AND ARAMID COMPOSITES

Carbon fibres, which were originally developed for the aerospace industry, combine strength and stiffness with low weight, but have poor impact resistance. They are produced from polyacrylonitrile fibres by controlled oxidation at 250°C followed by carbonisation at 2600°C in an inert atmosphere. Three grades, high strength (HS), high modulus (HM) and intermediate modulus (IM) are produced. Carbon fibres, like glass fibres, are available as woven material, chopped strand or continuous filament. Carbon fibres have a small negative coefficient of expansion along the fibre axis, thus composite materials of zero thermal expansion can be produced.

Aramids, are aromatic polyamide liquid crystalline polymers, with high strength to weight ratio in tension, but poorer properties under compression or bending. Impact resistance is greater than for carbon fibres. Aramid fibres, typically *Kevlar*, are produced by spinning the continuous fibre from solution. A variety of products are available with a range of modulus, elongation and impact resistance properties. Aramid composites exhibit good abrasion resistance. Carbon and aramid fibres may be combined in a composite material where strength, stiffness and impact resistance are all required.

Although polyester resins may be used as the matrix material for either carbon or aramid fibres, usually these more expensive fibres are incorporated into higher-performance epoxy resins. In addition to the standard GRP laying-up production process, pultrusion and preimpregnation are used for manufacturing carbon fibre-reinforced components. Pultrusion, a combination of extruding and pulling (Fig. 11.4), is used for making continuous profiles which may be either solid or hollow. Preimpregnation involves coating the continuous fibre or woven fibre fabric with a mixture of resin and curing agent, which can be stored frozen at this stage, then thawed and moulded into shape when required. By using low temperature moulding processes large and complex structures can be fabricated for the construction industry.

For externally bonded reinforcement to concrete structures, pultruded carbon fibre-reinforced plates may be bonded to the concrete with a thixotropic epoxy resin. Alternatively, woven carbon fibre mat is wrapped around the concrete and pasted on with epoxy resin. The required level of reinforcement can be achieved by building up the appropriate thickness of epoxy resin saturated carbon fibre mat. The technique can also be applied to reinforcing steel, masonry, timber or cast iron. Although usually used for remedial work, this type of reinforcement could also be considered for new-build elements.

RECYCLING FIBRE-REINFORCED POLYMERS

Currently, the majority of waste fibre-reinforced polymers (FRPs) are disposed into landfill sites. One

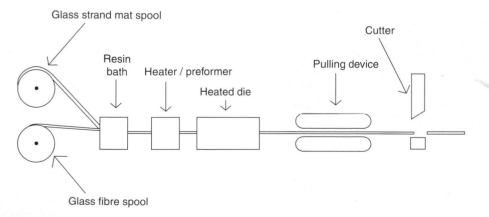

Fig. 11.4 Pultrusion

alternative is to grind the material into powder and use this ground GRP in conjunction with other binders; however, this process is difficult where embedded metal fixings were incorporated into the original components.

The recyclate powder can be blended with other recycled plastics to produce GRP/plastics lumber, which can be used for lightly loaded piles, decking, fencing and similar applications. This material can be cut and worked like the natural timber which it replaces. Alternatively, ground GRP can be incorporated into particleboard to make GRP-reinforced wood particleboard, which has enhanced mechanical properties compared to the standard grade (P5) of particleboard used for domestic flooring. However, when energy costs, transportation and other factors are considered, the ecological balance towards recycling fibre-reinforced polymers may be dependant on future considerations of recycling at the initial design stage.

Glass-fibre reinforced cement

Glass-fibre reinforced cement is a material that was developed in the early 1970s by the Building Research Establishment. The standard material is produced from a mixture of alkali-resistant glass fibres with Portland cement, sand aggregate and water. Admixtures such as pozzolanas, superplasticisers and polymers are usually incorporated into the mix to give the required fabrication or casting properties. The breakthrough in the development of the material was the production of the alkali-resistant (AR) glass fibres, as the standard E-glass fibres, which are used in GRP and GRG, corrode rapidly in the highly alkaline environment of hydrated cement. Alkali-resistant glass, in addition to the sodium, silicon and calcium oxide components of standard E-glass, contains zirconium oxide. Alkali-resistant glass fibres, which have been improved by progressive development, are manufactured under the trade name *Cem-FIL*. The addition to GRC mixes of metakaolin, a pozzolanic material produced by calcining china clay at 750–800°C, prevents the development of lime crystals around the glass fibres. In the unblended GRC this leads to some gradual loss of strength. Standard grey GRC has the appearance of sheet cement and is non-combustible.

MANUFACTURE OF ALKALI-RESISTANT GLASS FIBRES

Silica, limestone and zircon are melted in a furnace, the alkali-resistant glass produced is drawn into fibres of 14 or 20 microns diameter and rolled into cakes for subsequent use as continuous rovings or for conversion into chopped strand. The process is comparable to that for standard E-glass fibres. *Cem-FIL* glass tissue with a fine texture is also available.

CEMENT MATRIX

Portland cement, either strength class 42.5 or 42.5R (rapid early strength), is normally used. Portland cement will produce a grey finish, but white Portland cement or added pigments may be used to give different effects. However, with the use of pigments care must be taken to ensure uniformity of colour. Washed sand and fly ash (pulverised-fuel ash) are the usual aggregates, but crushed marble, limestone or granite can be used when a particular exposed aggregate finish is required.

FABRICATION PROCESSES

Fibre-reinforced cement components may be formed either by using a spray-gun, which mixes the glass fibres with a slurry of cement as it sprays directly into the mould, or by premixing a blend of cement, sand, water, admixtures and glass fibres before casting. Moulds similar to those required for the production of GRP components are used. Extrusion and injection moulding techniques are applicable for linear or small components, and bagged pre-blended mixes can be used for on-site applications.

Sprayed glass-fibre reinforced cement

Spray techniques, which may be manual (Fig. 11.5) or robotic, are used to build up the required thickness, usually between 10 and 20 mm. During spraying, the gun chops the fibres into 25–40 mm lengths, depositing a uniform felt of fibres and mortar into the mould. A typical sprayed mix would contain 5% glass fibres, 36% Portland cement, 36% washed sand, 11% additives/ polymer and 12% water. The curing of GRC is relatively slow, with 95% strength developed after seven days.

Premixed glass-fibre reinforced cement

It is normal to premix the cement, sand, water and admixtures then add the chopped fibres. A typical mix would contain up to 3.5% of 12 mm fibres in a

Fig. 11.5 Spraying glass-fibre reinforced cement. Photograph: Courtesy of Trent Concrete Ltd

sand : cement mix of 0.5 : 1 with a water/cement ratio of 0.35. The mix is then cast and vibrated, or pressed into form for smaller components. For renderings, a glass fibre content of between 1–2% is appropriate. A recent development involves the direct spraying of the premixed material.

PROPERTIES OF GLASS-FIBRE REINFORCED CEMENT

Appearance

While standard GRC has the appearance of cement, a wide diversity of colours, textures and simulated materials can be manufactured. A gloss finish should be avoided as it tends to craze and show any defects or variations. The use of specific aggregates followed by grinding can simulate marble, granite, terracotta, etc., while reconstructed stone with either a smooth or tooled effect can be produced by the action of acid etching. An exposed aggregate finish is achieved by the use of retardants within the mould, followed by washing and brushing. Applied finishes, which are usually water-based synthetic latex emulsions, can be applied to clean, dust-free surfaces.

Fig. 11.6 GRC components. Photograph: Courtesy of Trent Concrete Ltd

Moisture and thermal movement

GRC exhibits an initial irreversible shrinkage followed by a reversible moisture movement of approximately 0.2%. The coefficient of thermal expansion is within the range $7-20 \times 10^{-6}$ deg C^{-1}, typical for cementitious materials.

Thermal conductivity

The thermal conductivity of GRC is within the range 0.21–1.0 W/m K. Double-skin GRC cladding panel units usually incorporate expanded polystyrene, mineral wool or foamed plastic insulation. Cold bridging should be avoided where it may cause shadowing effects.

Durability

GRC is less permeable to moisture than normal concrete, so it has good resistance to chemical attack; however, unless manufactured from sulfate-resisting cement, it is attacked by soluble sulfates. GRC is unaffected by freeze/thaw cycling.

Impact resistance

GRC exhibits a high impact resistance but toughness and strength does decrease over long periods of time. However, the incorporation of metakaolin $(2SiO_2.Al_2O_3)$ into the mix appears to improve the long-term performance of the material.

USES OF GLASS-FIBRE REINFORCED CEMENT

GRC is used extensively for the manufacture of cladding and soffit panels because it is lightweight and easily moulded (Fig. 11.6). It is used in conservation work as a replacement for natural stone and in architectural mouldings, including sophisticated decorative screens within countries of the Middle East. It is used as permanent formwork for concrete, fire-resistant partitioning and in the manufacture of small components including slates, tiles and decorative ridge tiles. Glass-fibre reinforced cement slates are manufactured to simulate the texture and colour of natural slate. Some manufacturers incorporate blends of other non-asbestos natural and synthetic fibres, together with pigments and fillers, to produce a range of coloured products with glossy, matt or simulated riven finishes.

Glass-fibre reinforced gypsum (GRG)

Glass-fibre reinforced gypsum combines the non-combustibility of gypsum plaster with the reinforcing strength of glass fibres. Products contain typically 5% of the standard E-glass fibres, which considerably improve impact as well as fire resistance. Commercial GRG products are available as standard panels, encasement systems for the fire protection of steel and decorative wall panels. As with all gypsum products, GRG should not be used in damp conditions or at temperatures regularly over 50°C.

GLASS-FIBRE REINFORCED GYPSUM BOARDS

The standard boards, available in a range of thicknesses from 4 mm to 12.5 mm, are manufactured with a glass-fibre reinforced gypsum core and glass-fibre tissue immediately below the gypsum faces. The material is suitable for a wide range of applications including wall linings, ceilings and protected external positions such as roof soffits. The material can be easily cut on site and fixed with nails or screws; in addition, owing to the effect of the glass-fibre reinforcement, it can be curved to fit, for example, barrel-vault ceilings. The minimum radius of curvature depends upon the

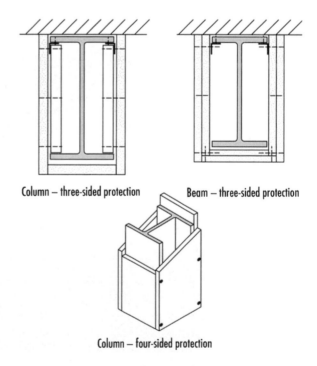

Column – three-sided protection   Beam – three-sided protection

Column – four-sided protection

Fig. 11.7 Fire protection with GRG panels

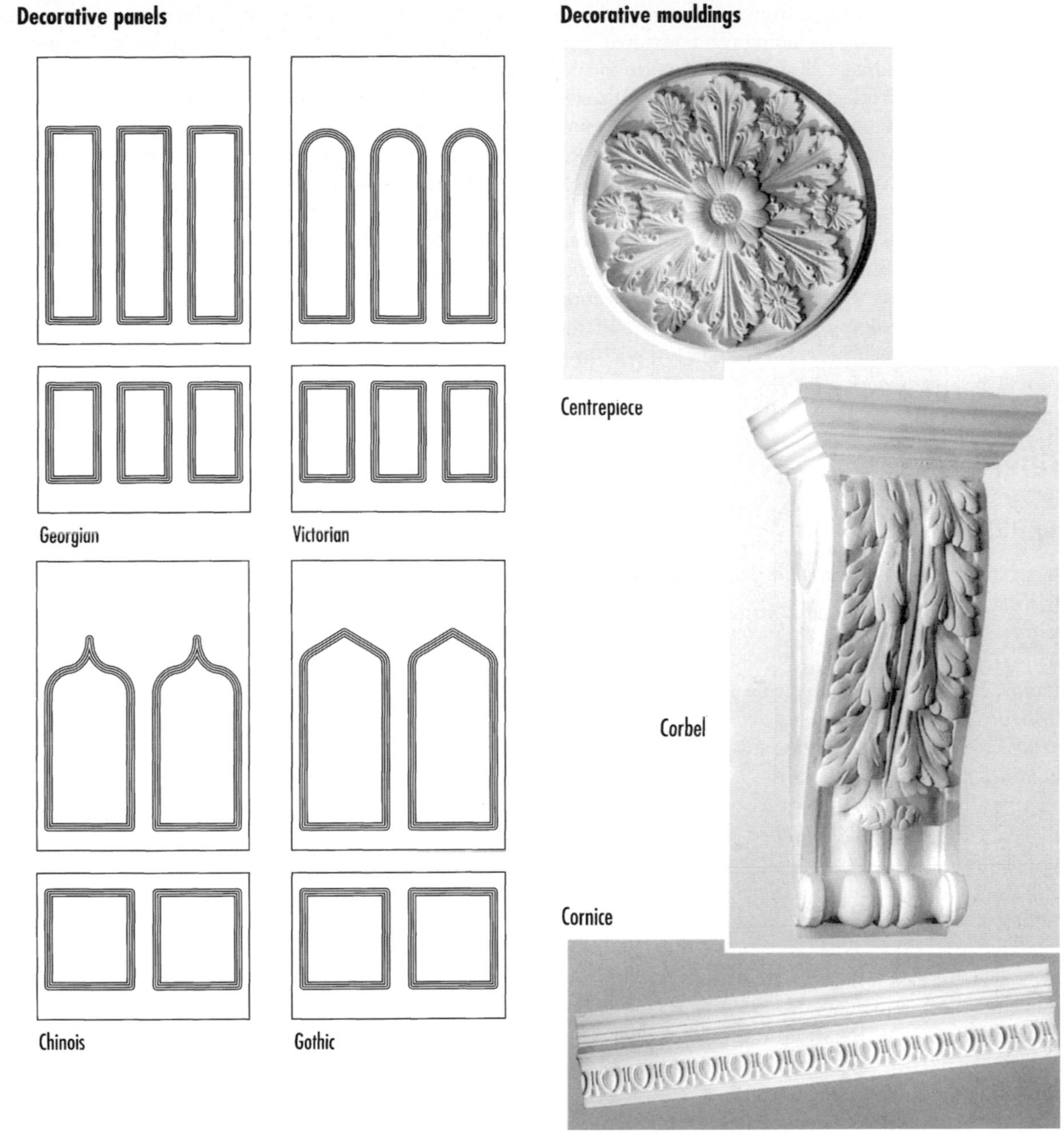

Decorative panels

Georgian   Victorian

Chinois   Gothic

Decorative mouldings

Centrepiece

Corbel

Cornice

Fig. 11.8 Decorative GRG plaster components

board thickness. The material has a smooth off-white finish; joints should taped before finishing board plaster is applied.

For steelwork, protection thicknesses of 15, 20 and 25 mm are available. Depending upon the steel section factor (Hp/A m^{-1}), with double layers and staggered joints, up to 120 minutes' fire resistance can be achieved (Fig. 11.7).

Decorative glass-fibre reinforced gypsum boards and ceiling tiles

Decorative boards manufactured with a range of motifs can be used as dado or wall panels (Fig. 11.8). Dabs of sealant are used initially to fix panels to existing walls and to allow adjustment to a flush finish. Panels may be painted after jointing.

Ceiling tiles manufactured from GRG are available to a wide range of designs, including plain, textured, patterned, open or closed-cell surface and with square, tapered or bevelled edges. The standard size is usually 600×600 mm, although some manufactures produce units at 300×600 or 1200 and 600 $\times$ 1200 mm. GRG has good fire-resistant properties; it is non-combustible to BS 476 Part 4: 1970, Class 1 surface spread of flame to BS 476 Part 6: 1989, Class 0 to Building Regulations Section E15, and does not emit smoke or noxious fumes in fire. Acoustic tiles with enhanced sound absorption and attenuation properties are normally part of the standard range, which may also include Imperial sizes for refurbishment work.

References

FURTHER READING

British Gypsum. 2005: *The white book.* Loughborough: British Gypsum Ltd.

British Gypsum. 2004: *Glasroc fire book.* Loughborough: British Gypsum Ltd.

Cripps, A. 2002: *Fibre-reinforced polymer composites in construction.* London: CIRIA,

Fordyce, M.W. and Wodehouse, R.G. 1983: *GRC and buildings.* London: Butterworth.

Glassfibre Reinforced Cement Association. 1986: *This is GRC.* Newport: The Glassfibre Reinforced Cement Association.

Holloway, L. 1994: *Handbook of polymer composites for engineers.* Cambridge: Woodhead Publishing Ltd.

Leggatt, A.J. 1984: *GRP and buildings.* London: Butterworth.

Majumdar, A.J. and Laws, V. 1990: *Glass-fibre reinforced cement.* Oxford: BSP Professional.

Swamy, R.N. (ed.) 1992: *Fibre-reinforced cement and concrete.* London: Spon Press.

Ture, G. 1986: *Glass fibre reinforced cement: Production and uses.* London: Spon Press.

STANDARDS

BS 476 Fire tests on building materials and structures:
Part 6: 1989 Methods of test for fire propagation for products.
Part 7: 1987 Method for classification of the surface spread of flame of products.

BS 5544: 1978 Specification for anti-bandit glazing (glazing resistant to manual attack).

BS 6206: 1981 Specification for impact requirements for flat safety glass and safety plastics for use in buildings.

BS EN 492: 2004 Fibre-cement slates and fittings – Product specification and test methods.

BS EN 494: 2004 Fibre-cement profiled sheets and fittings – Product specification and test methods.

BS EN 1013–2: 1999 Light transmitting profiled sheet for single-skin roofing (GRP).

BS EN 1169: 1999 Precast concrete products – factory production control of glass-fibre reinforced cement.

BS EN 1170 Precast concrete products – test method for GRC:
Part 1: 1998 Measuring the consistency – slump test method.
Part 2: 1998 Measuring the fibre content in fresh GRC.
Part 3: 1998 Measuring the fibre content in sprayed GRC.
Part 4: 1998 Measuring bending strength – simplified bending.
Part 5: 1998 Measuring bending strength – complete bending.
Part 6: 1998 Determination of the absorption of water.
Part 7: 1998 Measurement of extremes of dimensional variations.

BS EN 13280: 2001 Specification for glass fibre reinforced cisterns.

BS ISO 22314: 2006 Plastics, glass-fibre reinforced products – determination of fibre length.

BUILDING RESEARCH ESTABLISHMENT PUBLICATIONS

BRE Digests

BRE Digest 161: 1974 Reinforced plastics cladding panels.

BRE Digest 405: 2000 Polymer composites in construction.

BRE Digest 442: 1999 Architectural use of polymer composites.

BRE Good repair guide

BRE GRG 34: 2003 Repair and maintenance of FRP structures.

BRE Information papers

BRE IP 5/84 The use of glass-reinforced cement in cladding panels.
BRE IP 10/87 Polymer modified GRC.
BRE IP 1/91 Durability of non-asbestos fibre-reinforced cement.
BRE IP 7/99 Advanced polymer composites in construction.
BRE IP 19/01 The performance of fibre cement slates.
BRE IP 10/03 Fibre-reinforced polymers in construction: durability.
BRE IP 11/03 Fibre-reinforced polymers in construction: predicting weathering.
BRE IP 2/04 Wood plastic composites: market drivers and opportunities in Europe.
BRE IP 4/04 Recycling fibre-reinforced polymers in the construction industry.
BRE IP 5/04 Fibre-reinforced polymers in construction.

BRE Reports

BR 49: 1984 The use of glass-reinforced cement in cladding panels.
BR 405: 2000 Polymer composites in construction.
BR 461: 2003 Fibre-reinforced polymers in construction: long-term performance in service.
BR 467: 2004 Recycling fibre-reinforced polymers in construction: a guide to best practicable environmental option.

ADVISORY ORGANISATIONS

British Plastics Federation, 6 Bath Place, Rivington Street, London EC2A 3JE (020 7457 5000).
Glassfibre Reinforced Concrete Association, Concrete Society, 4 Meadows Business Park, Station Approach, Camberley, Surrey GU17 9AB (01276 607140).

12

PLASTER AND BOARD MATERIALS

—

Introduction

Plastering, based on lime, was brought to Britain by the Romans. In Britain it was originally used to strengthen and seal surfaces and in the case of combustible materials to afford some fire protection, but by the eighteenth century its value as a decorative finish had been appreciated. The use of gypsum plaster both as a sealant and as a decorative material by the Minoan civilisation is well documented, and current UK practice is now based on gypsum (hydrated calcium sulfate), rather than lime. Gypsum is mined from geological deposits produced by the gradual evaporation of lakes containing the mineral; there are extensive reserves within the UK, mainly in the North of England, but also in the East Midlands.

Historically, fibrous materials have been used to re-inforce plaster and particularly to control shrinkage in lime plaster. Traditionally ox, horse and goat hair were the standard materials; however, straw, hemp and jute have also been used. The earliest lightweight support for plasters was interwoven hazel twigs, but by the fifteenth century split timber laths were common. The modern equivalent is the use of galvanised and stainless steel expanded metal.

Gypsum plaster

MANUFACTURE OF GYPSUM PLASTER

Rock gypsum is mined, crushed and ground to a fine powder. The natural mineral may be white or discoloured pale pink, grey or brown due to small quantities of impurities which do not otherwise affect the product. On heating to temperatures in the range 130°–170°C, water is driven off the hydrated gypsum; the type of plaster produced is largely dependent upon the extent of this dehydration process.

$$CaSO_4.2H_2O \xrightarrow{130°C} CaSO_4.\tfrac{1}{2}H_2O \xrightarrow{170°C} CaSO_4$$

hydrated gypsum — hemihydrate — anhydrous gypsum

The standard classes of plaster are defined in the standard BS 1191–1: 1973. Lightweight plasters based on gypsum and lightweight aggregate are defined in BS 1191–2: 1973.

CLASSES OF PLASTER

Class A – Plaster of Paris

Plaster of Paris is produced by driving off three-quarters of the water content from natural hydrated gypsum. Plaster of Paris sets very quickly on the addition of water, and is therefore often used as a moulding material.

Class B – Retarded hemi-hydrate gypsum plaster

The majority of plasters in current use within construction are based on retarded hemi-hydrate gypsum. The addition of different quantities of a retarding agent, usually keratin, is used to adjust the setting time for different products.

Undercoat and one-coat plasters
The main constituents of undercoat and one-coat plasters are retarded hemi-hydrate gypsum, with expanded perlite or exfoliated vermiculite for the lightweight products, together with small quantities of limestone,

anhydrite (anhydrous gypsum), clay and sand. In addition, other materials are incorporated to adjust the product specification and setting time, which normally ranges between one and two hours. Thus, lime is added to undercoat plaster, and for backgrounds of high suction, a water retention agent is also required. For example, *browning* is suitable for use on backgrounds with moderate or high suction and a good mechanical key. For higher impact resistance, cement and granulated blastfurnace slag are incorporated, and for a one-coat plaster, limestone is added. Typical applications would be 11 mm for undercoats with a finish coat of 2 mm, or a single one-coat application of 13 mm.

Finish-coat plasters

For finish-coat plasters, like undercoat plasters, the main constituent is retarded hemi-hydrate gypsum, but with a small addition of lime to accelerate the set. The lightweight products contain exfoliated vermiculite. Finish coats on masonry substrates are usually 2 mm in thickness, and board finish plaster is normally applied to 2–3 mm.

Class C – Anhydrous gypsum plaster

When natural gypsum is heated at over 160°C, most of the water is driven off leaving anhydrous calcium sulfate or anhydrite. The proportion of the hemi-hydrate remaining is dependent upon the heating time and temperature. Anhydrous gypsum plaster sets very slowly, so an accelerator such as alum is added. The plaster has an initial set, after which it can be smoothed with the addition of more water to the surface. The material has been superseded by the Class B plasters.

Class D – Keene's plaster

Anhydrous gypsum with an accelerator sets slowly to a very hard surface, which can be worked to a high-quality glass-like finish. It is difficult to paint owing to its glassy surface and therefore requires a special primer to provide a key. The material has been superseded by cement-based products.

BACKGROUNDS FOR PLASTER

Plaster bonds to the background by a combination of mechanical key and adhesion. Backgrounds should be clean, dry and free from other contamination, and the specification of the plaster should be appropriate to the suction of the background surface. Where possible, as in the case of brickwork, a good mechanical key

should be obtained by raking out the joints. On hard low-suction materials such as smooth concrete and ceramic tiles, a PVA (polyvinyl acetate) bonding agent should be applied. Similarly, to control the high suction in substrates such as aerated concrete blocks, a PVA bonding agent can be applied or the substrate wetted prior to the application of plaster. Plaster can, however, be applied directly to dense aggregate concrete blocks without prior wetting. Where two or more coats of plaster are applied, the undercoats should be scratched to ensure good subsequent bonding. Gypsum plasters, if applied correctly, do not shrink or crack on drying out and subsequent coats can be applied in quick succession.

PLASTERBOARD

Plasterboard consists of an aerated gypsum core bonded to strong paper liners. Most wallboards have one light surface for direct decoration and one grey surface, which may be plastered with a skim of board plaster. The decorative surface may be bevelled or tapered at its edge, whilst the grey surface is square for plastering. Plasterboard may be cut with a saw or scored and snapped. Nail fixings should be driven in straight, leaving a shallow depression but without fracturing the paper surface. Alternatively boards may be screwed. Standard thickness are 9.5, 12.5, 15 and 19 mm. Only the moisture-resistant grades of plasterboard normally require the application of a PVA bonding agent before plastering. These have a water-resistant core and treated liners, so may be used behind external finishes such as vertical tiling and weatherboarding or in external sheltered positions protected from direct rain. Boards are available finished with PVC, backed with aluminium foil or laminated to insulation (expanded polystyrene, extruded polystyrene, rigid polyurethane foam or mineral wool) for increased thermal properties. (The thermal conductivity of standard plasterboard is 0.19 W/m K.)

Plasterboard systems

Plasterboard non-load-bearing internal walls may be constructed using proprietary metal stud systems or as traditional timber stud walls. Where appropriate, acoustic insulation should be inserted within the void spaces. Dry-lining to masonry may be fixed with dabs of adhesive, alternatively with metal or timber framing. Plasterboard suspended ceiling systems are usually supported on a lightweight steel framework fixed directly to either concrete or timber. Convex and concave surfaces can be achieved. Sound transmission through existing upper-storey timber-joist floors can be reduced by a combination of resiliently mounted plasterboard and mineral wool insulation.

Fibre-reinforced gypsum boards

Fibre-reinforced gypsum boards are manufactured with either natural or glass fibres. Glass-fibre reinforced gypsum (GRG) is described in Chapter 11.

Natural fibre-reinforced gypsum boards are manufactured from cellulose fibres, frequently from recycled paper, within a matrix of gypsum. The panel boards are either uniform or laminated with a perlite and gypsum core, encased in a hard layer of fibre-reinforced gypsum. Boards are impact and fire-resistant and easily fixed by nails, screws, staples or adhesive as a dry-lining system to timber, metal framing or masonry. Standard boards are 1200 × 2400 mm with thicknesses in the range 9.5–18 mm. Joints are filled or taped and corners beaded as for standard plasterboard products. A composite board of fibre-reinforced gypsum and expanded polystyrene offers enhanced insulation properties. (The thermal conductivity of fibre-reinforced gypsum board is typically 0.36 W/m K.)

ACCESSORIES FOR PLASTERING

Beads

Angle and stop beads are manufactured from galvanised or stainless perforated steel strip or expanded metal. They provide a protected, true straight arris or edge with traditional plastering to masonry or for thin-coat plasterboard. Proprietary systems are manufactured similarly from perforated galvanised or stainless steel to form movement joints in dry-lining systems (Fig. 12.1).

Scrim

Scrim, an open-weave material, is used across joints between plasterboards and in junctions between plaster and plasterboard. Both self-adhesive glass-fibre mesh and traditional jute scrim are available. For the prevention of thermal movement cracking at plasterboard butt joints, paper tape bedded into the plaster skim is often more effective than the use of self-adhesive scrim.

Coves and cornices

Decorative coves and cornices are manufactured from gypsum plaster encased in a paper liner. In some cases the gypsum is reinforced with glass fibres. The components (Fig. 12.2) can be cut to size with a saw, and are normally fixed with proprietary adhesives.

SPECIAL PLASTERS

Renovating plaster

Renovating plaster is used where walls have been stripped of existing plaster during the successful installation of a new damp-proof course. Renovating plasters contain aggregates which promote surface drying when they are applied to structures with residual moisture, but they should not be used in permanently damp locations below ground level. Renovating plaster should also not be used where masonry is heavily contaminated with salts, such as in buildings not originally built with damp-proof courses, and on the brickwork of chimney breasts. Renovating plasters contain a fungicide to inhibit mould growth during the drying out process.

Projection plaster

Projection plaster is sprayed onto the background from a plaster projection machine. The plaster should be built up to the required thickness, ruled to an even surface then flattened and trowelled to a flat surface. As with all plastering the process should not be carried out under freezing, excessively hot or dry conditions. A typical application to masonry would be 13 mm and should not exceed 25 mm.

Acoustic plaster

Acoustic plaster has a higher level of sound absorption than standard gypsum plasters owing to its porosity and surface texture. Aluminium powder is added to the wet plaster mix to produce fine bubbles of hydrogen gas which remain trapped as the plaster sets giving it a honeycomb structure. One form of acoustic plasterboard consists of a perforated gypsum plasterboard which may be backed with a 100 mm glass wool sound-absorbing felt.

X-ray plaster

X-ray plaster is retarded hemihydrate plaster containing barium sulfate (barytes) aggregate. It is used as an

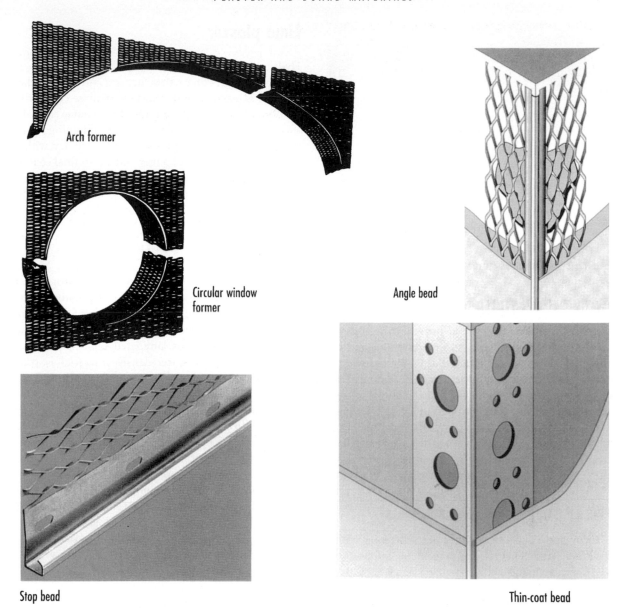

Arch former

Circular window former

Angle bead

Stop bead

Thin-coat bead

Fig. 12.1 Plastering beads and arch formers

undercoat plaster in hospitals, etc., where protection from X-rays is required. Typically, a 20 mm layer of X-ray plaster affords the same level of protection as a 2 mm sheet of lead, providing that it is free of cracks.

Textured plaster

Textured plaster is frequently applied to plasterboard ceilings. A variety of different patterns and textures can be achieved. The textured surface may be left as a natural white finish or painted as required.

Fibrous plaster

Fibrous plaster is plaster of Paris reinforced with jute, sisal, hessian, glass fibres, wire mesh or wood laths. It is used for casting in moulds, ornate plasterwork such as fire surrounds, decorative cornices, dados, friezes, panel mouldings, corbels and centrepieces for ceilings in both restoration and new work. The reinforcement material may be elementary in the form of random fibres or sheet material, or complementary as soft-wood laths or lightweight steel sections.

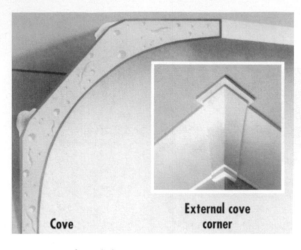

Fig. 12.2 Preformed plaster coves

GYPSUM FLOOR SCREED

Gypsum interior floor screed, manufactured from a mixture of hemi-hydrate gypsum, limestone and less than 2% cement, may be used as an alternative to a traditional sand and cement screed, providing that a floor covering is to be used. The material is self-smoothing and may be pumped. It is laid on a polythene membrane to a minimum thickness of 35 mm for floating screeds, and may be used over underfloor heating systems. When set, the hard plaster has a minimum 28-day compressive strength of 30 MPa.

FIRE RESISTANCE OF PLASTER MATERIALS

Gypsum products afford good fire protection within buildings due to their basic chemical composition. Gypsum, hydrated calcium sulfate ($CaSO_4.2H_2O$) as present in plaster and plasterboard, contains nearly 21% water of crystallisation. When exposed to a fire this chemically combined water is gradually expelled in the form of vapour. It is this process which absorbs the incident heat energy from the fire, considerably reducing the transmission of heat through the plaster, thus protecting the underlying materials. The process of dehydrating the gypsum commences on the face adjacent to the fire, and immediately the dehydrated material, because it adheres to the unaffected gypsum, acts as an insulating layer slowing down further dehydration. Even when all the water of crystallisation has been expelled, the remaining anhydrous gypsum continues to act as an insulating layer whilst it retains its integrity. The inclusion of glass fibres into gypsum plasterboards increases the cohesiveness of the material within fires.

Lime plaster

Hydraulic lime plaster is suitable for interior application, particularly on earth structures and unfired clay walls. It is usually applied in two or three coats – the best quality work requiring the three-coat system. In this case a 13 mm coat of *coarse stuff* containing 5 mm sand (lime : sand, 1 : 2½) is followed when dry, with a similar thickness of a 1 : 3 mix and a thin final coat of between 1 : 1 and 1 : 2 lime to sand. Other additions, including horse-hair and cow dung, may be added to improve the setting properties of the lime plaster.

Calcium silicate boards

Calcium silicate boards are manufactured from silica with lime and/or cement, usually incorporating cellulose fibres or softwood pulp and mica or exfoliated vermiculite filler, to produce a range of densities. The high-density material is laminated under steam and pressure, while the lower-density material is produced by rolling followed by curing in an autoclave. Calcium silicate boards like gypsum boards are non-combustible. The material is grey or off-white in colour, easily worked and nailed. Calcium silicate boards are durable, moisture-, chemical- and impact-resistant with dimensional stability and a good strength to weight ratio. They are available with a range of smooth or textured factory finishes for interior or exterior use and also laminated to extruded polystyrene for enhanced insulation properties. Standard thickness include 4.5, 6.0, 9.0 and 12.0 mm, although thicknesses up to 60 mm are available in the vermiculite lightweight boards used for fire protection, giving up to 240 minutes' resistance. Typical applications include wall, roof and partition linings, suspended ceilings, fasciae, soffits, weatherboarding and fire protection to structural steelwork. External cladding boards may be finished with a sprayed or trowelled render to produce a seamless finish. (The thermal conductivities of calcium silicate boards are usually within the range 0.13 to 0.29 W/m K depending upon their composition.)

References

FURTHER READING

Ashurst, J. and Ashurst, N. 1988: *Practical building conservation. Vol. 3. Mortars, plasters and renders.* Aldershot: Gower Technical Press.

British Gypsum. 2005: *The white book.* Loughborough: British Gypsum Ltd.

British Gypsum. 2002: *Site book.* Loughborough: British Gypsum Ltd.

Cape Boards. 1995: *The fire protection handbook.* Uxbridge: Cape Boards Ltd.

STANDARDS

BS 1191 Gypsum building plasters:

Part 1: 1973 Excluding premixed lightweight plasters.

Part 2: 1973 Premixed lightweight plasters.

BS 1230 Gypsum plasterboard:

Part 1: 1985 Specification for plasterboard excluding materials submitted to secondary operations.

BS 1369 Steel lathing for internal plastering and external rendering.

Part 1: 1987 Specification for expanded metal and ribbed lathing.

BS 4022: 1970 Prefabricated gypsum wallboard panels.

BS 5270 Bonding agents for use with gypsum plasters and cement:

Part 1: 1989 Specification for polyvinyl acetate (PVAC) emulsion bonding agents for indoor use with gypsum building plasters.

BS 5492: 1990 Code of practice for internal plastering.

BS 6100 Glossary of building and civil engineering terms:

Part 6 Concrete and plaster.

Subsec. 6.6.2: 1990 Plaster.

BS 6452 Beads for internal plastering and dry-lining:

Part 1: 1984 Specification for galvanised steel beads.

BS 7364: 1990 Galvanised steel studs and channels for stud and sheet partitions and linings using screw fixed gypsum wallboards.

BS 8000 Workmanship on building sites:

Part 8: 1994 Code of practice for plasterboard partitions and dry-linings.

Part 10: 1995 Code of practice for plastering and rendering.

BS 8212: 1995 Code of practice for dry-lining and partitioning using gypsum plasterboard.

BS EN 998 Specification for mortar for masonry:

Part 1: 2003 Rendering and plastering mortar.

BS EN 12859: 2001 Gypsum blocks – definitions, requirements and test methods.

BS EN 12860: 2001 Gypsum based adhesives for gypsum blocks – definitions.

pr EN 13815: 2000 Fibrous gypsum plaster products – definitions, requirements and test methods.

pr 14246: 2001 Gypsum elements for suspended ceilings – definitions, requirements and test methods.

pr EN 15319: 2005 General principles of design of fibrous gypsum plaster works.

BUILDING RESEARCH ESTABLISHMENT PUBLICATIONS

BRE Defect action sheets

BRE DAS 81: 1986 Plasterboard ceiling for direct decoration: nogging and fixing – specification.

BRE DAS 82: 1986 Plasterboard ceiling for direct decoration: nogging and fixing – site work.

BRE DAS 86: 1986 Brick walls: replastering following dpc injection.

BRE Good building guide

BRE GBG 7: 1991 Replacing failed plasterwork.

BRE GBG 65: 2005 Plastering and internal rendering (Parts 1 and 2).

BRE Good repair guide

BRE GRG 18: 1998 Replacing plasterwork.

ADVISORY ORGANISATION

Gypsum Products Development Association, PO Box 35084, London NW1 4XE (020 7935 8532).

13

INSULATION MATERIALS

—

Introduction

With the increasing emphasis on energy-conscious design and the broader environmental impact of buildings, greater attention is necessarily being focussed upon the appropriate use of thermal and sound insulation materials.

Thermal and sound insulation materials

The Approved Document of the Building Regulations gives guidance on minimum thermal performance criteria for buildings based on standards for their individual elements, or the overall energy efficiency of the whole building. To consider the relative efficiency of insulating materials, the thermal conductivities (W/m K) are quoted at the standard 10°C to allow direct comparisons. U-values would not illustrate direct comparability owing to the varying thicknesses used, and the wide variety of combinations of materials typically used in construction.

In considering acoustic control, distinction is made between the reduction of sound transmitted directly through the building components and the attenuation of reflected sound by the surfaces within a particular enclosure. Furthermore, transmitted sound is considered in terms of both impact and airborne sound. Impact sound is caused by direct impact onto the building fabric which then vibrates, transmitting the sound through the structure; it is particularly significant in the case of intermediate floors. Airborne sound waves, from the human voice and sound-generating equipment, cause the building fabric to vibrate, thus transmitting the sound. Airborne sound is particularly critical in relation to separating walls and is significantly increased by leakage at discontinuities within the building fabric, particularly around unsealed openings. The reduction in sound energy passing through a building element is expressed in decibels (dB). The doubling of the mass of a building component reduces the sound transmission by approximately 5 dB, thus sound insulating materials are generally heavy structural elements. However, the judicious use of dissipative absorbers within walls can reduce the reliance for sound absorption on mass alone. Noise may be transmitted through services installations, so consideration should be given to the use of acoustic sleeves and linings as appropriate.

The absorption of sound at surfaces is related to the porosity of the material. Generally, light materials with fibrous or open surfaces are good absorbers, reducing ambient noise levels and reverberation times, whereas smooth hard surfaces are highly reflective to sound (Table 13.1). Sound absorption is measured on a 0 to 1 scale with 1 representing total absorption of the sound.

FORMS OF INSULATION MATERIALS

Thermal and sound insulation materials may be categorised variously according to their appropriate uses in construction, their physical forms or their material origin. Many insulating materials are available in different physical forms each with their appropriate uses in building. Broadly, the key forms of material could be divided into:

- structural insulation materials;
- rigid and semi-rigid sheets and slabs;
- loose fill, blanket materials and applied finishes;
- aluminium foil.

Table 13.1 Typical sound absorption coefficients at 125, 500 and 2000 Hz for various building materials

Material	Absorption coefficient		
	125 Hz	500 Hz	2000 Hz
Concrete	0.02	0.02	0.05
Brickwork	0.05	0.02	0.05
Plastered solid wall	0.03	0.02	0.04
Glass 6 mm	0.1	0.04	0.02
Timber boarding, 19 mm over air space against solid backing	0.3	0.1	0.1
Wood wool slabs, 25 mm, on solid backing, unplastered	0.1	0.4	0.6
Fibreboard, 12 mm on solid backing	0.05	0.15	0.3
Fibreboard, 12 mm over 25mm air space	0.3	0.3	0.3
Mineral wool, 25 mm with 5% perforated hardboard over	0.1	0.85	0.35
Expanded polystyrene board, 25 mm over 50 mm airspace	0.1	0.55	0.1
Flexible polyurethane foam, 50 mm on solid backing	0.25	0.85	0.9

However, within this grouping, it is clear that certain materials spread over two or three categories. Insulation materials are therefore categorised according to their composition, with descriptions of their various forms, typical uses in construction and, where appropriate, fire protection properties. Materials are initially divided into those of inorganic and organic origin respectively.

The broad range of non-combustible insulating materials is manufactured from ceramics and inorganic minerals including natural rock, glass, calcium silicate and cements. Some organic products are manufactured from natural cork or wood fibres but materials developed by the plastics industry predominate. In some cases these organic materials offer the higher thermal insulation properties but many are either inflammable or decompose within fire. Cellular plastics include open and closed-cell materials. Generally the closed-cell products are more rigid and have better thermal insulation properties and resistance to moisture, whereas the open-cell materials are more flexible and permeable. Aluminium foil is considered as a particular case as its thermal insulation properties relate to the transmission of radiant rather than conducted heat. Typical thermal conductivity values are indicated in Table 13.2.

Inorganic insulation materials

FOAMED CONCRETE

The manufacture of foamed concrete is described in Chapter 3. Foamed concrete with an air content in the range 30–80% is a fire- and frost-resistant material. Foamed concrete can be easily placed without the need for compaction but it does exhibit a higher drying shrinkage than dense concrete. It is suitable for insulating under floors and on flat roofs where it may be laid to a fall of up to 1 in 100. (Thermal conductivity ranges from 0.10 W/m K at a density of 400 kg/m^3 to 0.63 W/m K at a density of 1600 kg/m^3.)

LIGHTWEIGHT AGGREGATE CONCRETE

Lightweight concrete blocks and *in-situ* concrete are discussed in Chapters 2 and 3 respectively. Lightweight concrete materials offer a range of insulating and load-bearing properties, starting from 0.10 W/m K at a crushing strength of 2.8 MPa. Resistance to airborne sound in masonry walls is closely related to the mass of the wall. However, any unfilled mortar joints which create air paths will allow significant leakage of sound. In cavity walls again mass is significant, but additionally to reduce sound transmissions the two leaves should be physically isolated, with the exception of any necessary wall ties, to comply with the Building Regulations.

GYPSUM PLASTER

Plasterboard thermal linings will increase the thermal response in infrequently heated accommodation; the effect can be enhanced with metallised polyester-backed boards which reduce radiant as well as transmitted heat loss. The addition of such linings for either new or upgrading existing buildings reduces the risk of thermal bridging at lintels, etc. (The thermal conductivity of gypsum plaster is typically 0.16 W/m K.)

Table 13.2 Typical thermal conductivity values for various building materials

Material	Thermal conductivity (W/m K)
Aerogel	0.018
Phenolic foam	0.018–0.031
Polyurethane foam (rigid)	0.019–0.023
Foil-faced foam	0.020
Polyisocyanurate foam	0.023–0.025
Extruded polystyrene	0.025–0.027
Expanded PVC	0.030
Mineral wool	0.031–0.040
Glass wool	0.031–0.040
Expanded polystyrene	0.033–0.040
Cellulose (recycled paper)	0.035–0.040
Flax	0.037
Sheep's wool	0.037–0.039
Rigid foamed glass	0.037–0.048
Urea-formaldehyde foam	0.038
Hemp	0.040
Corkboard	0.042
Coconut fibre boards	0.045
Fibre insulation board	0.050
Perlite board	0.050
Straw bales	0.050
Exfoliated vermiculite	0.062
Thatch	0.072
Wood wool slabs	0.077
Medium density fibreboard (MDF)	0.10
Foamed concrete (low density)	0.10
Lightweight to dense concrete	0.10–1.7
Compressed straw slabs	0.10
Softwood	0.13
Oriented strand board (OSB)	0.13
Hardboard	0.13
Particleboard/plywood	0.14
Gypsum plasterboard	0.19
Bituminous roofing sheet	0.19
Cement bonded particleboard	0.23
Unfired clay blocks	0.24
Calcium silicate boards	0.29
GRC – lightweight	0.21–0.5
GRC – standard density	0.5–1.0
Mastic asphalt	0.5
Calcium silicate brickwork	0.67–1.24
Clay brickwork	0.65–1.95
Glass – sheet	1.05

Notes:

Individual manufacturers' products may differ from these typical figures.
Additional data is available in BS 5250: 2002 and BS EN 12624: 2000.

Sound transmission through lightweight walls can be reduced by the use of two layers of differing thicknesses of gypsum plasterboard (e.g. 12.5 and 19 mm) as these resonate at different frequencies. The addition of an extra layer of plasterboard attached to existing ceilings with resilient fixings can reduce sound transmission from upper floors particularly if an acoustic quilt can also be incorporated.

WOOD WOOL SLABS

Wood wool slabs manufactured from wood fibres and cement (Chapter 4) are both fire- and rot- resistant. With their combined load-bearing and insulating properties, wood wool slabs are suitable as a roof decking material, which may be exposed, painted or plastered to the exposed lower face. Wood wool slabs offer good sound absorption properties due to their open textured surface, and this is largely unaffected by the application of sprayed emulsion paint. Acoustic insulation for a pre-screeded 50 mm slab is typically 30 dB. (The thermal conductivity of wood wool is typically 0.077 W/m K.)

MINERAL WOOL

Mineral wool is manufactured from volcanic rock (predominantly silica, with alumina and magnesium oxide) which is blended with coke and limestone and fused at 1500°C in a furnace. The melt runs onto a series of rotating wheels which spin the droplets into fibres; they are then coated with resin binder and water-repellent mineral oil. The fibres fall onto a conveyor belt, where the loose mat is compressed to the required thickness and density, then passed into an oven where the binder is cured; finally, the product is cut into rolls or slabs. Mineral wool is non-combustible, water-repellent, rot-proof and contains no CFCs or HCFCs.

Mineral wool is available in a range of forms dependent on its degree of compression during manufacture and its required use:

- loose for blown cavity insulation;
- mats for insulating lofts, lightweight structures and within timber-framed construction;
- batts (slabs) for complete cavity fill of new masonry;
- semi-rigid slabs for partial cavity fill of new masonry;
- rigid slabs for warm pitched roof and flat roof insulation;

- rigid resin-bonded slabs for floor insulation;
- weather-resistant boards for inverted roofing systems;
- dense pre-painted boards for exterior cladding;
- ceiling tiles.

The mats and board materials may be faced with aluminium foil to enhance their thermal properties. Roof slabs may be factory cut to falls or bitumen-faced for torch-on sheet-roofing systems. Floor units are coated with paper when they are to be directly screeded. A resilient floor can be constructed with floor units manufactured from mineral wool slabs, with the fibres orientated vertically rather than horizontally, bonded directly to tongued and grooved flooring-grade chipboard.

(The thermal conductivity of mineral wool products for internal use ranges typically between 0.031 and 0.039 W/m K at 10°C, although products for external use have higher conductivities.)

Mineral wool can be used effectively to attenuate transmitted sound. In lightweight construction, acoustic absorbent quilts are effective for reducing transmitted sound through separating walls when combined with double plasterboard surfaces and a wide airspace, as well as in traditional timber joist floors when combined with a resilient layer between joists and floor finish. Pelletised mineral wool can be used for *pugging* between floor joists to reduce sound transmission, and is particularly appropriate for upgrading acoustic insulation during refurbishment.

Mineral wool, due to its non-combustibility, is used for manufacture of fire stops to prevent fire spread through voids and cavities, giving fire resistance ratings between 30 and 120 minutes. Mineral wool slabs give typically between 60 minutes' and 4 hours' fire protection to steel. Similar levels of protection can be achieved with sprayed-on mineral wool which may then be coated with a decorative finish.

Ceiling tiles for suspended ceilings manufactured from mineral wool typically provide Class 1 Spread of Flame to BS 476 Part 7 (1997) and Class 0 to Part 6 (1989) on both their decorative and back surfaces. The thermal conductivity of mineral wool suspended ceiling tiles is typically within the range 0.052–0.057 W/m K. Sound attenuation of mineral wool ceiling tiles usually lies within the range 34–6 dB, but depending upon the openness of the tile surface, the sound absorption coefficient may range from 0.1 for smooth tiles, through 0.5 for fissured finishes to 0.95 for open-cell tiles overlaid with 20 mm mineral wool.

GLASS WOOL

Glass wool is made by the Crown process (Fig. 13.1), which is similar to that used for mineral wool. A thick stream of glass flows from a furnace into a forehearth and by gravity into a rapidly rotating steel alloy dish, punctured by hundreds of fine holes around its perimeter. The centrifugal force expels the filaments which are further extended into fine fibres by a blast of hot air. The fibres are sprayed with a bonding agent and then sucked onto a conveyor to produce a mat of the appropriate thickness. This is cured in an oven to set the bonding agent, then finally cut, trimmed and packaged.

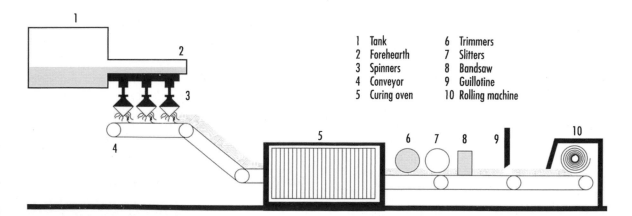

1	Tank	6	Trimmers
2	Forehearth	7	Slitters
3	Spinners	8	Bandsaw
4	Conveyor	9	Guillotine
5	Curing oven	10	Rolling machine

Fig. 13.1 Crown process for the manufacture of glass wool

Glass wool is non-combustible, water-repellent, rot proof and contains no CFCs or HCFCs. It is available in a range of product forms:

- loose for blown cavity wall insulation;
- rolls, either unfaced or laminated between kraft paper and polythene, for roofs, within timber-frame construction, internal walls and within floors;
- semi-rigid batts with water-repellent silicone for complete cavity fill of new masonry;
- rigid batts for partial cavity fill within new masonry;
- compression-resistant slabs for solid concrete or beam and slab floors;
- a laminate of rigid glass wool and plasterboard for dry-linings;
- PVC-coated rigid panels for exposed factory roof linings.

(The thermal conductivity of glass wool products ranges typically between 0.031 and 0.040 W/m K at 10°C.)

The sound and fire-resistant properties of glass wool are similar to those of mineral wool. Glass wool sound-deadening quilts, which have overlaps to seal between adjacent units, are used to reduce impact sound in concrete and timber floating floors. Standard quilts are appropriate for use in lightweight partitions and over suspended ceilings.

Resin-bonded glass wool treated with water-repellent is used to manufacture some ceiling tiles which meet the Class 0 fire spread requirements of the Building Regulations (BS 476: Parts 6 and 7) and also offer sound absorption to reduce reverberant noise levels.

CELLULAR OR FOAMED GLASS BLOCKS

Cellular or foamed glass (CG) is manufactured from a mixture of crushed glass and fine carbon powder, which on heating to 1000°C, causes the carbon to oxidise creating bubbles within the molten glass. The glass is annealed, cooled and finally cut to size. The black material is durable, non-combustible, easily worked and has a high compressive strength. It is water-resistant due to its closed cell structure, impervious to water vapour and contains no CFCs.

Cellular glass slabs are appropriate for roof insulation, including green roofs and roof-top car parks owing to their high compressive strength. The slabs are usually bonded in hot bitumen to either concrete screeds, profile metal decking or bitumen-felt-coated

timber roofing. Foamed glass is suitable for floor insulation under the screed and may be used internally, externally or within the cavity of external walls. Externally it may be rendered or tile hung and internally finished with plasterboard or expanded metal and conventional plaster. (The thermal conductivity of cellular glass is within the range 0.037–0.048 W/m K at 10°C, depending upon the grade.)

EXFOLIATED VERMICULITE

Exfoliated vermiculite, which contains up to 90% air by volume, is used as a loose fill for loft insulation and within a cementitious spray produces a hard fire protection coating for exposed structural steelwork. Where thicknesses over 30 mm are required, application should be in two coats. The product has a textured surface finish which may be exposed internally or painted in external applications. Depending upon the thickness of application and the ratio (Hp/A) between exposed surface area and steel cross-section (Chapter 5), up to 240 minutes' fire protection may be obtained. Vermiculite is used for certain demountable fire stop seals where services penetrate through fire compartment walls. (The thermal conductivity of exfoliated vermiculite is 0.062 W/m K. Within lightweight aggregate concrete a thermal conductivity of typically 0.11 W/m K can be achieved.)

EXPANDED PERLITE

Expanded perlite is manufactured from natural volcanic rock minerals. It is used for loose and bonded *in-situ* insulation for roofs, ceilings, walls and floors also as preformed boards. (The thermal conductivity of expanded perlite boards is 0.05 W/m K.)

CALCIUM SILICATE

Calcium silicate, which is described in Chapter 12, has the advantage of good impact resistance and is very durable. Various wallboards are manufactured with calcium silicate boards laminated to extruded polystyrene. (Calcium silicate typically has a thermal conductivity of 0.29 W/m K.)

GLASS AND MULTIPLE GLAZING

The thermal and sound insulation effects of double and triple glazing and the use of low-emissivity glass are described in Chapter 7.

AEROGEL

Aerogels are extremely lightweight, hydrophobic amorphous silica materials with densities as low as 3 mg/l (ρ air = 1.2 mg/l). They are manufactured by solvent evaporation from silica gel under reduced pressure. Aerogels are highly porous with typically 95–7% and even 99.8% air space, but significantly the pore size of 20 nm is so small that it is less than the mean free path of nitrogen and oxygen in the air. This prevents the air particles moving and colliding with each other which would normally give rise to gas phase heat conduction. With only 3–5% solid material, heat conduction in the solid phase is very limited. When used to fully fill the cavity in glazing units, 0.5–4 mm aerogel granules prevent the movement of air, thus reducing heat transfer by convection currents. Limited heat transfer can therefore only occur across the glazing unit by radiation.

Light transmission through aerogel is approximately 80% per 10 mm thickness, giving a diffuse light and eliminating the transmission of UV. Airborne sound transmission is reduced particularly for lower frequencies of less than 500 Hz. The material is hydrophobic and therefore resistant to mould growth.

Polycarbonate glazing units filled with aerogel are available as 10 and 16 mm panels, over a range of sizes to fit profiled glass trough sections (Chapter 7) or as translucent rooflights or wall panels (e.g. 1220 × 3660 mm). (The thermal conductivity of aerogel silica is typically 0.018 W/m K.)

Organic insulation materials

The use of straw bales is described in Chapter 15 with other recycled products.

CORK PRODUCTS

Cork is harvested from the Cork Oak (*Quercus suber*) on a nine-year (or more) cycle and is therefore considered to be an environmentally-friendly material. For conversion into boards, typically used for roof insulation, cork granules are expanded, then formed under heat and pressure into blocks using the natural resin within the cork. The blocks are trimmed to standard thicknesses or to a taper to produce falls for flat roofs (Fig. 13.2). For increased thermal insulation properties, the cork may be bonded to closed-cell

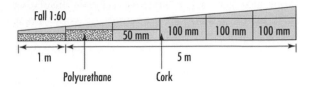

Fig. 13.2 Cork insulation to falls for flat roofs

polyurethane or polyisocyanurate foam. In this case the laminate should be laid with the cork uppermost. The overall thickness of the insulation required will be dependent on the structural substrate material and the target thermal properties of the roof. Cork products are unaffected by the application of hot bitumen in flat roofing systems. (The thermal conductivity of corkboard is 0.042 W/m K.)

SHEEP'S WOOL

Sheep's wool is a very efficient renewable resource insulation material, with a low conductivity that compares favourably to other fibrous insulants. It is available in grey batts ranging in thickness from 50- and 75- to 100-mm-thick. Wool is a hygroscopic material, that is, it reversibly absorbs and releases water vapour, and this effect is advantageous when it is used for thermal insulation. When the building temperature rises, wool releases its moisture causing a cooling effect in the fibre and thus a reduced flow of heat into the building. In winter the absorption of moisture warms the material. This evolution of heat helps to prevent interstitial condensation in construction cavities by maintaining the temperature of the fibres above the dew point, and also effectively reduces the heat loss from the building.

Wool is safe to handle, only requiring gloves and a dust mask as minimal protection. It causes no irritation except in the rare cases of people with a specific wool allergy. Wool batts, which contain 85% wool and 15% polyester to maintain their form can easily be cut with a sharp knife or torn to size. Wool is potentially susceptible to rodents, which may use it as a nesting material if it is accessible, but the batts are treated with an insecticide to prevent moth or beetle attack and with an inorganic fire retardant.

Wool batt insulation is suitable for ventilated loft applications between joists or rafters and for timber-frame construction. It should be installed with a vapour permeable breather membrane on the cold side, and kept clear of any metal chimneys or flues.

Wool also acts as an effective acoustic insulating material. (The thermal conductivity of wool batts is 0.039 W/m K.)

Sheep's wool has also been used experimentally as loose fill insulation for lofts, sloping ceilings, timber-frame walls and timber floors. Natural wool from sheep that have not been dipped is washed several times to remove the natural oil, lanolin, then opened out to the required density. It is sprayed with borax as a fire retardant and insect repellent. Supplied loose as hanks, wool is only suitable for locations where it will not get wet, which would cause it to sag, thus reducing its thermal efficiency. Wool insulation is a renewable source with low embodied energy, but it is currently more expensive than the standard mineral wool alternative. (The thermal conductivity of loose wool is 0.037 W/m K.)

CELLULOSE INSULATION

Cellulose insulation is manufactured from shredded recycled paper. It is treated with borax for flammability and smouldering resistance; this also makes it unattractive to vermin and resistant to insects, fungus and dry rot. Unlike mineral fibre and glass-fibre insulation it does not cause skin irritation during installation. Recycled cellulose has a low embodied energy compared to mineral and glass-fibre insulation, and when removed from a building it may be recycled again, or disposed of safely without creating toxic waste. (Treatment with the inorganic salt, borax, ensures that cellulose insulation conforms to BS 5803 Part 4: 1985 – Fire Test Class 1 and Smoulder Test Class B2.)

Cellulose insulation may be used directly from bags for internal floors and also lofts where the required eaves ventilation gap must be maintained. For other cavities, including sloping roof voids, the material is dry injected under pressure completely filling all spaces to prevent air circulation. In breathing walls, cellulose insulation is filled inside a breathing membrane, which allows the passage of water vapour through to the outer leaf of the construction. Cellulose may be damp-sprayed in-between wall studs before the wall is closed. Cellulose is a hygroscopic material, which under conditions of high humidity absorbs water vapour and then releases it again under dry conditions. Cellulose is an effective absorber of airborne sound. (The thermal conductivity of cellulose is 0.035 W/m K in horizontal applications and 0.038 to 0.040 W/m K in walls.)

FLAX, HEMP AND COCONUT FIBRE

As the demand for sustainable insulating materials increases, products derived from renewable flax, hemp and coconut fibres are becoming available. Flax insulation is suitable for ventilated or breathing constructions. The batts may be used in ceilings and walls, and rolls in lofts, suspended floors and walls. Flax is treated with borax for fire and insect resistance and bonded with potato starch, giving a moisture-absorbing, non-toxic product with good thermal and acoustic insulation properties. (The thermal conductivity of flax is 0.037 W/m K.)

Chopped hemp fibres, treated with borax for fire resistance, are used to produce insulation batts, and also as loose fill for floors and roofs. Hemp, a very tough material, is used in the manufacture of certain particleboards in Germany and generally for paper production. *Hempcrete* is described in Chapter 3. (The thermal conductivity of hemp is 0.040 W/m K.)

Coconut fibre thermal- and acoustic-insulation boards have the advantage of natural rot resistance. They are available in a range of thicknesses from 10 to 25 mm, and typical uses include ceiling and floor insulation including under-screed applications. (The thermal conductivity of coconut fibre is 0.045 W/m K. The sound reduction for a typical 18 mm under screed application is 26 dB.)

FIBRE INSULATION BOARD

The manufacture of insulation board (WF) or softboard, which is a low density wood fibre building board, is described in Chapter 4. Standard grades of insulation board should only be used in situations where they are not in contact with moisture, or at risk from the effects of condensation. Insulating board is used for wall linings and may be backed with aluminium foil for increased thermal insulation.

Insulating board may be impregnated with inorganic fire retardants to give a Class 1 Surface Spread of Flame to BS 476 Part 7 or finished with plasterboard to give a smooth Class 0 fire rated surface. The Euroclass fire performance rating under the conditions specified in BS EN 13986:2004 for 9 mm untreated high-density medium board of 600 kg/m^3 is Class D-s2, d0 for non-floor use. For untreated low-density medium board of 400 kg/m^3 the equivalent rating is Class E, pass for non-floor use and for 9 mm untreated softboard of 250 kg/m^3 the rating is Class E, pass for non-floor use.

Exposed insulation board has good sound-absorbing properties due to its surface characteristics. Standard 12 mm lining softboard has a noise reduction coefficient of 0.42, although this is increased to 0.60 for the 24 mm board.

Bitumen-impregnated insulating board, with its enhanced water-resistant properties, is used as a thermal insulation layer on concrete floors. The concrete floor slab is overlaid with polythene, followed by bitumen-impregnated insulation board and the required floor finish such as particleboard. In the upgrading of existing suspended timber floors, a loose-laid layer of bitumen-impregnated insulating board under a new floor finish can typically reduce both impact and airborne sound transmission by 10 dB. Bitumen-impregnated insulating board is frequently used in flat-roofing systems as a heat protective layer to polyurethane, polystyrene, or phenolic foams prior to the application of the hot bitumen waterproof membrane. It is also used for sarking in pitched roofs. (The thermal conductivity of insulating board is typically 0.050 W/m K.)

EXPANDED POLYSTYRENE

Expanded polystyrene is a combustible material which, in fire, produces large quantities of noxious black smoke, although Type A with a flame retardant additive is not easily ignitable. Expanded polystyrene, a closed-cell product, is unaffected by water, dilute acids and alkalis but is readily dissolved by most organic solvents. It is rot- and vermin-proof, and CFC and HCFC-free.

Polystyrene beads

Expanded polystyrene beads are used as loose fill for cavity insulation. To prevent subsequent slippage and escape through voids, one system bonds the polystyrene beads by spraying them with atomised PVA adhesive during the injection process, although other processes leave the material loose. Walls up to 12 m in height can be insulated by this type of system. Polystyrene bead insulation should not be used where electrical wiring is present in the cavity, as the polystyrene gradually leaches the plasticiser out from plastic cables causing their embrittlement which could lead to problems later if the cables are subsequently moved. Polystyrene bead aggregate cement is used to form an insulating sandwich core in concrete cladding panel systems.

Expanded polystyrene boards

Expanded polystyrene rigid lightweight boards are used for thermal insulation and four standard grades are available (Table 13.3). The standard material is classified as Euroclass F in relation to fire, but certain flame retardant modified boards are classified as Euroclass E. (Grades range from A1 and A2 through to F). Load-bearing expanded polystyrene for impact sound insulation properties is designated type EPS T to BS EN 13163: 2001.

The boards, which are manufactured by fusing together pre-foamed beads under heat and pressure, can easily be cut, sawn or melted with a hot wire. Polystyrene boards provide thermal insulation for walls, roofs and floors. In addition, polystyrene may be cast into reinforced concrete, from which it is easily removed to create voids for fixings.

In cavity wall insulation, a 50 mm cavity may be retained to prevent the risk of water penetration, with proprietary wall ties fixing the boards against the inner leaf. Alternatively, with a full-fill cavity system, the boards may be slightly moulded on the outer surface to shed any water back onto the inside of the external masonry leaf. Interlocking joints prevent cold bridging, air leakage and water penetration at the board joints. In upgrading existing walls, external expanded polystyrene insulation should be protected

Table 13.3 Standard grades of polystyrene (BS 3837–1: 2004 and BS EN 13163: 2001)

Grade		Description	Typical density (kg/m³)	Thermal conductivity (W/m K)
BS 3837	BS EN 13163			
SD	EPS 70	standard duty	15	0.038
HD	EPS 100	high duty	20	0.036
EHD	EPS 150	extra high duty	25	0.035
UHD	EPS 200	ultra high duty	30	0.034

by suitably supported rendering or tile hanging. For internal wall insulation, expanded polystyrene can be used in conjunction with 12.5 mm plasterboard either separately or as a laminate. Expanded polystyrene is used to give thermal insulation in ground floors. It may be laid below or above the oversite slab; in the latter case it may be screeded or finished with chipboard. Composite floor panels manufactured from expanded polystyrene and oriented strand board are suitable for beam and block floors while proprietary systems offer thermal insulation to prestressed concrete beam and reinforced concrete screed floors. Expanded polystyrene boards reduce impact and airborne sound transmission through intermediate floors.

Expanded polystyrene is suitable for thermal insulation in flat and pitched roofs. For flat roofs it may be cut to falls. Where hot bitumen products are to be applied, the expanded polystyrene boards must be protected by an appropriate layer of bitumen-impregnated fibreboard, perlite board or corkboard. In metal deck applications the insulating layer may be above or below the purlins, whereas in traditional pitched roofs expanded polystyrene panels are normally installed over the rafters. Expanded polystyrene, although a closed-cell material, acts as a sound-absorber, providing it is installed with an air gap between it and the backing surface. It particularly absorbs sound at low frequencies and may be used in floors and ceilings. It is, however, less effective than the open-cell materials such as flexible polyurethane foam. (The thermal conductivity of expanded polystyrene is in the range 0.033–0.040 W/m K depending upon the grade.)

EXTRUDED POLYSTYRENE

Extruded polystyrene is normally manufactured by a vacuum process, although some is blown with CFCs. It is slightly denser and therefore slightly stronger in compression than expanded polystyrene, but has a lower thermal conductivity. It has a closed-cell structure with very low water-absorption and vapour-transmission properties. Extruded polystyrene is available with densities ranging from 20–40 kg/m^3. Extruded polystyrene is widely used for cavity wall and pitched roof insulation. Because of its high resistance to water absorption, extruded polystyrene may be used for floor insulation below the concrete slab and on inverted roofs where its resistance to mechanical damage from foot traffic is advantageous. Extruded polystyrene is also available laminated to

tongued and grooved moisture-resistant flooring grade particleboard for direct application to concrete floor slabs, and laminated to plasterboard as a wallboard. (The thermal conductivity of extruded polystyrene is typically 0.025–0.027 W/m K.)

EXPANDED PVC

Plasticised PVC open, partially open and closed-cell foams are manufactured as flexible or rigid products within the density range of 24–72 kg/m^3. The rigid closed-cell products provide low water permeability and are self-extinguishing in fire. Expanded PVC boards are used in sandwich panels and for wall linings. The low density open-cell material has particularly good acoustic absorbency and can be used to reduce sound transmission through unbridged cavities and floating floors. (The thermal conductivity of expanded PVC is typically 0.030 W/m K.)

POLYISOCYANURATE FOAM

Polyisocyanurate foam (PIR), usually blown with HCFCs is available in two grades: PIR1 and PIR2. It is used as a roof insulation material since it is more heat-resistant than other organic insulation foams, which cannot be directly hot-bitumen bonded. Polyisocyanurate is also appropriate for use in wall and floor insulation. PIR is combustible (BS 476 Part 4) with a Class 1 Surface Spread of Flame (BS 476 Part 7) but is more fire-resistant than polyurethane foam. Polyisocyanurate tends to be rather friable and brittle. Certain proprietary systems for insulated cavity closers use PVC-U-coated polyisocyanurate insulation. Such systems offer a damp-proof barrier and can assist in the elimination of cold bridging, which sometimes causes condensation and mould growth around door and window openings. (The thermal conductivity of polyisocyanurate foam is usually in the range 0.023–0.025 W/m K.)

POLYURETHANE FOAM

Rigid polyurethane (PUR) is closed-cell foam currently manufactured using CFCs (chlorofluorocarbons) or HCFCs (hydrochlorofluorocarbons); the latter being slightly less damaging to the atmospheric ozone layer. The CFCs and HCFCs remain trapped in the closed cells enhancing the thermal performance. Certain polyurethanes are modified with polyisocyanurates.

Rigid polyurethane is a combustible material producing copious noxious fumes and smoke in fire, although a flame-resistant material is available. It is used to enhance the thermal insulation properties of concrete blocks either by filling the void spaces in hollow blocks or by direct bonding onto the cavity face. Roofboards, in certain systems pre-bonded to bitumen roofing sheet, are suitable for mastic asphalt and built-up roofing systems. Owing to the temperature stability of polyurethane, no additional protection from the effects of hot bitumen application is required; the durability of the material also makes it suitable for use in inverted roofs. Laminates with foil or kraft paper are available. Factory-manufactured double-layer profiled-metal sheeting units are frequently filled with rigid polyurethane foam due to its good adhesive and thermal insulation properties. Polyurethane laminated to plasterboard is used as a wallboard. When injected as a pre-mixed two-component system into cavity walls, polyurethane adheres well to the masonry, foaming and expanding *in-situ* to completely fill the void space. It has been used in situations where the cavity ties have suffered serious corrosion, and where additional bonding between the two leaves of masonry is required. However, polyurethane foam is not now widely available as a cavity insulation material.

Flexible polyurethane foam is an open-cell material offering good noise absorption properties. It is therefore used in unbridged timber-frame partitions, floating floors and duct linings to reduce noise transmission. Polyurethane foams are resistant to fungal growth, aqueous solutions and oils, but not to organic solvents. (The thermal conductivity of rigid polyurethane foam is usually in the range 0.019–0.023 W/m K at a nominal density of 32 kg/m^3. Flexible polyurethane foam typically has a thermal conductivity of 0.048 W/m K.)

UREA-FORMALDEHYDE FOAM

Urea-formaldehyde foam was used extensively in the 1980s for cavity wall insulation, but it can shrink after installation, creating fissures which link the outer and inner leaves. Occasionally, in conditions of high exposure, this had led to rainwater penetration. After installation the urea-formaldehyde foam emits formaldehyde fumes, which have in certain cases entered buildings, causing occupants to suffer from eye and nose irritation. The problem normally arises only if the inner leaf is permeable and a cavity greater than 100 mm is being filled. Recent advances claim to have reduced formaldehyde emissions but all installations must be undertaken to the stringent British Standard BS 5618: 1985. (The thermal conductivity of urea-formaldehyde foam is typically 0.038 W/m K.)

PHENOLIC FOAM

Phenolic foams, which have very low thermal conductivities, are used as alternatives to rigid polyurethane and polyisocyanurate foams, where a self-extinguishing low smoke emission material is required. Phenolic foams are produced with densities in the range 35 kg/m^3 to 200 kg/m^3, but some material is still blown with CFCs or HCFCs. Wallboard laminates with plasterboard offer good thermal insulation properties due to the very low thermal conductivity of phenol foam, compared to polyurethane or extruded polystyrene. Phenolic foams are stable up to a continuous temperature of 120°C. (The thermal conductivity of phenolic foam in the density range 35–60 kg/m^3 is typically 0.020 W/m K, although the open-cell material has a thermal conductivity of 0.031 W/m K.)

Aluminium foil

Aluminium foil is frequently used as an insulation material in conjunction with organic foam or insulating gypsum products. It acts by a combination of two physical effects. First, it reflects back incident heat due to its highly reflecting surface. Second, owing to its low emissivity, the re-radiation of any heat that is absorbed is reduced.

THERMO-REFLECTIVE INSULATION PRODUCTS

Proprietary quilt systems incorporating multilayers of aluminium foil, fibrous materials and cellular plastics act as insulation by reducing conduction, convection and radiation (Fig. 13.3). A range of these thermo-reflective insulation products are manufactured using different combinations of thin plastic foam, plastic bubble sheet, and non-woven fibrous wadding with plain and reinforced aluminium foil. Thus a 10 mm sandwich system consisting of four layers of aluminium foil alternating with three layers of polyethylene bubble sheet can achieve the thermal insulation effect equivalent to that of approximately 100 mm mineral wool. A 25-mm-thick system, composed

Fig. 13.3 Multi-layer aluminium foil insulation system

of 14 intermixed layers of aluminium foil, foam and wadding, can achieve the same thermal efficiency as roof insulation equivalent to that of 200 mm of mineral wool. Higher levels of insulation are achieved by systems composed of larger numbers of layers and increased overall thicknesses, e.g. a 30 mm system with 19 layers can achieve a U-valve of 0.19 W/m² K. (The thermal conductivity of foil-faced foam is typically 0.020 W/m K.)

Chlorofluorocarbons in foamed plastics

Until recently, rigid polyurethane and polyisocyanurate foams were blown with chlorofluorocarbons (CFCs). However, due to the worldwide concern over the effects of these gases on the ozone layer, the use of CFCs is being rapidly phased out in favour of reduced ozone depletion potential (ODP) blowing agents such as the partially halogenated alkanes (PHAs), usually hydrochlorofluorocarbons (HCFCs). HCFCs themselves are due to be phased out early this century, and it is likely that hydrofluorocarbons (HFCs) with zero ozone depletion potential will quickly become the standard blowing agent. Carbon dioxide can be used as a blowing agent, but it produces less dimensionally stable products with higher thermal conductivities. Generally, CFC-blown polyurethane foam has better insulating properties (thermal conductivity 0.019 W/m K) than the equivalent foam blown by non-CFCs (thermal conductivity 0.022 W/m K).

References

FURTHER READING

Bynum, R. and Rubino, D. 2000: *Insulation handbook.* Maidenhead: McGraw.

CIRIA. 1986: *Sound control for homes – A design manual.* CIRIA Report No.114. London: Construction Industry Research and Information Association.

Johnson, S. 1993: *Greener buildings – Environmental impact of property.* Basingstoke: Macmillan.

Kefford, V.L. 1993: *Plastics in thermal and acoustic building insulation.* Review Report No. 67, **6**(7): Shrewsbury: RAPRA Technology Ltd.

TIMSA. 1994: *Handbook – The specifiers insulation guide.* Aldershot: Thermal Insulation Manufacturers & Suppliers Association.

STANDARDS

BS 476 Fire tests on building materials and structures:
  Part 6: 1989 Method of test for fire propagation of materials.
  Part 7: 1997 Method for classification of the surface spread of flame of products.
BS 874: 1973 Methods for determining thermal insulating properties.
BS 874–1: 1986 Methods for determining thermal insulating properties.
BS 2750 Acoustics – measurement of sound insulation in buildings and of building elements:
  Part 2: 1993 Determination, verification and application of precision data.
  Part 3: 1995 Laboratory measurement of airborne sound insulation of building elements.
  Part 9: 1987 Laboratory measurement of room-to-room airborne sound insulation of a suspended ceiling with a plenum above it.
BS 3379: 2005 Flexible polyurethane foam materials for load-bearing applications.
BS 3533: 1981 Glossary of thermal insulation terms.
BS 3837 Expanded polystyrene boards:
  Part 1: 2004 Boards and blocks manufactured from expandable beads.
BS 4023: 1975 Flexible cellular PVC sheeting.
BS 4841 Rigid polyurethane (PUR) and polyisocyanurate foam for building applications:
  Part 1: 1993 Laminated board for general purposes.
  Part 2: 1975 Laminated board for use as a wall and ceiling insulation.
  Part 3: 1994 Specification for two types of laminated board (roofboards).
BS 5241 Rigid polyurethane (PUR) and polyisocyanurate (PIR) foam when dispensed or sprayed on a construction site:
  Part 1: 1994 Specification for sprayed foam thermal insulation applied externally.
  Part 2: 1991 Specification for dispensed foam for thermal insulation or buoyancy applications.
BS 5250: 2002 Code of practice for control of condensation in buildings.
BS 5422: 2001 Thermal insulating materials on pipes, ductwork and equipment.
BS 5608: 1993 Specification for preformed rigid polyurethane (PUR) and polyisocyanurate (PIR) foams for thermal insulation of pipework and equipment.
BS 5617: 1985 Specification for urea-formaldehyde (UF) foam systems suitable for thermal insulation of cavity walls with masonry or concrete inner and outer leaves.
BS 5618: 1985 Code of practice for thermal insulation of cavity walls by filling with urea-formaldehyde (UF) foam systems.
BS 5803 Thermal insulation for use in pitched roof spaces in dwellings:
  Part 2: 1985 Specification for man-made mineral fibre thermal insulation in pelleted or granular form for application by blowing.
  Part 3: 1985 Specification for cellulose fibre thermal insulation for application by blowing.
  Part 4: 1985 Methods for determining flammability and resistance to smouldering.
  Part 5: 1985 Specifications for installations of man-made mineral fibre and cellulose fibre insulation.
BS 5821 Methods for rating the sound insulation in buildings and of building elements:
  Part 3: 1984 Airborne sound insulation of façade elements and facades.
BS 6203: 2003 Guide to the fire characteristics and fire performance of expanded polystyrene materials (EPS and XPS) used in building applications.
BS 7021: 1989 Code of practice for thermal insulation of roofs externally by means of sprayed rigid polyurethane (PUR) or polyisocyanurate (PIR) foam.
BS 7456: 1991 Code of practice for stabilization and insulation of cavity walls by filling with polyurethane (PUR) foam systems.
BS 7457: 1994 Specification for polyurethane (PUR) foam systems suitable for stabilisation and thermal

insulation of cavity walls with masonry or concrete inner and outer leaves.

BS 8207: 1985 Code of practice for energy efficiency in buildings – design guide.

BS 8216: 1991 Code of practice for use of sprayed lightweight mineral coatings used for thermal insulation and sound absorption in buildings.

BS 8233: 1999 Sound insulation and noise reduction for buildings.

BS EN ISO 717 Acoustics – rating of sound insulation in buildings:

    Part 1: 1997 Airborne sound insulation.

    Part 2: 1997 Impact sound insulation.

BS EN ISO 5999: 2004 Polymeric materials, cellular flexible – polyurethane foam for load-bearing applications excluding carpet underlay.

BS EN ISO 6946: 1997 Building components and building elements – thermal resistance and thermal transmittance – calculation method.

BS EN ISO 10077 Thermal performance of windows, doors and shutters:

    Part 1: 2000 Calculation of transmittance.

    Part 2: 2003 Calculation of transmittance – frames.

BS EN ISO 12241: 1998 Thermal insulation for building equipment and industrial installations.

BS EN 12354 Building acoustics – estimation of acoustic performance of buildings.

    Part 1: 2000 Airborne sound insulation between rooms.

    Part 2: 2000 Impact sound insulation between rooms.

    Part 3: 2000 Airborne sound insulation against outdoor sound.

    Part 4: 2000 Transmission of indoor sound to the outside.

    Part 6: 2003 Sound absorption in enclosed spaces.

BS EN 12524: 2000 Building materials and products – hygrothermal properties – tabulated design values.

BS EN ISO 12567 Thermal performance of windows and doors:

    Part 1: 2000 Complete windows and doors.

BS EN 12758: 2002 Glass in building – Glazing and airborne sound insulation.

BS EN 13162: 2001 Thermal insulation products for building – factory made mineral wool (MW) products.

BS EN 13163: 2001 Thermal insulation products for building – factory made products of expanded polystyrene (EPS).

BS EN 13164:2001 Thermal insulation products for building – factory made products of extruded polystyrene foam (XPS).

BS EN 13165: 2001 Thermal insulation products for building – factory made rigid polyurethane foam (PUR) products.

BS EN 13166: 2001 Thermal insulation products for building – factory made products of phenolic foam (PF).

BS EN 13167: 2001 Thermal insulation products for building – factory made cellular glass (CG) products.

BS EN 13168: 2001 Thermal insulation products for building – factory made wood wool (WW) products.

BS EN 13169: 2001 Thermal insulation products for building – factory made products of expanded perlite (EPB).

BS EN 13170: 2001 Thermal insulation products for building – factory made products of expanded cork (ICB).

BS EN 13171: 2001 Thermal insulation products for building – factory made wood fibre (WF) products.

BS EN 13172: 2001 Thermal insulating products.

BS EN ISO 13370: 1998 Thermal performance of buildings – heat transfer via the ground.

BS EN 13467: 2001 Thermal insulation for building equipment and industrial installations.

BS EN 13468: 2001 Thermal insulation for building equipment and industrial installations.

BS EN 13469: 2001 Thermal insulation for building equipment and industrial installations.

BS EN 13470: 2001 Thermal insulation for building equipment and industrial installations.

BS EN 13471: 2001 Thermal insulation for building equipment and industrial installations.

BS EN 13472: 2001 Thermal insulation for building equipment and industrial installations.

BS EN 13494: 2002 Thermal insulation products for building applications.

BS EN 13495: 2002 Thermal insulation products for building applications.

BS EN 13496: 2004 Thermal insulation products for building applications.

BS EN 13497: 2002 Thermal insulation products for building applications. External thermal insulation.

BS EN 13498: 2002 Thermal insulation products for building applications. External thermal insulation.

BS EN 13499: 2003 Thermal insulation products for building applications. External thermal insulation.

BS EN ISO 13788: 2002 Hygrothermal performance of building components and building elements.

BS EN ISO 13789: 1999 Thermal performance of buildings. Transmission heat loss coefficient.

BS EN 13791: 2004 Thermal performance of buildings – Calculation of internal temperatures of a room in summer without mechanical cooling.

pr EN 14064–1: 2000 Thermal insulation for buildings – *in situ* formed loose-fill mineral wool products.

pr EN 14064: 2000 Thermal insulation in buildings – *in situ* formed loose-fill mineral wool products.

pr EN 14303: 2002 Thermal insulation products for building equipment and industrial installations. MW.

pr EN 14304: 2005 Thermal insulation products for building equipment and industrial installations. FEF.

pr EN 14305: 2002 Thermal insulation products for building equipment and industrial installations. CG.

pr EN 14306: 2002 Thermal insulation products for building equipment and industrial installations. Calcium silicate CS.

pr EN 14307: 2005 Thermal insulation products for building equipment and industrial installations. XPS.

pr EN 14308: 2005 Thermal insulation products for building equipment and industrial installations. PUR.

pr EN 14309: 2005 Thermal insulation products for building equipment and industrial installations. EPS.

pr EN 14313: 2002 Thermal insulation products for building equipment and industrial installations. PEF.

pr EN 14314: 2005 Thermal insulation products for building equipment and industrial installations. PF.

pr EN 14315: 2002 Thermal insulating products for buildings. PUR.

pr EN 14316: 2002 Thermal insulation products for buildings. EP.

BS EN 14317: 2004 Thermal insulation products for buildings. EV.

pr EN 14318: 2002 Thermal insulating products for buildings. PUR.

pr EN 14319: 2002 Thermal insulating products for building equipment and industrial installations. PUR.

pr EN 14320: 2002 Thermal insulating products for building equipment and industrial installations. PUR.

pr EN 15100: 2004 Thermal insulating products for buildings. UF.

PD CEN/TR 15131: 2006 Thermal performance of building materials.

BS EN 20140 Acoustics – measurement of sound insulation in buildings and of building elements:

    Part 2: 1993 Determination, verification and application of precision data.

    Part 9: 1994 Room to room airborne sound insulation of a suspended ceiling.

    Part 10: 1992 Airborne sound insulation of small building elements.

pr EN ISO 23993: 2004 Thermal insulation for building equipment and industrial installations.

PD 6680: 2002 Guidance on the new European Standards for thermal insulation materials.

BUILDING RESEARCH ESTABLISHMENT PUBLICATIONS

BRE Digests

BRE Digest 293: 1985 Improving the sound insulation of separating walls and floors.

BRE Digest 294: 1985 Fire risk from combustible cavity insulation.

BRE Digest 295: 1985 Stability under wind load of loose-laid external roof insulation boards.

BRE Digest 324: 1987 Flat roof design: thermal insulation.

BRE Digest 336: 1989 Swimming pool roofs: minimising the risk of condensation.

BRE Digest 337: 1988 Sound insulation: basic principles.

BRE Digest 338: 1988 Insulation against external noise.

BRE Digest 347: 1989 Sound insulation of lightweight buildings.

BRE Digest 358: 1992 CFCs in buildings.

BRE Digest 369: 1992 Interstitial condensation and fabric degradation.

BRE Digest 379: 1993 Double glazing for heat and sound insulation.

BRE Digest 453: 2000 Insulating glazing units.

BRE Information papers

BRE IP 6/88 Methods for improving the sound insulation between converted flats.

BRE IP 12/89 Insulation of dwellings against external noise.

BRE IP 3/90 U-value of ground floors; application to building regulations.

BRE IP 12/91 Fibre building boards: types and uses.

BRE IP 18/92 Sound insulation and the 1992 edition of Approved Document E.

BRE IP 7/93 U-value of solid ground floors with edge insulation.

BRE IP 21/93 Noise climate around our homes.

BRE IP 6/94 The sound insulation provided by windows.

BRE IP 12/94 Assessing condensation risk and heat loss at thermal bridges around openings.

BRE IP 14/94 U-values for basements.

BRE IP 3/95 Comfort, control and energy efficiency in offices.

BRE IP 9/98 Energy efficient concrete walls using EPS permanent formwork.

BRE IP 1/00 Airtightness in UK dwellings.

BRE IP 14/01 Reducing impact and structure-borne sound in buildings.

BRE IP 14/02 Dealing with poor sound insulation between new dwellings.

BRE IP 3/03 Dynamic insulation for energy saving and comfort.

BRE IP 2/05 Modelling and controlling interstitial condensation in buildings.

BRE IP 4/06 Airtightness of ceilings. Energy loss and condensation risk.

BRE IP 5/06 Modelling condensation and airflow in pitched roofs.

BRE Defect action sheets

BRE DAS 77: 1986 Cavity external walls: cold bridges around windows and doors.

BRE DAS 79: 1986 External masonry walls : partial cavity fill insulation – resisting rain penetration.

BRE DAS 104: 1987 Masonry separating walls: airborne sound insulation in new-build housing.

BRE DAS 105: 1987 Masonry separating walls: improving airborne sound insulation between existing dwellings.

BRE DAS 119: 1988 Slated or tiled pitched roofs – conversion to accommodate rooms: installing quilted insulation at rafter level.

BRE DAS 131: 1989 External walls: combustible external plastics insulation – horizontal fire barriers.

BRE DAS 132: 1989 External walls: external combustible plastics insulation – fixings.

BRE DAS 133: 1989 Solid external walls: internal dry-lining – preventing summer condensation.

BRE Good building guides

BRE GBG 5: 1990 Choosing between cavity, internal and external wall insulation.

BRE GBG 22: 1999 Improving sound insulation (Parts 1 and 2).

BRE GBG 28: 1997 Domestic floors: construction, insulation and damp-proofing.

BRE GBG 31: 1999 Insulated external cladding systems.

BRE GBG 37: 2000 Insulating roofs at rafter level: sarking insulation.

BRE GBG 43: 2000 Insulating profiled metal roofs.

BRE GBG 44: 2001 Insulating masonry cavity walls (Parts 1 and 2).

BRE GBG 45: 2001 Insulating ground floors.

BRE GBG 50: 2002 Insulating solid masonry walls.

BRE Good repair guides

BRE GRG 7: 1997 Treating condensation in houses.

BRE GRG 22: 1999 Improving sound insulation.

BRE GRG 26: 1999 Improving energy efficiency: Part 1 – thermal insulation.

BRE GRG 30: 2001 Remedying condensation in domestic pitched tiled roofs.

BRE Reports

BR 238: 1993 Sound control for homes.

BR 262: 2002 Thermal insulation: avoiding risks.

BR 347: 1998 Energy efficient *in situ* concrete housing using EPS permanent formwork.

BR 358: 1998 Quiet homes; a guide to good practice.

BR 406: 2000 Specifying dwellings with enhanced sound insulation: a guide.

ADVISORY ORGANISATIONS

British Rigid Urethane Foam Manufacturers Association Ltd., Second Floor, Portland Tower, Portland Street, Manchester M1 3LF (0161 236 7575).

Cork Industry Federation, 13 Felton Lea, Sidcup, Kent DA14 6BA (020 8302 4801).

Eurisol-UK Mineral Wool Association, PO Box 35084, London NW1 4XE (020 7935 8532).

Expanded Polystyrene Insulation Board Information Service, The British Plastics Federation, EPS Construction Group, PO Box 72, Billingshurst, West Sussex RH14 0FD (01403 701167).

Insulated Render and Cladding Association Ltd., PO Box 12, Haslemere, Surrey GU27 3AH (01428 654011).

National Cavity Insulation Association, PO Box 12, Haslemere, Surrey GU27 3AH (01428 654011).

Thermal Insulation Manufacturers & Suppliers Association, Association House, 99 West Street, Farnham, Surrey GU9 7EN (01252 739154).

14

ENERGY-SAVING MATERIALS AND COMPONENTS

—

Introduction

The trend towards increasingly energy-conscious design has resulted in a greater focus on energy-saving materials and components. These include photovoltaics (PVs) and solar collectors, which turn the sun's energy into electricity and hot water respectively. Light tubes and wind catchers are energy-saving devices which can make modest reductions in the energy consumption of buildings in the context of a holistic energy-efficient strategy.

Photovoltaics

Photovoltaics are silicon-based devices which, under sunlight, generate a low-voltage direct electric current. The quantity of electricity produced is directly related to the intensity of incident solar radiation or irradiance (W/m^2). Both direct and diffuse sunlight are effective, although the intensity of direct sunlight is typically ten-fold that of an overcast sky, and the efficiency of energy conversion is around 15%. Photovoltaic cells are connected in series to generate a higher voltage. The supply is then passed through an inverter to convert the direct current into more useable alternating current at the standard voltage. The electricity generated may then be used within the building or sold back into the national supply if generation exceeds the demand.

Photovoltaic units are manufactured from a sandwich of at least two variants of mono- or poly-crystalline silicon (Fig. 14.1). These n and p-type (negative & positive) silicon crystals generate electricity at their interface under solar (photon) radiation. Cells are arranged in rectangular modules ranging from 0.3 to 1.5 m^2. A typical unit of one square metre could produce 150 watts of electrical power under bright sunshine of 1000 W/m^2.

The cells are usually laminated with a protective layer of glass, backed with metal sheeting and mounted on a steel frame. However, translucent systems built into glass double-glazing units or flexible units faced with a plastic cover are available. Mono- or poly-crystalline silicon modules are usually blue or black in colour. The mono-crystalline modules are a uniform colour, whereas the polycrystalline units have a sparkling surface. Other colours can be achieved, but with a reduced level of efficiency. The alternative amorphous thin-film silicon (TFS) cell modules are matt red, orange, yellow, green, blue or black in colour, and can be laminated into glass or mounted on a flexible plastic backing. These systems have significantly lower levels of efficiency than the crystalline cells in good light conditions, but are the more efficient in poor light conditions. Hybrid units combining mono-crystalline and thin film technologies give good output over the range of light conditions. Photovoltaic systems are normally supplied as panels, but PV slates and glazing systems are also available.

The location and tilt angle of PV installations is critical in respect of maximising output. The maximum efficiency in the northern hemisphere is gained from a south orientation with a tilt from the horizontal equal to the geographic latitude $-20°$ (Fig. 14.2). Thus for London, at a latitude of 51°, the optimum tilt is 31° from the horizontal. However, in urban areas, the effects of the immediate environment must be taken into account when assessing the available solar energy. Shadowing and inter-reflection from adjacent buildings, together with local regular weather patterns, affect the total annual solar energy, which ultimately determines the electrical output.

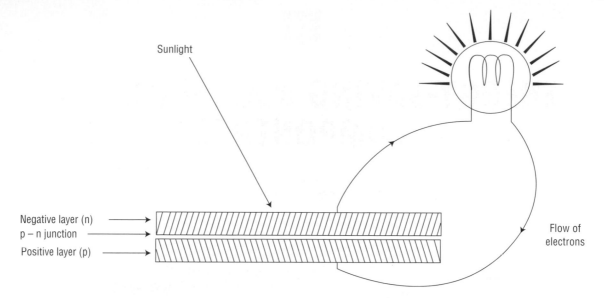

Fig. 14.1 Photovoltaic cell

Fig. 14.2 Photovoltaic array. Photograph: solarcentury.com

PV units, particularly those manufactured from crystalline silicon, decrease in performance with increased temperature, and any roof, curtain walling or rainscreen cladding system should be naturally ventilated to maintain efficiency. The use of PVs should be an integral part of the energy strategy for the building.

A 1 m square crystalline PV unit operating efficiently will generate about 100 kWh per year. The pay-back time for such a unit would be of the order of 10 years including the energy of manufacture; most installations will last between 20 and 30 years.

PHOTOVOLTAIC SLATES AND TILES

Photovoltaic slates (Fig. 14.3) and tiles, which have the general appearance of fibre-cement slates and

Fig. 14.3 Photovoltaic slates. Photograph: solarcentury.com

shiny plain tiles respectively, can be used on suitably orientated roofs as an ecological alternative to standard roofing, subject to appropriate planning consent. The lower half of the individual slates consist of photvoltaic cells which are connected into a standard photovoltaic system. The tiles are 2.1 m strips of material, marked in units to resemble conventional roof tiles and overlapped to give the required visual effect. To obtain 1 kW of power under optimum conditions, 10 m^2 of slate or 16 m^2 of tiling are required.

PHOTOVOLTAIC GLAZING

Amorphous silicon photovoltaic cells integrated into the glass of laminated or double-glazing systems can be a source of electrical energy. The glazing may be semi-transparent, retaining visual contact with the exterior or opaque in various colours. The electrical connections are made at the perimeter of the units within the frame system.

Solar collectors

The two standard types of solar collectors are the flat-plate and the evacuated-tube systems. Flat-plate collectors consist of a metal heat-absorbing plate, closely bonded to copper water pipes which transport the heated water to a storage system. The maximum efficiency is achieved using a low-emissivity matt black absorbing plate, which limits the loss of energy through re-radiation from the hot surface. A low iron-content double-glazed cover, which admits the maximum quantity of short-wave energy, protects the absorbing plate and retains the entrapped heat. The underside of the pipework is insulated with fibreglass or polyisocyanurate foam to prevent heat loss to the aluminium casing and the underlying roof structure or support system.

Evacuated-tube collectors consist of a double layer glass tube, with a vacuum between the two layers. The outer glass is clear, admitting light and heat with minimal reflection. The inner tube is coated to absorb the maximum quantity of radiation. The heat from the inner tube is transferred in a sealed unit vaporising and condensing system to a heat exchanger within the main liquid flow to the heat storage system. Evacuated-tube collectors are substantially more expensive than flat-plate collectors, but are more efficient if angled correctly and will produce higher temperatures.

Flat-plate solar collectors may be located in any unshaded location, at ground level or attached to buildings. The best orientation is directly towards the midday sun, but a variation of up to 15° east or west will have little adverse effect. The optimum tilt from horizontal for solar hot water collectors for maximum all year round efficiency equals the location's latitude. However, for increased winter efficiency, when solar gains are at a premium, the tilt from horizontal should be increased by 10°, to pick up more energy at lower sun altitudes. Solar hot water systems are heavy and must be fixed securely to suitable substrates. On tiled or slated pitched roofs an air gap should allow for the clear passage of rainwater and melting snow.

Hot water from the solar collector is usually circulated through an indirect system to a solar storage tank (Fig. 14.4). This acts as a heat store of preheated water to be fed into a standard hot water cylinder system, where the temperature can then be boosted from a boiler to the required level. Circulation may either be a gravity thermosyphon system operated by hot water convection with the storage tank located above the collector, or through a pumped system, in which case the tank may be below the collector. The circulated water must contain antifreeze and a rust inhibitor. An alternative direct system feeds tap water directly into the solar collector, but scaling and corrosion of the pipework can be problematic. A 5 m² solar collector panel will heat 250 litres of hot water per day, which is a typical four person family demand.

Solar energy district-heating plants in Europe, including Scandinavia, contribute significantly to a direct reduction in the energy requirements for small town domestic hot water systems. Water is preheated by large arrays of solar collectors before the local conventional-fuel heating system tops up the temperature to the required domestic level. Furthermore, solar heating systems in conjunction with large underground heat storage tanks can significantly reduce winter energy consumption by preheating the water supplies during periods when direct solar gain is ineffective.

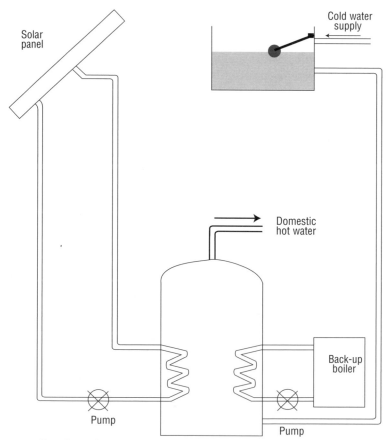

Note: Expansion tanks to heating systems not illustrated.

Fig. 14.4 Solar collector and domestic hot water system

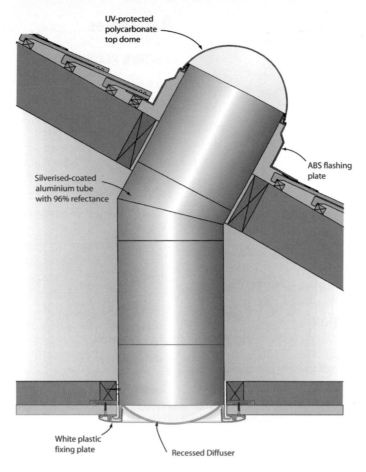

Fig. 14.5 Light pipe. Illustration: Courtesy of Monodraught

Light pipes

Light pipes or tubes transmit direct sunlight and natural daylight from roof level into the building space below (Fig. 14.5). At roof level an acrylic self-cleaning dome admits light into a highly reflective pipe, which transmits it down to a white translucent dome at ceiling level where the light is diffused into the space below. The mirror-finish aluminium tube can be of any length including offsets, although the quantity of light transmitted is typically reduced by 3% for each metre length and by 8% for each bend.

Standard light pipe diameters range from 200 mm to 600 mm, although larger sizes up to 1000 mm are available for commercial applications. The systems should be free of condensation and not cause winter heat loss or summer solar gain to the building enclosure. Rectangular units, similar in appearance to standard or flush-fitting conservation roof lights are also available, and the ceiling unit can be a square diffuser to integrate into suspended ceiling systems.

A 330 mm diameter system will typically deliver between 100 W from a winter overcast sky to 400 W under full summer sun, for a straight tube not exceeding three metres in length. Such systems can offer energy-saving solutions to existing buildings, and may be considered as one element within a fully-integrated lighting strategy for new-build.

A more sophisticated system combines the functions of both a light pipe and a wind catcher to admit natural daylight and ventilation into internal spaces poorly served by normal external glazing.

Wind catchers

Wind catchers (Fig. 14.6) have been standard architectural features on the roofs of buildings in hot dry climates for centuries. However, in order to reduce

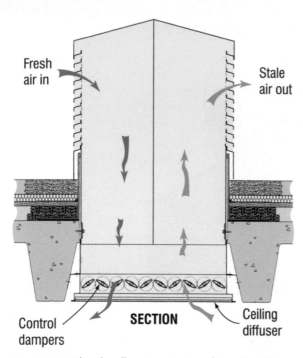

Fig. 14.6 Wind catcher. Illustration: Courtesy of Monodraught

energy costs associated with air-conditioning systems, this additional source of natural ventilation can now be designed into larger temperate-climate buildings to supplement other natural ventilation systems.

A wind catcher operates by capturing the air on the windward side of the shaft, and deflecting it down one quadrant by a series of vanes. The force of the wind drives it into the space below. As the entering air is cooler and more dense than that within the building, it displaces the warm vitiated air which rises by natural stack ventilation through the other quadrants of the shaft, leaving through the leeward side of the wind catcher. With a symmetrical system one quadrant will predominantly face the prevailing wind to act as the catcher and the opposite quadrant will provide the majority of the stack-ventilation effect. A glazed top to the wind catcher, which heats up further the vitiated air, can enhance the stack effect. Dampers can be used to reduce air flow during winter months. Wind catchers should be located near to the ridge on pitched roofs to maximise their efficiency.

For functioning in multistorey buildings, wind catchers require appropriate ducting and damper systems, and may incorporate heat exchangers from the central heating system to admit tempered fresh air in winter operation.

References

FURTHER READING

Addington, M. and Schodek, D. 2005: *Smart materials and technologies.* Oxford: Elsevier.

Baker, N. and Steemers, K. 2000: *Energy and the environment in architecture – a technical design guide.* London: E. & F.N. Spon.

Berge, B. 2001: *Ecology of building materials.* Oxford: Architectural Press.

Brown, G.Z. and DeKay, M. 2001: *Sun, wind and light: architectural design strategies.* 2nd ed. New York: John Wiley.

German Solar Energy Society. 2005: *Planning and installing photovoltaic systems: A guide for installers, architects and engineers.* London: Earthscan Publications.

Goetzberger, A and Hoffmann, V.U. 2005: *Photovoltaic solar energy generation.* Berlin: Springer-Verlag.

Gouldi, J.R. 1992: *Energy conscious design – a primer for architects.* Commission of the European Communities: Batsford.

Herzog, T. 1996: *Solar energy in architecture.* Munich: Prestel.

Herzog, T. (ed.) 1998: *Solar energy in architecture and urban planning.* Munich: Prestel.

Humm, O. and Toggweiler, P. 1993: *Photovoltaics in architecture.* Basle: Birkhäuser.

Kibert, C.J. (ed.) 2001: *Construction ecology: Nature as a basis for green buildings.* Spon.

Langston, C. (ed.) 2001: *Sustainable practices for the construction industry.* Oxford: Butterworth-Heinemann.

Lewis, J.O. (ed.) 1998: *European directory of sustainable and energy efficient building: components, services, materials.* London: James & James Science Publishers.

Luque, A. 2003: *Handbook of photovoltaic science and engineering.* London: John Wiley and Sons.

Roaf, S., Fuentes, M. and Thomas, S. 2001: *Ecohouse: a design guide.* Oxford: Architectural Press.

Roaf, S. and Walker, V. 1996: *Photovoltaics.* Oxford: Oxford Brookes University.

Sick, F. and Erge, T. (ed.) 1996: *Photovoltaics in buildings: a design handbook for architects and engineers.* London: James and James.

Smith, P.F. 2001: *Architecture in a climate of change – a guide to sustainable design.* London: Architectural Press.

Smith, P. and Pitts, C.A. 1997: *Concepts in practice – energy – building for the third millennium.* London: Batsford.

Thomas, R. (ed.) 2001: *Photovoltaics and architecture.* London: Spon.

Thomas, R. 1999: *Photovoltaics in buildings: a design guide.* London: ETSU Department of Trade and Industry.

Vale, B. and Vale, R. 2000: *The new autonomous house: design and planning for sustainability.* London: Thames and Hudson.

BUILDING RESEARCH ESTABLISHMENT PUBLICATIONS

BRE Digests

BRE Digest 355: 1991 Energy efficiency in dwellings.
BRE Digest 438: 1999 Photovoltaics integration into buildings.
BRE Digest 446: 2000 Assessing environmental impacts of construction: industry consensus, BREAM and UK ecopoints.
BRE Digest 452: 2000 Whole life-cycle costing and life-cycle assessment for sustainable building design.
BRE Digest 457: 2001 The carbon performance rating for offices.
BRE Digest 486: 2004 Reducing the effects of climate change by roof design.
BRE Digest 489: 2004 Wind loads on roof-based photovoltaic systems.
BRE Digest 495: 2005 Mechanical installation of roof-mounted photovoltaic systems.

BRE Information papers

BRE IP 2/90 Ecolabelling of building materials and building products.
BRE IP 11/93 Greenhouse-gas emissions and buildings in the United Kingdom.
BRE IP 15/98 Water conservation.
BRE IP 13/00 Green buildings revisited (Parts 1and 2).
BRE IP 17/00 Advanced technologies for 21st century building services.
BRE IP 5/01 Solar energy in urban areas.
BRE IP 3/03 Dynamic insulation for energy saving and comfort.
BRE IP 13/03 Sustainable buildings (Parts 1–4).
BRE IP 10/04 Whole life value: sustainable design in the built environment.
BRE IP 15/05 The scope for reducing carbon emissions from housing.
BRE IP 16/05 Domestic energy use and carbon emissions: Scenarios to 2050.

BRE Good practice guide

(Building Research Energy Conservation Support Unit [BRECSU])
GPG 287: 2000 Design teams guide to environmentally smart buildings: energy efficient options for new and refurbished offices.

BRE Reports

Report 370: 1999 BRE methodology for environmental profiles of construction materials, components and buildings.
Report 431: 2001 Cooling buildings in London.

ADVISORY ORGANISATIONS

British Photvoltaic Association, National Energy Centre, Davy Avenue, Knowhill, Milton Keynes, Buckinghamshire MK5 8NG (01908 442291).
Centre for Alternative Technologies, Machynlleth, Powys SY20 9AZ (01654 705950).

15

RECYCLED AND ECOLOGICAL MATERIALS

Introduction

Realisation of the finite nature of global resources and the greenhouse effect of ever-increasing carbon dioxide emissions has promoted consideration of the potential for recycling in construction of many mass produced waste products which are currently either burned or buried in landfill sites. Such materials include plastics, cardboard, straw, paper and tyres. Whilst some recycled products are only in the experimental stage, others are now becoming recognised as standard building materials. The re-use of building materials is well illustrated by the Earth Centre, Doncaster which is constructed using many recycled and reclaimed products including crushed concrete, telegraph poles, glass and radiators (Fig. 15.1).

Straw bales

Straw bales, a by-product of the mechanical harvesting and threshing of grain, are produced in large quantities in mechanised agricultural countries. The traditional rectangular bales, which are cheap and can be manhandled individually, are appropriate for building. The large cylindrical and very large rectangular bales, which require mechanical lifting, are less useful in construction and are not considered here. Standard bales (typically 330 × 530 × 1050 mm) are produced within the baler by compressing quantities of straw into flakes about 100 mm thick. These layers are built up along the length of the bale, which is then automatically tied, usually with two polypropylene strings. There is inevitably some variation in length, and the ends are slightly rounded. For construction,

the bales should be well compressed in manufacture, dry (maximum 20% moisture) to prevent the growth of moulds and fungi, and with the minimum amount of remaining grain, which might attract rodents.

In building construction, bales are stacked, large faces down, making the orientation of the straw fibres

Fig. 15.1 Recycled materials – Conference Centre at the Earth Centre, Doncaster. Architects: Bill Dunster Architects. Photograph: Courtesy of Nick Riley

predominantly horizontal. At ground level, straw bales must be protected from rising damp and from any risk of saturation from surface water. Additionally steel mesh protection from rodents is necessary. Adjacent bales must be firmly packed together to ensure stability and to reduce settling under load both during and after construction. Bales are normally secured with metal spikes or hazel rods from coppiced timber and may be sprayed with insecticide for added protection. Externally, lime render on wire mesh is appropriate as it is flexible, self-healing, and will breathe to prevent the build-up of trapped moisture. Alternatively a rainscreen, separated from the external face of the bales, may be used. Internally, straw bales are usually finished with gypsum plaster on wire mesh. Openings in straw bale construction may be formed with timber framing, but careful detailing is required to prevent water penetration at these locations. Roofs are normally set onto a timber wall plate fixed through the top bales for stability.

An alternative approach to using load-bearing straw bales is timber- or steel-frame construction with straw bales as the insulating infill (Fig. 15.2). While fire is a risk during straw bale construction, the non-combustible internal and external finishes and the compact nature of the straw make the completed construction resistant to fire. (The thermal conductivity of baled straw is approximately 0.050 W/m K.)

Cardboard

The cardboard classroom at Westborough School, Essex (Fig. 15.3) illustrates the potential of this largely recycled product as a useful construction material. A combination of flat composite panels and tubes forms the structure of this building which has an estimated life of 20 years.

In the recycling process, waste paper and cardboard are broken down and converted to pulp, which is a suspension of cellulose fibres in water. The pulp flows onto a conveyor belt, where it is drained of the excess water and compressed, causing the fibres to felt

Fig. 15.2 Straw bales in construction. Architects and Photograph: Sarah Wigglesworth Architects

Fig. 15.3 Cardboard Classroom – Westborough School, Essex. Engineers: Buro Happold. Architects: Cottrell and Vermeulen. Photograph: Copyright Adam Wilson/Buro Happold

together producing a long roll of paper. Flat cardboard sheets are formed by gluing together successive layers of paper. Tubes are manufactured from multiple layers of spirally wound paper plies, starting on a steel tube former of the appropriate size, the adhesive being starch or PVA glue. The first and last layers of paper can be of a different quality, for example, impregnated or coloured to create the required surface finish. For the Westborough School building the flat sections are composite panels consisting of multiple cardboard sheets and honeycomb cardboard interlayers, surrounded by a timber frame to facilitate ease of fixing between units. Adjacent roof and walls panels are articulated to ensure overall structural rigidity.

PROPERTIES

Cardboard, like timber, is combustible and can be treated to improve its fire performance, particularly in relation to the surface spread of flame test. However, some fire-retardant materials are environmentally unfriendly and should be avoided if the material is to be subsequently recycled.

The structural strength of cardboard is seriously affected by water. Cardboard, even if specially treated in its manufacture, is a hygroscopic material which will readily absorb moisture. It is therefore necessary to protect it from warm moist air within the building using an impermeable membrane, and externally from rain using a breathing membrane, the latter preventing the build-up of trapped interstitial moisture. In the Westborough School building, the inner plastic membrane is protected from physical damage with a further 1 mm cardboard layer, and the external membrane is covered by cement-bonded fibreboard for fire and rain protection.

Cardboard is potentially vulnerable to rot and insect attack. This could be prevented by treatment with boron products; however, this would adversely affect the potential for ultimate recycling of the cardboard. As cardboard is a recycled material it has a low embodied energy, and can legitimately be considered *green*.

Rammed-earth and cob construction

Earth construction is one of the oldest forms of building used by mankind. Rammed-earth buildings can be found in most countries, and many have survived hundreds of years. The ideal material is a well-graded mixture of gravel, sand, silt and clay fines. The clay content should be sufficient to act as an efficient binder, but not in excess to cause large moisture movement or cracking of the finished construction. In modern rammed-earth construction Portland cement is frequently incorporated as a binder to improve cohesion of the stabilised earth mix.

In rammed-earth construction the mix is placed in layers, typically 100 to 150 mm deep, within the rigid formwork and firmly tamped down, thus inevitably giving some variation in density between the top and bottom of each lift. The compaction should be sufficient to ensure good strength and a smooth finish.

Window and door openings should be limited to no more than one-third of the length of any wall to ensure structural stability. Lintels should be sufficiently robust to take static loading and the effects of ramming further lifts of earth. A wall plate of timber or poured reinforced concrete, which may be hidden within the top lift of the earth wall, is necessary to spread the loading from the roof structure. The eaves should be detailed to ensure appropriate shelter to rammed earth walling, which is usually protected with several coats of limewash finish.

Cob construction differs from rammed earth in that the clay is mixed with straw. In the traditional process, a fine tilth of clay is spread about 100 mm deep over a thin straw bed; water and a second thicker layer of straw is added. The mixture is well trodden to produce a reasonably uniform mix. Devon clay is ideal for this process as it is well graded with a range of particle size from coarse gravel through fine sand to coarse clay. Devon clay has only a low expansion and contraction which otherwise can cause cracking of the completed structure. In the construction process, the mixture of the straw and clay is tamped together, usually commencing from a minimum 450 mm stone plinth. Free-form designs may be achieved without the use of shuttering. This type of construction was common in many parts of the UK and many old cob buildings still exist in Devon. The thermal mass of cob construction stabilises seasonal variations, helping to keep the interior cool in summer and warm in winter. The external finish should be of lime plaster rather than a Portland cement render, which cracks or breaks away allowing rainwater to penetrate the wall. As with rammed-earth construction, the eaves should be deep enough to protect the walls from severe weathering.

The 'House for Stories' designed by Tono Mirai at the Bleddfa Centre for the Arts in Powys, Wales is constructed from mud and straw. The building, which is intended to be a quiet space for contemplation and imagination is partly subterranean, but spirals out of the ground into the delightful helical form shown in Figure 15.4.

Fig. 15.4 Mud and straw construction – The House for Stories, Bleddfa Centre for the Arts, Powys, Wales. Architect: Tono Mirai. Photograph: Courtesy of Richard Weston

Earth-sheltered buildings

Earth-sheltered buildings, including homes, are defined as those where the roof and some sides are covered by earth. Increased depth of earth cover improves the thermal performance, but this has to be balanced against consequential increased structural strength requirements. Typically, 400–50 mm of earth cover is appropriate, and the weight of this material is usually supported by concrete or masonry construction. Water exclusion is key to the design, requiring land drainage and the use of reinforced membranes.

One method of construction uses fibrous plaster shells to create the internal organic form. These are sprayed with a layer of lightweight aggregate concrete insulation, followed by 100 mm of structural concrete. After the concrete is structurally sound, the building may be fitted out and covered externally with soil and grass. An alternative construction system uses extruded polystyrene insulation between the structural concrete and the soil backfill. Either approach uses the temperature stabilising effect of the mass concrete and soil cover to significantly reduce energy consumption. In order to introduce sufficient light, at least one elevation is usually glazed and, in addition, interesting effects can be achieved with roof lights or light pipes. Ventilation may be mechanical but is normally provided through opening glazing. At the same time, unwanted cold air infiltration is eliminated by the earth enclosure.

The Hockerton Housing Project at Southwell, Nottinghamshire (Fig. 15.5) illustrates an ecological development of earth-sheltered housing in which the residents generate their own energy, harvest their own water, and recycle waste materials eliminating pollution and carbon dioxide emissions. Only the south elevation of the development overlooking the reed pond is visible, as grass covers the majority of the construction.

Clay products

Concern over the intrinsic energy in most manufactured building products has led to the further development of a range of clay-based products including clay boards and plasters. Unfired clay building products are hygroscopic and have the positive effect of controlling internal environments by absorbing odours and stabilising humidity and temperature.

CLAY BOARDS

Clay boards, which are an alternative to gypsum plasterboards, are manufactured from clay and layers of reed along and across the board length. Hessian on both faces acts as reinforcement and a key for a 2–3 mm finish of lime-based or earth plaster. Joints should be scrimmed before skimming, although clay boards if sealed can be painted directly. The 25 mm boards may be used for drywall construction and ceilings, where they should be screw-fixed at 600 mm or 400 mm centres respectively.

CLAY PLASTERS

Clay plasters, also known as earth plasters, are available in a range of self colours, which do not require paint decoration. Clay plaster, manufactured from a blend of clay and fine aggregate, may be applied in two layers of 10 mm and 3 mm respectively or as a single 10 mm coat. If necessary, an initial 1 mm bonding coat may be applied to the substrate. Clay plasters harden only by drying out without any chemical processes. As with all clay products, these plasters absorb moisture and are responsive to environmental conditions helping to control the internal relative humidity.

Recycled plastics

The increasing use of plastics in our everyday lives has led to a large waste problem, which can only be resolved by extensive recycling. Many plastics are slow to degrade in landfill sites and, as many are based on products from the petrochemical industry, this finite resource should not be wasted. One major problem in recycling plastics is their wide diversity (Chapter 10); thus separation into single recyclable products is difficult unless we are educated to do this within our own homes. However, recently it has been demonstrated that structural products can be manufactured from recycled mixed plastic waste.

Recycled plastic lumber

Mixed domestic plastic waste is cut up into small flakes, melted at 200°C into a grey viscous liquid and cast into moulds to produce structural components. The product, *polywood*, has been used to create a 7.5 m span lightweight bridge with a capacity of 30 tonnes in America. Recycled plastic was used to construct the

South elevation

North elevation

Fig. 15.5 Earth-sheltered housing – south and north elevations. The Hockerton Housing Project, Southwell, Nottinghamshire. Photographs: Copyright of Hockerton Housing Project

l-beams and other structural sections. Recycled plastic lumber has the advantage over timber that it requires no maintenance or treatment with noxious chemicals and its use reduces the demand on landfill sites. Polywood is light, although more dense than timber; however, it suffers from creep. Also, it has greater thermal movement and lower stiffness (modulus of elasticity) than timber. Prior to its recent structural use, polywood has been used for decking, fencing, garden furniture and various marine applications.

RECYCLED TYRES

Recycled tyres have been used to create the structural walls of new homes called *earthships* in Fife and near Brighton, UK. The tyres are laid in courses, filled with rammed earth, and finished internally with plaster and externally with solar tiles. To date only a few small housing units have been built, but subject to Building Regulations, there is no theoretical limitation on size. The UK produces 40 million used tyres per year,

Fig. 15.6 Sandbags in construction. Architects and Photograph: Sarah Wigglesworth Architects

enough to create 20 000 low carbon autonomous earthships annually.

Papercrete

Papercrete is made from recycled paper and/or cardboard with sand and Portland cement. Pulverised waste glass from recycled bottles may be used instead of sand and glossy magazines can be mixed in with standard newsprint. The material is made by dry mixing shredded paper with sand and Portland cement in the approximate ratio of 3 : 1 : 1. Water is added to make a paper-mâché slurry that can be cast into block units or into monolithic structures. Where papercrete blocks are used in construction, then the same material can be used as the mortar. The material dries to a grey colour. It is, however, very water-absorbent and must be protected from moisture and weather by appropriate detailing. Externally a stronger mix of 1 : 1 paper and cement may be used as a stucco layer, and internally papercrete plaster may be used to give a

textured or patterned finish. Papercrete is still an experimental material, but it has the potential to remove up to 20% of the waste material currently deposited in landfill sites. As a lightweight material it has good insulating properties and the cement content significantly increases its fire resistance.

Sandbags

In recently constructed office/home accommodation in Islington, London, sandbags were used as a sound-absorbing facade adjacent to a noisy main railway line (Fig. 15.6). The sandbags, filled with a sand-cement-lime mixture and exposed to the elements, will eventually set hard. Over a further period of time, possibly 30 years, the hessian bags will disintegrate leaving the undulating concrete exposed, imprinted with the texture of the hessian fabric. The wall is experimental in that the weathering effects cannot be predicted as with other more standard forms of construction.

References

FURTHER READING

Bee, B. 1998: *The cob builders handbook.* Vermont: Chelsea Green.

Davis, M. and Davis, L. 1993: *How to make low-cost building blocks: stabilised soil block technology.* London: ITDG Publishing.

Easton, D. 1995: *The rammed- earth house.* Vermont: Chelsea Green.

Fernandez, J. 2006: *Material architecture: emergent materials and issues for innovative buildings and ecological construction.* Oxford: Elsevier.

Guelberth, C.R. 2003: *The natural plaster book, earth, lime and gypsum plasters for natural homes.* USA: New Society Publishers.

Houben, H. and Guillaud, H. 1994: *Earth construction.* London: Intermediate Technology Publications.

Janssen, J. J. A., 1995: *Building with bamboo: a handbook.* Warwickshire: ITDG Publishing.

Jones, B. 2002: *Building with straw bales: a practical guide for the UK and Ireland.* Green Books.

Keable, J. 1996: *Rammed-earth structures: A code of practice.* Warwickshire: ITDG Publishing.

Khalili, N. 2002: *Ceramic houses and earth architecture.* Vermont: Chelsea Green Publishing.

King, B. 1996: *Buildings of earth and straw.* California: Ecological Design Press.

Lacinski, P. and Bergeron, M. 2000: *Serious straw bale: a home construction guide for all climates.* Totnes: Chelsea Green Publishing.

Magwood, C. and Mack, P. 2005: *More straw bale building: a complete guide to designing and building with straw.* USA: New Society Publishers.

McCann, J. 2004: *Clay and cob buildings.* Princes Risborough: Shire Publications.

Minke, G. 2000: *Earth construction handbook.* Southampton: WIT Press.

Minke, G. and Friedemann, M. 2005: *Building with straw: design and technology of a sustainable architecture.* Basle: Birkhäuser.

Norton, J. 1997: *Building with earth: a handbook.* London: ITDG Publishing.

Schofield, J. and Smallcombe, J. 2004: *Cob buildings: A practical guide.* South Carolina: Black Dog Press.

Smith, M.G. and Bednar, D. 1999: *Cobber's companion: How to build your own earthen home.* USA: Cob Cottage.

Wells, M. 1998: *Earth sheltered house: an architect's sketchbook.* Totnes: Chelsea Green Publishing.

William-Ellis, C. 1999: *Building with cob, pisé and stabilised earth.* London: Donhead.

Woolley, T. and Kimmins, S. 2000: *Green building handbook.* vol. 2. E. & F.N. Spon.

Woolley, T., Kimmins, S., Harrison P. and R. 1997: *Green building handbook.* vol. 1. E. & F.N. Spon.

BUILDING RESEARCH ESTABLISHMENT PUBLICATIONS

BRE Digests

Digest 446: 2000 Assessing environmental impacts of construction.

Digest 452: 2000 Whole life costing and life-cycle assessment for sustainable building design.

Digest 470: 2002 Life-cycle impacts of timber. A review of the environmental impacts of wood products in construction.

BRE Information papers

BRE IP 3/97 Demonstrating reuse and recycling materials: BRE energy efficient office of the future.

BRE IP 12/97 Plastics recycling in the construction industry.

BRE IP 7/00 Reclamation and recycling of building materials.

BRE IP 3/03 Sustainable buildings (Parts 1–4).

ADVISORY ORGANISATION

Earth Centre, Denaby Main, Doncaster, South Yorkshire DN12 4EA (01709 512000).

16

SEALANTS, GASKETS AND ADHESIVES

—

Introduction

Although used in relatively small quantities compared with the load-bearing construction materials, sealants, gaskets and adhesives play a significant role in the perceived success or failure of buildings. A combination of correct detailing and appropriate use of these materials is necessary to prevent the need for expensive remedial work.

Sealants

Sealants are designed to seal the joints between adjacent building components while remaining sufficiently flexible to accommodate any relative movement. They may be required to exclude wind, rain and airborne sound. A wide range of products is available matching the performance characteristics of the sealant to the requirements of the joint. Incorrect specification or application, poor joint design or preparation, are likely to lead to premature failure of the sealant. The standard (BS EN ISO 11600: 2003) classifies sealants into type G for glazing applications and type F (facade) for other construction joints. For both types, classes are defined by movement capability, modulus and elastic recovery (Fig. 16.1).

Key factors in specifying the appropriate sealant are:

- understanding the cause and nature of the relative movement;
- matching the nature and extent of movement to an appropriate sealant;
- appropriate joint design, surface preparation and sealant application;
- the service life of the sealant.

Relative movement within buildings

The most common causes of movement in buildings are associated with settlement, dead and live load effects including wind loading, fluctuations in temperature, changes in moisture content and, in some cases, the deteriorative effects of chemical or electrolytic action. Depending upon the prevailing conditions, the various effects may be additive or compensatory.

SETTLEMENT

Settlement is primarily associated with changes in loadings on the foundations during the construction process although it may continue for some time, frequently up to five years, after the construction is complete. Subsequent modifications to a building or its contents may cause further relative movement. Settlement is usually slow and in one direction, creating a shearing effect on sealants used across the boundaries.

THERMAL MOVEMENT

All building materials expand and contract to some degree with changes in temperature. For timber the movement is low, but for glass, steel, brick, stone and concrete it is moderate, and relatively high for plastics and aluminium. The effects of colour, insulation and the thickness of the material accentuate thermal movements. Dark materials absorb solar radiation and heat more quickly than light reflective materials. Also, well-insulated claddings respond quickly to changes in solar radiation, producing rapid cyclical expansion movements, whereas heavy construction materials respond more slowly but will still exhibit

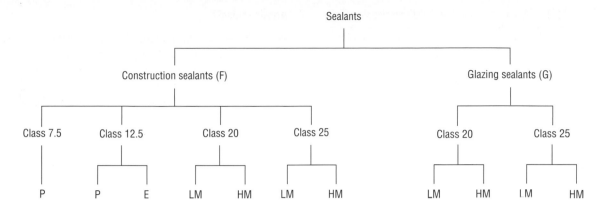

Notes: F refers to facade, G to glazing.
Class number indicates the movement accommodation as a percentage.
P refers to plastic, E to elastic, LM to low modulus and HM to high modulus.

Fig. 16.1 Classification of sealants in construction

considerable movements over an annual cycle. Typical thermal movements are shown in Table 16.1.

MOISTURE MOVEMENT

Moisture movement falls into two categories: irreversible movements as new materials acclimatise to the environment; and reversible cyclical movements due to climatic variations. Many building materials, especially concrete and mortars, exhibit an initial contraction during the drying-out process. Incorrectly seasoned timber will also shrink but new bricks used too quickly after manufacture will expand. After these initial effects, all materials which absorb moisture will expand and contract to varying degrees in response to changes in their moisture content. Depending upon climatic conditions, moisture and thermal movements may oppose or reinforce each other. Typical irreversible and reversible moisture movements are shown in Table 16.2.

LOADING AND DETERIORATION

Movements associated with live loads such as machinery, traffic and wind can cause rapid cyclical movements within building components. The deterioration of materials, such as the corrosion of steel or sulfate attack on concrete, is often associated with irreversible expansion, causing movement of adjacent components. Concrete structures may exhibit creep, which is gradual permanent deformation under load, over many years.

Types of sealant

There are three distinct types of sealant – plastic, elastoplastic and elastic – each of which exhibits significantly different properties which must be matched to the appropriate application (BS 6213: 2000).

PLASTIC SEALANTS

Plastic sealants, which include general-purpose mastics, allow only a limited amount of movement, but when held in a deformed state they stress-relax. Elastic recovery is limited to a maximum of 40%. Plastic sealants dry by the formation of a surface skin, leaving liquid material encased to retain flexibility. However, with time the plastic core continues to harden; thus durability is related to the thickness of the material used.

Oil-based mastics

For oil-based mastics a 10 mm depth is required for optimum durability with a typical life expectancy of 2 to 10 years. The effects of ultraviolet degradation are reduced by painting. Typical uses include sealing around window and door frames in traditional low rise building. (The typical movement accommodation for oil-based mastics is 10%.)

Butyl sealants

Butyl sealants are plastic but with a slightly rubbery texture. They are used in small joints as a gap filler

Table 16.1   Thermal movements of building materials

Typical thermal movements of building materials in use calculated for a temperature variation of 85°C (e.g. −15 to + 70°C) (measured in mm per metre)

Material	Typical thermal movement (mm/m for 85°C change)	Coefficient of linear expansion °C × 10^{−6}
Masonry		
Concrete – standard aggregates	1.2	10–14
Calcium silicate brickwork	1.2	8–14
Concrete blockwork	1.0	6–12
Concrete – aerated	0.7	8
Concrete – limestone aggregate	0.6	7–8
Clay brickwork	0.5–0.7	5–8
GRC	0.8–1.7	10–20
Metals		
Zinc (along roll)	2.7	32 (23 across roll)
Lead	2.5	29
Aluminium	2.0	23
Titanium zinc	1.8	20–22
Copper	1.4	17
Stainless steel	1.4	17
Terne coated stainless steel	1.4	17
Structural steel	1.0	12
Stone and glass		
Glass	0.9	9–11
Slate	0.9	9–11
Granite	0.8	8–10
Sandstone	0.8	7–12
Marble	0.4	4–6
Limestone	0.3	3–4
Plastics		
ABS	8.0	83–95
PVC	6.0	40–80
GRP	3.0	20–35
Timber		
Wood (along grain)	0.5	4–6

and general-purpose sealant where oil-based mastics would dry too rapidly. Life expectancy is between 10 and 20 years if they are protected from sunlight by painting, but only up to 5 years in exposed situations. (The typical movement accommodation for butyl sealants is 10%.)

Acrylic sealants

Water-based acrylic sealants are frequently used for internal sealing such as between plaster and new windows. The solvent-based acrylic sealants are durable for up to 20 years, with good adhesion to slightly contaminated surfaces. They accommodate only limited movement but produce a good external seal around windows, both for new and remedial work. (The typical movement accommodation for water-based and solvent-based acrylic sealants is 15% and 20% respectively.)

Polymer/bitumen sealants

Solvent-based bitumen sealants are generally suitable for low-movement joints in gutters and flashings. Hot-poured bitumen is used for sealing movement joints in asphalt and concrete floor slabs, although compatibility with any subsequent floor coverings should be verified.

Linseed oil putty

Traditional putty contains a mixture of linseed oil and inorganic fillers (BS 544: 1969), which sets by a combination of aerial oxidation of the oil and some absorption into the timber. A skin is produced initially, but the mass ultimately sets to a semi-rigid material. Application is with a putty knife onto primed timber. For application to steel window frames, non-absorbent hardwoods and water-repellent preservative treated softwoods, non-linseed oil putty is appropriate. Linseed oil putty should be painted within two weeks, whereas metal casement putty may be left three months before painting.

ELASTOPLASTIC SEALANTS

Elastoplastic sealants will accommodate both slow cyclical movements and permanent deformations. A range of products offer appropriately balanced strength, plastic flow and elastic properties for various applications.

Polysulfide sealants

Polysulfide sealants are available as one- or two-component systems. The one-component systems have the advantage that they are ready for immediate use. They cure relatively slowly by absorption of

Table 16.2 Moisture movements of typical building materials

Typical reversible and irreversible moisture movements of building materials in use (measured in mm per metre)

Material	Reversible (mm/m)	Irreversible (mm/m)
Concrete	0.2–0.6	0.3–0.8 (shrinkage)
Aerated concrete	0.2–0.3	0.7–0.9 (shrinkage)
Brickwork — clay	0.2	0.2–1.0 (expansion)
Brickwork — calcium silicate	0.1–0.5	0.1–0.4 (shrinkage)
Blockwork — dense	0.2–0.4	0.2–0.6 (shrinkage)
Blockwork — aerated	0.2–0.3	0.5–0.9 (shrinkage)
Glass-fibre reinforced cement	1.5	0.7 (shrinkage)
Softwood	5–25 (60–90% relative humidity)	
Hardwood	7–32 (60–90% relative humidity)	
Plywood	2–3 (60–90% relative humidity)	

(Seasoned timber has no irreversible movement)

moisture from the atmosphere, initially forming a skin and fully curing within 2–5 weeks. One-component systems are limited in their application to joints up to 25 mm in width, but their ultimate performance is comparable to that of the two-component materials. Typical uses include structural movement joints in masonry, joints between precast concrete or stone cladding panels and sealing around windows.

The two-component polysulfide sealants require mixing immediately before use and fully cure within 24–48 hours. They are more suitable than one-component systems for sealing joints which are wider than 25 mm, have large movements, or are subject to vandalism during setting. Uses include sealing joints within concrete and brickwork cladding systems and also within poorly insulated lightweight cladding panels. Polysulfides have a life expectancy of 20 to 25 years. (The typical movement accommodation for polysulfide sealants is up to 25% for one-part systems and up to 30% for two-part systems.)

ELASTIC SEALANTS

Elastic sealants are appropriate for sealing dynamic joints where rapid cyclic movement occurs. They are often sub-classified as low- or high-modulus depending upon their stiffness. Low modulus sealants should be used where joints are exposed to long periods of compression or extension and where the substrate material is weak.

Polyurethane sealants

Polyurethane sealants are available as one- or two-component systems. The products are highly elastic but surfaces should be carefully prepared and usually primed to ensure good adhesion. Durability is good, ranging from 20 to 25 years. Typical applications are joints within glazing, curtain walling and lightweight cladding panels. (The typical movement accommodation for polyurethane sealants is between 10% and 30% depending on the modulus.)

Silicone sealants

Silicone sealants are usually one-component systems which cure relatively quickly in air, frequently with the evolution of characteristic smells such as acetic acid. Generally, silicone sealants adhere well to metals and glass, but primers may be necessary on friable or porous surfaces such as concrete or stone. High-modulus silicone sealants are resilient. Typical applications include glazing and curtain-wall systems, movement joints in ceramic tiling and around sanitary ware. Low-modulus silicone sealants are very extensible and are appropriate for use in joints subject to substantial thermal or moisture movement. Typical applications are the perimeter sealing of PVC-U and aluminium windows, and also cladding systems. Silicone sealants are durable with life expectancies within the range 25 to 30 years. (The typical movement accommodation for silicone sealants ranges from 20% to 70% depending upon the modulus.)

Epoxy sealants

Epoxy sealants are appropriate for stress-relieving joints where larger movements in compression than tension are anticipated. Typical applications include floor joints and the water-sealing of tiling joints within swimming pools. Epoxy sealants have a life expectancy of 10 to 20 years. (The typical movement accommodation of epoxy sealants is within the range 5% to 15%.)

Joint design

There are three forms of joint: butt, lap and fillet (Fig. 16.2). However, only butt and lap joints will accommodate movement. Generally, lap joints in which the sealant is stressed in shear will accommodate double the movement of butt joints in which the sealant is under tension or compression. Furthermore, lap joints tend to be more durable as the sealant is partially protected from the effects of weathering.

However, lap joints are generally more difficult to seal than butt joints. Frequently, joints are made too narrow, either for aesthetic reasons or due to miscalculation of component tolerances. The effect is that extent of movement is excessive in proportion to the width of sealant, causing rapid failure.

To correctly control the depth of the sealant and to prevent it adhering to the back of the joint, a compressible back-up material, usually rectangular or round closed-cell polyethylene, is inserted (Fig. 16.3). The polyethylene acts as a bond-breaker by not adhering to the sealant. Where the joint is filled with a filler board, such as impregnated fibreboard or corkboard, a plastic bond-breaker tape or closed-cell polyethylene strip should be inserted. Normally the depth of the sealant should be half the width of the joint for elastic and elastoplastic sealants and equal to the width of the joint for plastic sealants, the minimum width of the joint being calculated from the maximum movement to be accommodated and the movement accommodation factor (MAF), i.e. the extensibility of the sealant. Where insufficient depth is available to insert a polyethylene foam strip, a tape bond-breaker should be inserted at the back of the joint.

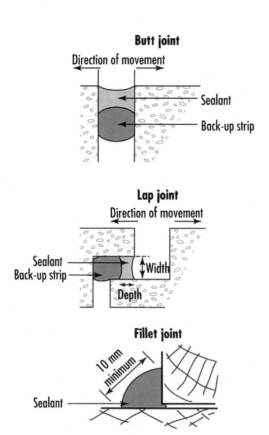

Fig. 16.2 Butt, lap and fillet joints

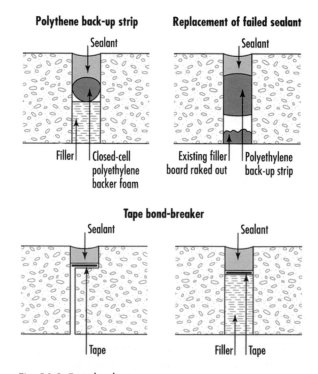

Fig. 16.3 Typical sealant systems

Minimum joint width calculation:

Total movement	$= 5$ mm
Movement accommodation factor (MAF)	$= 25\%$
Width of sealant to accommodate movement	$= {}^{5}/_{0.25} = 20$ mm

In order to obtain good adhesion, the joint surfaces should be prepared by the removal of contaminants, loose material or grease and by the application of a primer if specified by the sealant manufacturer. Most sealants are applied directly by gun application, although tooled, poured and tape/strip sealants are also used. Tooling helps to remove air bubbles entrained in two-component mixes; if left, air bubbles would reduce the durability of the seal. Externally, recessed cladding joints show less staining than flush joints, although the usual finish is a slightly concave surface. Where stonework is being sealed, non-staining silicone sealants must be used to prevent the migration of plasticiser into the stone which could cause discolouration. Sealants to floor joints need to be tough, therefore wider to accommodate the necessary movements and recessed to prevent mechanical damage. Alternatively proprietary mechanical jointing systems should be used.

COLOUR MATCHING

While most sealants, except the black bituminous products, are available in white, translucent, greys and browns, the silicone sealants appropriate for use around kitchen and bathroom units are available in a wide range of colours. For these purposes, fungicides are often included within the formulation.

Fire-resistant sealants

Many fire-resistant sealants are based on the use of intumescent materials which expand copiously in fire. The intumescent components commonly used are either ammonium phosphate, hydrated sodium silicate or intercalated graphite (layers of water and carbon), and these are incorporated into the appropriate sealant. Intumescent oil-based mastics and acrylic sealants are suitable for sealing low-movement joints around fire check doors. For the fire-resistant sealing of structural movement joints, fire-resistant grades of low-modulus silicone, two-part polysulfide and

acrylic sealants are available. Maximum fire resistance is obtained if the sealant is applied to both faces of the joint, with mineral wool or glass-fibre insulation in the void space. Four hours of fire resistance with respect to both integrity and insulation can be achieved for a 20 mm wide movement joint within 150 mm concrete (BS 476–20: 1987). The low-modulus silicone is appropriate for sealing fire-resisting screens, curtain walls, claddings and masonry subject to movement. The two-part polysulfide is designed for use in concrete and masonry fire-resisting joints. Acrylic sealants are appropriate over a wide range of materials but where timber is involved an allowance must be made for its loss by charring.

Intumescent fillers manufactured from acrylic emulsions with inert fillers and fire-retardant additives can be applied either by gun or trowel to fill voids created around service ducts within fire-resistant walls. Four hours of fire resistance can be achieved with these materials. Intumescent tapes are appropriate for application within structural movement joints. Most intumescent sealants are now *low-smoke* and evolve no halogenated products of combustion in fire situations. (The typical movement accommodation for intumescent acrylic sealants is 15%.)

Foam sealants

Compressible strips of closed-cell PVC and polyethylene, or open-cell polyurethane foams, coated on one or both edges with pressure-sensitive adhesive are used in air-conditioning ductwork and to seal thermal movement and differential settlement joints, gaps around window and door frames. Strips may be uniform in section or profiled for particular applications. Aerosol-dispensed polyurethane foam is widely used as an all-purpose filler. It is available either as foam or as expanding foam, and acts as an adhesive, sealant, filler and insulator.

Concrete joint fillers and sealants

Concrete joint fillers for use in pavements are specified by the standards BS EN 14188: 2004, Parts 1 and 2 for hot and cold application sealants respectively. Sealants for cold application are classified as single-component systems (S) or multi0component systems (M) and sub-divided into self-levelling (sl) or non-sag (ns) types. An additional classification A, B, C or D relates to increasing level of resistance to chemicals. Standard hot applied joint sealants are classified as elastic (high extension) Type N1, and normal (low extension) Type N2. Where fuel resistance is also required, the higher specification grades F1 and F2 are necessary.

Gaskets

Gaskets are lengths of flexible components of various profiles, which may be solid or hollow and manufactured from either cellular or non-cellular materials. They are held in place either by compression or encapsulation into the adjacent building materials and maintain a seal by pushing against the two surfaces (Fig. 16.4). Typical applications include the weather sealing of precast cladding units and facade systems. Within precast concrete, GRP (glass-fibre reinforced polyester) or GRC (glass-fibre reinforced cement) cladding units, the gaskets are typically inserted into recessed open-drained joints. The gaskets therefore act as a rain barrier, but because they

do not necessarily fit tightly along their full length, they can be backed up by a compressed cellular foam wind penetration seals. Gaskets should not be either stretched or crammed in during insertion as they will subsequently shrink leaving gaps or pop out causing failure.

In glazing and related curtain walling systems, gaskets may be applied as capping seals, retained by appropriate profiles within the mullions and transoms; alternatively, the gaskets may be recessed within the joints of the glazing system to give narrower visual effect to the joint. Some glazing gaskets of H- or U-sections are sealed with a zipper or filler strip which is inserted in the profile, compressing the material into an air- and watertight seal. Gaskets and weather-stripping for use on doors, windows and curtain walling are classified by a letter and digit code which defines the use and key physical properties of the particular product, enabling appropriate specification (Table 16.3).

The standard materials for gaskets used in construction are neoprene which is highly elastic; EPDM (ethylene propylene diene monomer) which has better weathering characteristics than neoprene; and silicone rubbers which are highly resistant to ultraviolet light, operate over a wide range of temperatures, and are available in almost any colour. Cruciform section gaskets of polychloroprene rubber are suitable for vertical joints between precast concrete panels.

Waterstops for embedding into *in-situ* concrete for sealing movement and construction joints are manufactured in PVC or rubber according to the required movement (Fig. 16.5). Sections are available in long extruded lengths and factory-produced intersections. Applications include water-containing structures and water exclusion from basements. Waterstops placed centrally within concrete will resist water pressure from either side, but externally positioned waterstops, not encased below the concrete slab or within permanent concrete shuttering, will only resist water pressure from the outer face.

Proprietary systems offer watertight expansion jointing for horizontal surfaces such as roof car parks and pedestrian areas. Systems usually combine complex aluminium or stainless steel profiles with extruded synthetic rubber inserts. Materials can withstand high loads, with good resistance to bitumen and salt water.

Dry glazing strips are based on elastomeric polymers, typically EPDM or butyl rubber. Usually

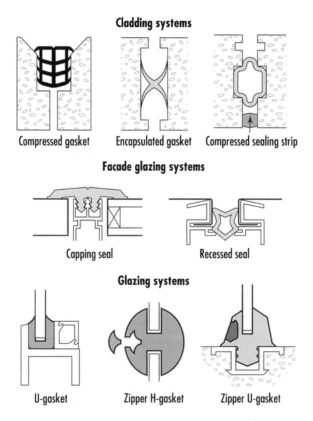

Cladding systems

Compressed gasket   Encapsulated gasket   Compressed sealing strip

Facade glazing systems

Capping seal   Recessed seal

Glazing systems

U-gasket   Zipper H-gasket   Zipper U-gasket

Fig. 16.4 Typical gaskets for cladding and glazing systems

Table 16.3   Classification of gaskets and weatherstripping to BS EN 12365–1: 2003

Letter and five number code					
letter (G or W)	digit 2	digit 3	digit 4	digit 5	digit 6
category	working range mm	compression force KPa	working temperature range °C	deflection recovery %	recovery after ageing %
gasket (G) weatherstripping (W)	9 grades identified (1 − 9)	9 grades identified (1 − 9)	6 grades identified (1 − 6)	8 grades identified (1 − 8)	8 grades identified (1 − 8)

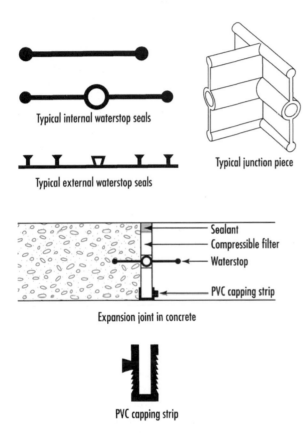

Typical internal waterstop seals

Typical external waterstop seals

Typical junction piece

Sealant
Compressible filter
Waterstop
PVC capping strip

Expansion joint in concrete

PVC capping strip

Fig. 16.5 Concrete waterstop seals

the synthetic rubber strip has a self-adhesive backing which adheres to the rebate upstand. With external beading, the dry glazing strip can also be applied to each bead, which is then fixed with suitable compression to ensure a good seal to the glass. The performance requirements and classification for gaskets and weather-stripping for doors, windows and curtain walling are described in the standard BS EN 12365–1: 2003.

Adhesives

TYPES OF ADHESIVE

The traditional adhesives based on animal and vegetable products have largely been superseded by synthetic products manufactured by the polymer industry, except for casein, manufactured from skimmed milk, which is currently used as a timber adhesive (BS EN 12436: 2002). The range of adhesives is under constant development and particular applications should always be matched to manufacturers' specifications. Special notice should be taken of exclusions where materials and adhesives are incompatible, also to safety warnings relating to handling and the evolution of noxious fumes or flammable vapours. Adhesives are more efficient when bonding components are subject to shear forces rather than direct tension. They are least efficient against the peeling stresses. Most adhesives have a *shelf life* of 12 months when stored unopened under appropriate conditions. The *pot life* after mixing the two-component systems ranges from a few minutes to several hours.

Tile adhesives

The standard BS EN 12004: 2001 classifies adhesives for tiles into three types: cementitious (C); dispersion (D); and reaction resin (R). Each of these types may have further characteristics defined by classes relating to enhanced adhesive properties, faster setting, reduced slip or extended open time (the time between spreading the adhesive and applying the tiles) (Table 16.4). Dispersion adhesives are the ready-for-use aqueous polymer dispersions, whilst the reaction resin adhesives are one- or two-component systems which set by chemical reaction.

Table 16.4 Classification of tile adhesives by composition and properties

Classification	Composition and properties
Type C	cementitious adhesive – hydraulic binding resin
Type D	dispersion adhesive – aqueous organic polymer resin
Type R	reaction resin adhesive – one or two component synthetic resin
Class 1	normal adhesive
Class 2	improved adhesive
Class F	fast setting adhesive
Class T	reduced slip adhesive
Class E	extended open time adhesive

Ceramic wall tile adhesives

Wall tile adhesives are usually PVA (polyvinyl acetate), acrylic or cement-based compositions. The standard PVA thin-bed adhesives, typically to 3 mm, will only tolerate moisture, whereas the thin-bed water-resistant acrylic-based adhesives are suitable for fixing wall tiles and mosaics in damp and wet conditions associated for example with domestic showers. Some acrylic-based products evolve ammonia on setting. The water-resistant cements and polymer-modified cement products are appropriate both for internal and external use and can usually be applied with either thin or thick bedding. The polymer-modified cement adhesives are also suitable for fixing marble, granite and slate tiles up to 15 mm thick. For chemical resistance thin-bed epoxy-resin-based adhesives are available. In all cases the substrate must be sound with new plaster, brickwork, concrete, fully dried out for 2–6 weeks. Plasterboard and timber products must be adequately fixed at 300 mm centres horizontally and vertically to ensure rigidity. In refurbishment work, flaking or multi-layered paint should be removed and glazed surfaces made good. Where the tile adhesive is classified as waterproof, either acrylic or cement-based, it may be used as the grouting medium. Alternatively, equivalent waterproof grouting is available in a wide range of colours to blend or contrast with the wall tiles. Epoxy-resin tile grout is available for very wet conditions.

Ceramic floor tile adhesives

The majority of ceramic floor tile adhesives are cement-based, used either as thick bed (up to 25 mm) or thin bed according to the quality of the substrate. Standard products are suitable for fixing ceramic tiles, quarries, brick slips, stone and terrazzo to well-dried-out concrete or cement/sand screed. Where suspended timber floors are to be tiled, they must be well ventilated and strong enough to support the additional dead load. An overlay of 12 mm exterior grade plywood, primed with bonding agent and screwed at 200 mm centres may be necessary. In refurbishment work, it is better to remove all old floor finishes, but ceramic floor tiles may be fixed over cleaned ceramic or possibly primed vinyl tiles, providing all loose material is first removed.

Cement-based grouting can be pigmented to the required colour, but care must be taken to ensure that excess grout is removed from the surface of the tiles before staining occurs. Thin-bed two-component epoxy-based adhesives are more water- and chemical-resistant than the standard cement-based products and are appropriate for use where repeated spillage is likely from industrial processes. Where there is likely movement of the substrate, two-component rubber-based adhesives are generally appropriate.

Contact adhesives

Contact adhesives based on polychloroprene rubber, either in organic solvents or aqueous emulsions, are normally suitable for bonding decorative laminates and other rigid plastics such as PVC and ABS to timber, timber products and metals. The adhesive is usually applied to both surfaces, the solvent or emulsion allowed to become touch dry, prior to bringing the two surfaces into contact when an immediate strong bond is produced. The aqueous emulsion products can also be suitable for fixing sealed cork and expanded polystyrene and have the advantage that no fumes are evolved. Expanded polystyrene tiles may be adversely affected by solvent-based formulations.

Vinyl floor tile and wood block adhesives

Most vinyl floor tile and wood block adhesives are based on either rubber/bitumen rubber/resin or modified bitumen emulsions. In all cases it is essential that the sub-floor is dry, sound, smooth and free from any contamination which would affect the adhesion. Where necessary cement/acrylic or cement/latex floor levelling compound should be applied to concrete, asphalt or old ceramic tiled floors. Some cement/latex materials evolve ammonia during application.

Wood adhesives

Wood joints generally should be close contact with a gap of less than 0.15 mm, but so-called *gap-filling* adhesives satisfactorily bond up to 1.3 mm. Polyvinyl acetate (PVA) wood glues are widely used for most on-site work and in the factory assembly of mortice and tenon joints for doors, windows and furniture. The white emulsion sets to a colourless translucent thermoplastic film, giving a bond of similar strength to the timber itself, but insufficient for bonding load-bearing structural members. Components should be clamped in position for up to 12 hours to ensure maximum bonding, although increasing the temperature may reduce this. Waterproof PVA adhesives which partially cross-link on curing are suitable for protected external use but not immersion in water. PVA adhesives generally retain their strength up to 60°C and do not discolour the timber, except by contact with ferrous metals.

The thermosetting wood resins are mainly two-component systems based on phenolic compounds such as urea, melamine, resorcinol or phenol which cure with formaldehyde to produce load-bearing adhesives (BS EN 301: 1992). Most formulations require the mixing of the resin and hardener, but a premixed dry powder to which water is added is also available. Structural resin-based adhesives are designated for exterior (Type 1) or protected (Type 2) use. Melamine formaldehyde adhesives will not resist prolonged exposure to weathering. Urea formaldehyde adhesives are generally moisture-resistant or for interior use only. Certain timber fire-retardant and preservative treatments reduce the efficiency of timber adhesives, although generally those based on phenol formaldehyde/resorcinol formaldehyde are unaffected.

Wallpaper adhesives

Standard wallpaper adhesives are based on methyl cellulose, a white powder which is water soluble giving a colourless solution. For fixing the heavier papers and decorative dado strips, polyvinyl acetate (PVA) is an added component. Cold water starch is also available as both a wall sizing agent and wallpaper adhesive. Most wallpaper pastes contain fungicide to inhibit mould growth. The standard BS 3046: 1981 describes five types of adhesive ranging from low solids to high wet and dry strength with added fungicide.

Epoxy resin adhesives

Epoxy resins are two-component cold-curing adhesives which produce high strength durable bonds. Most require equal quantities of the resin and hardener to be mixed and various formulations are available giving curing times ranging from minutes to hours. Strong bonds can be obtained to timber, metal, glass, concrete, ceramics and rigid plastics. Epoxy resins may be used internally or externally and they are resistant to oils, water, dilute acids, alkalis, and most solvents except chlorinated hydrocarbons. Epoxy resins are frequently used for attaching stainless steel fixings into stone and brick slips prior to their casting into concrete cladding panels. Epoxy flooring adhesives may be used for bonding vinyl floor finishes in wet service areas and to metal surfaces.

Cyanoacrylate adhesives

Cyanoacrylates are single-component adhesives which bond components held in tight contact within seconds. A high tensile bond is produced between metals, ceramics, most plastics and rubber. The curing is activated by adsorbed moisture on the material surfaces, and only small quantities of the clear adhesive are required. The bond is resistant to oil, water, solvents, acid and alkalis but does not exhibit high impact resistance. A range of adhesive viscosities is manufactured to match to particular applications.

Hot-melt adhesives

Hot-melt adhesives for application by glue-gun are usually based on the thermoplastic copolymer, ethylene vinyl acetate (EVA). Formulations are available for joining materials to either flexible or rigid substrates. Generally, the adhesive should be applied to the less easily bonded surface first (e.g. the harder or smoother surface) and then the two components should be pressed together for at least one minute. Where metals are to be bonded they should be pre-warmed to prevent rapid dissipation of the heat. Similar adhesives are used in iron-on edging veneers for plastic- and wood-faced particleboard.

Bitumen sheet roofing adhesives

Bitumen adhesives are available for hot application, emulsion or in hydrocarbon solvent for the cold-bonding bituminous sheet roofing. The adhesives should be poured and spread by trowel to avoid air pockets, which may cause premature delamination of the sheet from the substrate. Excess bitumen should be removed as it may stain adjacent materials.

Plastic pipe adhesives

Solvent-based vinyl resin adhesives are used for bonding PVC-U and ABS pipes and fittings. The adhesive is brush-applied to both components which are then united and slightly rotated to complete the seal. Curing is rapid but in cold water supply systems water pressure should not be applied for several hours.

Gap-filling adhesive

Gun-grade gap-filling adhesives, usually based on solvent-borne rubber or synthetic rubber resins with filler reinforcement, are versatile in their applications. They are generally formulated to bond timber, timber products, decorative laminates, sheet metals, PVC-U and rigid insulating materials (except polystyrene), to themselves and also to brickwork, blockwork, concrete, plaster and GRP. Typical applications include the fixing of decorative wall panels, dado rails, architraves and skirting boards without nailing or screwing. Surfaces to be bonded must be sound and clean, but the gap-filling properties of the products can allow fixing to uneven surfaces. The materials have good immediate adhesion and can allow the components to be adjusted into position.

PVA bonding agent and sealant

PVA (polyvinyl acetate) is a versatile material which will not only act as an adhesive as described, but also as a bonding agent or surface sealant. As a bonding agent it will bond cement screeds, rendering and plaster to suitable sound surfaces without the requirement for a good mechanical key. PVA will seal porous concrete surfaces to prevent dusting.

References

FURTHER READING

BASA 2001: *The BASA guide to the British Standard BS 6213*. Stevenage: The British Adhesives and Sealants Association.

BASA. 1999: *The BASA guide to the ISO 11600 classification of sealants for building construction*. Stevenage: The British Adhesives and Sealants Association.

CIRIA. 1991: *Manual of good practice in sealant application*. Special Publication 80, London: Construction Industry Research and Information Association/ British Adhesives and Sealants Association.

Cognard, P. 2005: *Handbook of adhesives and sealants: basic concepts and high tech bonding*. Netherlands: Elsevier.

Dunn, D.J. 2003: *Handbook of adhesives and sealants: applications and markets*. Shrewsbury: RAPRA Technology.

Hussey, B. and Wilson, J. 1996: *Structural adhesives directory and data book*. London: Chapman and Hall.

Intumescent Fire Seals Association. 1999: *Sealing apertures and service penetrations to maintain fire resistance*. Princes Risborough: IFSA.

Ledbetter, S.R., Hurley, S. and Sheehan, A. 1998: *Sealant joints in the external envelope of buildings: a guide to design*. Report R178. London: Construction Industry Research and Information Association.

Panek, J.R. 1991: *Construction sealants and adhesives*. 3rd ed. New York: John Wiley & Son Inc.

Petrie, E.M. 2003: *Handbook of adhesives and sealants*. USA: McGraw-Hill Education.

Wolf, A.T. 1999: *Durability of building sealants*. London: Spon.

Woolman, R. and Hutchinson, A. (eds) 1994: *Resealing of buildings: A guide to good practice*. Oxford: Butterworth-Heinemann.

STANDARDS

BS 544: 1969 Linseed oil putty for use in wooden frames.

BS 1203: 2001 Hot-setting phenolic and aminoplastic wood adhesives.

BS 2499 Hot-applied joint sealant systems for concrete pavements:
    Part 2: 1992 Code of practice for application and use of joint sealants.
    Part 3: 1993 Methods of test.

BS 3046: 1981 Specification for adhesives for hanging flexible wallcoverings.

BS 3712 Building and construction sealants:
    Part 1: 1991 Method of test of homogeneity, relative density and penetration.
    Part 2: 1973 Methods of test for seepage, staining, shrinkage, shelf-life and paintability.
    Part 3: 1974 Methods of test for application life, skinning properties and tack-free time.

Part 4: 1991 Methods of test for adhesion in peel.

BS 4071: 1966 Polyvinyl acetate (PVA) emulsion adhesives for wood.

BS 4254: 1983 Specification for two-part polysulphide-based sealants.

BS 4255 Rubber used in preformed gaskets for weather exclusion from buildings:

Part 1: 1986 Specification for non-cellular gaskets.

BS 4346 Joints and fittings for use with unplasticized PVC pressure pipes:

Part 3: 1982 Specification for solvent cement.

BS 5212 Cold applied joint sealants for concrete pavements:

Part 1: 1990 Specification for joint sealants.

Part 2: 1990 Code of practice for application and use of joint sealants.

Part 3: 1990 Methods of test.

BS 5270 Bonding agents for use with gypsum plaster and cement:

Part 1: 1989 Specification for polyvinyl acetate (PVAC) emulsion bonding agents for indoor use with gypsum building plasters.

BS 5385 Wall and floor tiling:

Part 1: 1995 Code of practice for the design and installation of internal ceramic wall tiling and mosaics in normal conditions.

Part 2: 1991 Code of practice for the design and installation of external ceramic wall tiling and mosaics (including terra cotta and faience tiles).

Part 3: 1989 Code of practice for the design and installation of ceramic floor tiles and mosaics.

Part 4: 1992 Code of practice for tiling and mosaics in specific conditions.

Part 5: 1994 Code of practice for the design and installation of terrazzo tile and slab, natural stone and composition block floorings.

BS 5442 Adhesives for construction:

Part 1: 1989 Classification of adhesives for use with flooring materials.

Part 2: 1989 Classification of adhesives for use with interior wall and ceiling coverings (excluding decorative flexible material in roll form).

BS 6093: 1993 Code of practice for design of joints and jointing in building construction.

BS 6209: 1982 Specification for solvent-cement for non-pressure thermoplastics pipe systems.

BS 6213: 2000 Selection of constructional sealants – guide.

BS 6446: 1997 Specification for manufacture of glued structural components of timber and wood-based panel products.

BS 6576: 1985 Code of practice for installation of chemical damp-proof courses.

BS 8000 Workmanship on building sites:

Part 11: 1989 Code of practice for wall and floor tiling.

Part 12: 1989 Code of practice for decorative wall coverings and painting.

Part 16: 1997 Code of practice for sealing joints in buildings using sealants.

BS 8203: 2001 Code of practice for installation of resilient floor coverings.

BS EN 204: 2001 Classification of thermoplastic wood adhesives for non-structural applications.

BS EN 205: 1991 Test methods for wood adhesives for non-structural applications determination of tensile shear strength of lap joints.

BS EN 301: 1992 Adhesives, phenolic and aminoplastic, for load-bearing timber structures: classification and performance requirements.

BS EN 302 Adhesives for load-bearing timber structures:

Part 1: 2004 Determination of bond strength in longitudinal shear.

Part 2: 2004 Determination of resistance to delamination.

Part 3: 2004 Determination of effect of acid damage to wood fibres.

Part 4: 2004 Determination of the effects of wood shrinkage on shear strength.

Part 6: 2004 Determination of conventional pressing time.

Part 7: 2004 Determination of the conventional working life.

BS EN 1965: 2001 Structural adhesives – corrosion.

BS EN ISO 9047: 2003 Building construction – jointing products – determination of adhesion/cohesion properties of sealants at variable temperatures.

BS EN ISO 9664: 1995 Adhesives – test methods for fatigue properties of structural adhesives in tensile shear.

BS EN ISO 11431: 2003 Building construction – jointing products – determination of adhesion/cohesion properties of sealants after exposure to heat, water and artificial light through glass.

BS EN ISO 11600: 2003 Building construction – jointing products – classification and requirements for sealants.

BS EN 12004: 2001 Adhesives for tiles – definitions and specifications.

BS EN 12365 Building hardware – gaskets and weather stripping for doors, windows shutters and curtain walling:

Part 1: 2003 Performance requirements and classification.

Part 2: 2003 Linear compression force test methods.

Part 3: 2003 Deflection recovery test method.

Part 4: 2003 Recovery after accelerated ageing test method.

BS EN 12436: 2002 Adhesives for load-bearing timber structures – casein adhesives.

BS EN 12765: 2001 Classification of thermosetting wood adhesives for non-structural applications.

BS EN 12860: 2001 Gypsum based adhesives for gypsum blocks.

BS EN 13415: 2002 Adhesives – test of adhesives of floor coverings.

BS EN 14187-9: 2006 Cold applied joint sealants.

BS EN 14188 Joint fillers and sealants:

Part 1: 2004 Specification for hot applied sealants.

Part 2: 2004 Specification for cold applied sealants.

Part 3: 2006 Specification for preformed joint sealants.

pr Part 4: 2006 Specifications for primers to be used with joint sealants.

BS EN 14496: 2005 Gypsum based adhesives for thermal/acoustic insulation composite panels and plasterboards.

pr EN 14815: 2003 Adhesives, phenolic and amino-plastic for finger-joints in lameliae for load-bearing timber structures.

pr EN 15416: 2005 Adhesives for load-bearing timber structures.

pr EN 15425: 2005 Adhesives, one component polyurethane, for load bearing timber structures.

pr EN ISO 17087: 2003 Specification for adhesives used for finger joints.

BS EN 26927: 1991 Building construction – Jointing products – Sealants vocabulary.

BS EN 27389: 1991 Building construction – Jointing products – Determination of elastic recovery.

BS EN 27390: 1991 Building construction – Jointing products – Determination of resistance to flow.

BS EN 28339: 1991 Building construction – Jointing products – Sealants – Determination of tensile properties.

BS EN 28340: 1991 Building construction – Jointing products – Sealants – Determination of tensile properties or maintained extension.

BS EN 28394: 1991 Building construction – Jointing products – Determination of extrudability of one-component sealants.

BS EN 29046: 1991 Building construction – Determination of adhesion/cohesion properties at constant temperature.

BS EN 29048: 1991 Building construction – Jointing products – Determination of extrudability of sealants under standardized apparatus.

BUILDING RESEARCH ESTABLISHMENT PUBLICATIONS

BRE Digests

BRE Digest 227: 1979 Estimation of thermal and moisture movements and stresses: Part 1.

BRE Digest 228: 1979 Estimation of thermal and moisture movements and stresses: Part 2.

BRE Digest 245: 1986 Rising damp in walls: diagnosis and treatment.

BRE Digest 340: 1989 Choosing wood adhesives.

BRE Digest 346: 1992 The assessment of wind loads.

BRE Digest 463: 2002 Selecting building sealants with ISO 11600.

BRE Digest 469: 2002 Selecting gaskets for construction joints.

BRE Information papers

BRE IP 25/81 The selection and performance of sealants.

BRE IP 8/84 Ageing of wood adhesives – loss of strength with time.

BRE IP 12/86 Site-applied adhesives – failures and how to avoid them.

BRE IP 9/87 Joint primers and sealants: performance between porous claddings.

BRE IP 4/90 Joint sealants and primers: further studies of performance with porous surfaces.

BRE IP 12/03 VOC emissions from flooring adhesives.

ADVISORY ORGANISATION

British Adhesives and Sealants Association, 5 Alderson Road, Worksop, Nottinghamshire S80 1UZ (01909 480888).

17

PAINTS, WOOD STAINS, VARNISHES AND COLOUR

Introduction

As colour is an important factor in the description of paints, wood stains and varnishes, the key elements of the British Standards, the *Natural Color System*, RAL and *Pantone* systems and the *Colour Palette* notation are described, although other colour systems including Munsell are also used within the construction industry.

Colour

BRITISH STANDARDS SYSTEM

The British Standards BS 5252: 1976 and BS 4800: 1989 define colour for building purposes and paints respectively. A specific colour is defined by the framework with a three-part code consisting of hue (two digits, 00–24), greyness (letter, A to E) and weight (two further digits) (Fig. 17.1). Hue is the attribute of redness, yellowness, blueness, etc., and the framework consists of 12 rows of hue in spectral sequence plus one neutral row. Greyness is a measure of the grey content of the colour at five levels from the maximum greyness Group A, to clear Group E. The third attribute, weight, is a subjective term which incorporates both lightness (reflectivity to incident light) and greyness. Within a given column, colours have the same weight, but comparisons between columns in different greyness groups should only be made in respect of lightness. The framework has up to 8 columns of equal lightness in each greyness group commencing with the highest lightness. Thus any colour is defined through the system by its three-part code, e.g. Magnolia is yellow-red 08, nearly grey B, and low weight 15 (i.e. 08 B 15), Midnight 20 C 40 and Plum 02 C 39.

NATURAL COLOR SYSTEM

The *Natural Color System* (NCS) was developed by the Scandinavian Colour Institute in the 1980s. It is a colour language system, which can describe any colour by a notation, communicable in words without the need for visual matching. It has been used by architects, builders and designers who need to co-ordinate colour specification across a broad range of building products. A range of materials can be colour referenced using the system; these include wall, floor and ceiling tiles, carpets, fabrics, wall coverings, flexible floor finishes, paints, architectural ironmongery and metalwork, sanitary fittings, laminates and furniture.

The *Natural Color System* is based on the assumption that for people with normal vision there are six pure colours: yellow, red, blue, green, white and black. The four colours yellow, red, blue and green are arranged around the *colour circle*, which is then subdivided into 10% steps. For example, yellow changes to red through orange, which could be described as Y50R (yellow with 50% red) (Fig. 17.2). In order to superimpose the black/white variation and also intensity of colour, each of the forty 10% steps around the colour circle may be represented by colour triangles, with the pure colour at the perimeter apex and the vertical axis illustrating blackness/whiteness. A colour may be therefore described as having 10% blackness and 80% chromatic intensity. The full colour specification thus reads 1080-Y50R for an orange with 10% blackness, 80% chromatic intensity at yellow with 50% red.

BS4800:1989 Specification for paint colours for building purposes

Hue	Weight	A group grey (01 03 05 07 09 11 13)	B group nearly grey (15 17 19 21 23 25 29)	C group grey/clear (31 33 35 37 39 40)	D group nearly clear (41 42 43 44 45 46)	E group clear (49 50 51 53 55 56 57 58)
02 red-purple				Plum 02 C 39		
04 red					Tawny 04 D 44	Poppy 04 E 53
06 yellow-red					Cinnamon 06 D 43	Apricot 06 E 50
08 yellow-red			Magnolia 08 B 15	Bamboo 08 C 35		
10 yellow		Flake grey 10 A 03				Jonquil 10 E 49
12 green-yellow			Chive 12 B 25			
14 green				Conifer 14 C 40		
16 blue-green				Duck egg 16 C 33		
18 blue			Raven 18 B 29			
20 purple-blue				Midnight 20 C 40		Cornflower 20 E 51
22 violet			Dove 22 B 17	Heather 22 C 37		
24 purple				Mallow 24 C 33		
00 neutral		Portland 00 A 01				Black 00 E 53

Fig. 17.1 British Standards Colour System with some illustrative examples

The system allows for a finer subdivision of the colour circle, and this is necessary to define any colour and to make direct comparisons with colours defined within the British Standards system. Thus Magnolia (BS 08 B 15) is 0606-Y41R (6% blackness, 6% chromatic intensity on a yellow with 41% red). Plum (BS 02 C 39) is 5331-R21B and Midnight (BS 20 C 40) is 7415-R82B in the *Natural Color System*.

RAL COLOUR COLLECTION

The RAL colour collection is used significantly within the building industry for defining the colours of finishes, particularly to plastics and metals, but also materials such as glazed bricks. Typical applied finishes include acrylics, polyesters and polyurethane as well as some paints and lacquers. The RAL system, established in Germany in 1925, has developed through several phases. It commenced with 40 colours; subsequently many were added and others removed, leaving 170 standard colour shades. Because of it development, the RAL system (designated RAL 840-HR) does not have a systematic order of colours with equal steps between shades. Colours are defined by four digits, the first being the colour class (1 yellow, 2 orange, 3 red, 4 violet, 5 blue, 6 green, 7 grey, 8 brown and 9 black/white) and the further three digits relate only to the sequence in which the colours were filed. An official name is also applies to each standard RAL colour (e.g. RAL 1017 Saffron Yellow, RAL 5010 Gentian Blue, RAL 6003 Olive Green). Some additional colours have been added to the RAL classic

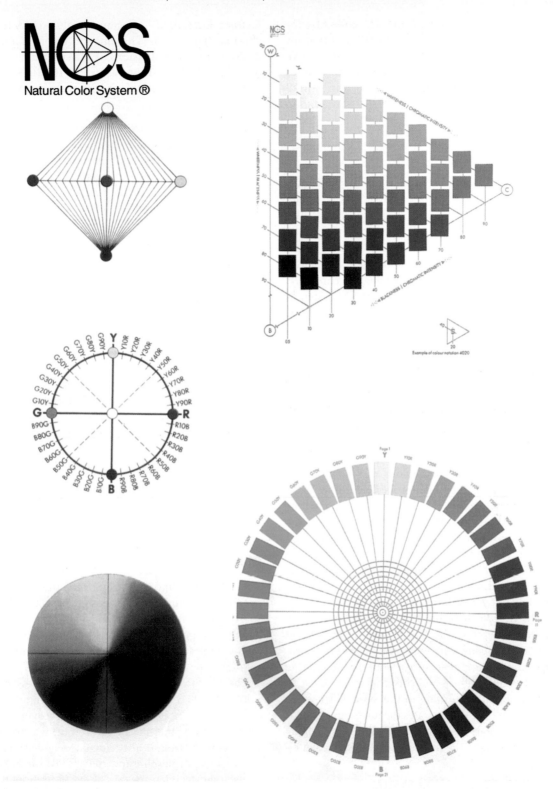

Fig. 17.2 *Natural Color System©* Images: Courtesy of the Scandinavian Colour Institute AB www.sci-sweden.se

collection giving over 200 colours. The collection for matt shades is designated RAL 840-HR and that for glossy shades is RAL 841-GL. A CD of 195 of the RAL classic colours is available for computer applications giving the colour specifications in RGB (red/green/blue), HLC (hue/lightness/chroma) and in offset printing format.

RAL DESIGN SYSTEM

Unlike the RAL colour collection which only has a limited selection of standard colours, the RAL design system has 1688 colours arranged in a colour atlas based on a three-dimensional colour space defined by the coordinates of Hue, Lightness and Chroma. Hue is the attribute of colour, e.g. red, blue or yellow. Lightness ranges from black to white, and chroma is the saturation or intensity of the colour. The system is equivalent to the HLS (Hue, Lightness, Saturation) system which is used alongside RGB (Red, Green, Blue) in many computer colour systems. The RAL design system is similar to the *Natural Color System*, except that it is based on a mathematical division of the whole visible wavelength spectrum, rather than the visually assessed four standard colours yellow, red, blue and green.

The colour spectrum is therefore divided into mostly 10° steps around a circle. Each step, illustrated on a page of the associated colour atlas, represents a particular hue. For each hue on the colour atlas page, samples illustrate lightness decreasing from top to bottom and intensity or saturation increasing from the inside to the outside. Any colour is therefore coded with the three numbers relating to hue, lightness and chroma, e.g. 70 75 55. The standard RAL colour collection numbers do not fit neatly to the RAL design system coding but any colour can be defined, thus Saffron Yellow (RAL 1017) becomes 69.9 75.6 56.5. However, as the number defining the hue is not exactly 70, the colour Saffron Yellow will not appear on the atlas page. Computer programs generating colour through the attributes of hue, lightness and chroma can immediately formulate colours according to this system. The electronic version of RAL illustrates the three-dimensional colour atlas, offers 1900 standard colours, and links proposed colours to the nearest standard RAL colour.

COLOUR PALETTE NOTATION SYSTEM

The *Colour Palette* (*Dulux*) notation system (also known as the *Master Palette*) is based on the three

factors: hue; light reflectance value (LRV); and chroma (Fig. 17.3). The hue or colour family is derived from eight divisions of the spectrum, each of which is sub-divided into a further 100 (0–99) divisions to give a precise colour within a particular hue.

Hue families:

RR	magenta through to red
YR	red through to orange
YY	orange through yellow to lime
GY	lime through to green
GG	green through to turquoise
BC	turquoise through to blue
BB	blue through to violet
RB	violet through to magenta

Thus hues are described by two digits and two letters (e.g. 50RR).

Light reflectance value (LRV) is a measure of lightness or darkness, with light colours having a high two-digit number. Thus, most pastel shades have a light reflectance value between 75 and 83 and the majority of colours fall within the range 04 (very dark) to 83 (very light).

Chroma is a measure of the saturation or strength of the colour measured in 1000 steps from 000 to 999 with high numbers indicating high saturation or intensity.

Thus a lavender would be specified as:

45RB	44	/	242
hue	LRV		chroma

(where LRV is light reflectance value)

PANTONE

The *Pantone Textile Colour System* is frequently used by architects and interior designers for specifying colours for plastics, fabrics and paint. The system is based on a cylindrical colour solid; the position of each colour within the cylinder is represented by a six-digit code and additional suffix letters.

The first two digits indicate the lightness of the colour with 11 as the lightest (nearest to white) and 19 as the darkest (nearest to black). The third and fourth digits represent the hue of the colour, which has 64 sub-divisions around the circular sequence yellow, orange, red, purple, blue and green. The fifth and sixth digits indicate intensity or colour saturation represented by the distance of the particular colour from the axis of the colour cylinder where 0 represents white, grey or black to 64 indicating the most intense

Example

30BB 08/263

HUE   LRV   CHROMA

Hue

The colour family

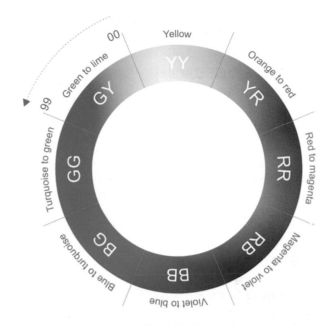

Light Reflectance Value (LRV)

The lightness or darkness of the colour.

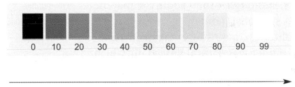

The higher the number the lighter the colour.

Chroma

The intensity of the colour.

The higher the number the more intense the colour.

Fig. 17.3 Dulux Colour Palette Notation. Images: Courtesy of Imperial Chemical Industries Plc (*Dulux, Colour Palette* and *Master Palette* are trademarks of ICI.)

colour. The suffix letters differentiate between the textile cotton (TC) and textile paper (TPX) editions (the X refers to the 2003 edition). The distinction between the two suffix notations is necessary as colours appear different according to the surface material to which they are applied. The designation for a particular scarlet red is thus – *Pantone* Scarlet 19–1760 TP.

VISUAL COMPARISON OF PAINT COLOURS

The European standard for the comparison of paint colours (BS EN ISO 3668: 2001) is based on observation under specified illumination and viewing conditions (either natural diffuse daylight or scientifically specified artificial light). The colour-matching process is based on an assessment of the differences in hue, chroma and lightness between test panels and reference colour standards.

Paints

COMPONENTS OF PAINTS

Paints consist of a blend of components, each with their specific function. Commonly these include the binder (or medium), solvent, base, extenders, pigments and driers, although other additives may be incorporated into specialist paints.

The binder solidifies to produce the paint film. Traditionally, the binder was natural linseed oil, which set by gradual oxidation on exposure to air. However, linseed oil has now largely been replaced by alkyd resins which oxidise in air, or vinyl and acrylic resins which solidify by drying. To ensure adequate fluidity of the paint during application by brushing or spraying, either water or organic solvents (hydrocarbons, ketones or esters) are incorporated; paint thinners have the same effect. The base material, usually white titanium dioxide, produces the required opacity, although the *body* of the paint may be increased by the incorporation of inert extenders such as silica, calcium carbonate, china clay or barytes. Colouring materials are frequently a mixture of organic and inorganic dyes and pigments. Driers which induce the polymerisation of the binder ensure a rapid drying process.

Changes in legislation and environmental concerns have led to the development of paints with reduced levels of volatile organic compounds (VOCs). Mainly this has been through the increased use of water-borne rather than solvent-borne paints. In some respects

water-borne paints have the advantage. They have low odour emissions, brushes can be cleaned in water, and they will tolerate damp surfaces. However, they are not ideal for use externally in cold and wet conditions. Other developments have been towards *high solids* paints, which have low solvent content and therefore very low VOC emissions. A further trend is towards the use of *natural paints* based on plant and mineral compositions, which incorporate a lower proportion of components of high embodied energy materials such as titanium dioxide and petrochemical products These new formulations consume considerably less energy in manufacture, are environmentally friendly in application and less problematic in waste management; however, they are not suitable for all applications.

PAINT SYSTEMS

Coats within a paint system perform specific tasks. Usually a complete system would require primer,

Table 17.1 Recommended primers for various substrates

Primer	Suitable substrates and conditions
Timber	
Preservative primer	Exterior use, contains fungicides
Wood primer	Softwoods and hardwoods, interior and exterior
Aluminium wood primer	Resinous softwoods and hardwoods
Acrylic primer	Softwoods and hardwoods
Plaster and masonry	
Alkali resisting primer	Plaster, cement and concrete
Acrylic primer sealer	Loose, friable surfaces
Ferrous metals	
Zinc phosphate	Steel, iron, galvanised steel. Good rust inhibitor
Red lead	Steel. Contains lead but excellent rust inhibitor
Metal primer	Steel and iron. Non-toxic alternative to red lead. Grey
Calcium plumbate	Galvanised steel. Contains lead but excellent rust inhibitor
Acrylated rubber	Steel, iron, galvanised steel. Must be a full acrylated rubber system
Zinc-rich primer	Steel. Two-component system
Non-ferrous metals	
Zinc phosphate	Aluminium
Acrylated rubber	Aluminium. Must be a full acrylated rubber system
Acrylic metal primer	Aluminium, copper, lead, brass. Quick-drying water-based primer

undercoat and finishing coat, although in the case of new external materials, four coats may be appropriate.

Primers

The primer must adhere well to the substrate, offer protection from deterioration or corrosion and provide a good base for the undercoat. To ensure adhesion, the substrate surface must be free of loose or degraded material. Appropriate systems are indicated in Table 17.1. For use on timber, primers may be oils, alkyd resins or acrylic emulsions, frequently with titanium oxide. Aluminium wood primer is recommended for resinous woods and to seal aged creosoted and bitumen-coated surfaces. For the corrosion protection of ferrous metals, primers incorporate zinc or lead-rich compounds within oils or alkyd resins. While lead-based paints such as red lead and calcium plumbate are considered environmentally less acceptable than the alternatives, they remain very efficient in the inhibition of steel corrosion. The newly developed low-VOC coatings offer temporary protection against the corrosion of structural steelwork either as prefabrication or post-fabrication primers. Alternatively, acrylated rubber paints, which form a physical barrier over steel, may be used as primers. For non-ferrous metals, zinc phosphate primers are frequently used. The application of primers suitable to ferrous metals may cause increased corrosion on non-ferrous substrates, particularly aluminium. Masonry paints are usually based on alkyd or acrylic resins with titanium oxide; where surfaces are likely to be alkaline, such as new plaster, brickwork or concrete, alkali-resisting primer should be used.

Undercoats

Undercoats provide cover and a good base for the finishing coat. Most undercoats are based on alkyd resins or acrylic emulsions.

Finishing coats

Finishing coats provide a durable and decorative surface. Some gloss, eggshell and satin finishes are still based on oils and alkyd resins, although increasingly water-borne products are becoming predominant. Some water-borne gloss finishes tend to be visually softer and are more moisture-permeable than the traditional solvent-borne hard glosses. However, they have the advantage of quick drying without the evolution of solvent odour; generally they are more durable and do not yellow on ageing. Matt and silk finishes are usually vinyl or acrylic emulsions.

Special paints

MULTICOLOUR PAINTS

Multicolour paints incorporating flecks give a hard wearing surface which may be glazed over to ease the removal of graffiti. Application is with a spray gun, which can be adjusted to change the pattern and texture of the fleck. This type of paint system may be applied to most dust- and grease-free internal surfaces.

Broken-colour paints

Broken colour effects, reflecting the traditional processes of graining, marbling, ragging and stippling, are once again popular. Most modern broken-colour effects require a base coat, applied by brush or roller, which is then overpainted, with a clear coloured glaze. The glaze is then patterned or distressed to create the desired effect. One proprietary system uses a special rag-roller, which flails the wet finish coat giving random partial exposure of the darker first coat. Alternative finishes include metallic, pearlescent and graining effects. An iridescent finish produces a two-tone shimmer effect by optical interference of the reflected light. Water-based acrylic glazes are virtually odour-free and are touch dry within two hours.

ACRYLATED RUBBER PAINTS

Acrylated rubber paints are suitable for internal and external applications exposed to chemical attack or wet and humid atmospheric conditions. Acrylated rubber paints are tending to replace chlorinated rubber coatings which rely on carbon tetrachloride solvent, now considered environmentally damaging. Acrylated rubber paints may be applied to metal or masonry by either brushing or spraying. Usually a film of dry thickness 100 microns is applied compared to 25–30 microns for most standard paint products.

HEAT-RESISTING PAINT

Aluminium paint, which has a lustrous metallic finish, is resistant to temperatures up to 230–60°C. A dry-film thickness of 15 microns is typical.

Acrylated rubber paints can usually be used satisfactorily to 100°C.

FLAME-RETARDANT PAINTS

Flame-retardant paints emit non-combustible gases when subject to fire, the usual active ingredient being antimony oxide. Combustible substrates such as plywood and particleboard can be raised to Class 1 (BS 476 Part 7) surface spread of flame. Products include matt, semi-gloss and gloss finishes, and may be applied by brush, roller or spray.

INTUMESCENT COATINGS

Thin-film intumescent coatings, typically 1 or 2 mm in thickness, offer fire protection to structural steel without noticeable visual effect. In the event of fire, the thin coating expands up to 50 times to form a layer of insulating foam. The carbonaceous material in the coating, typically starch, is charred, whilst the heat also causes the release of acids. These act to produce large volumes of non-inflammable gases which blow up the charring starch within the softened binder into an insulating cellular carbon layer. Coatings may be applied to give 30, 60 or 120 minutes' fire protection. Intumescent emulsion paints or clear varnishes are appropriate for use on timber, although where timber has been factory-impregnated with a flame-retardant salt, the compatibility of the intumescent coating and flame retardant must be verified.

FUNGICIDE PAINTS

Fungicide paints for application in areas where mould growth is a recurrent problem usually contain a blend of fungicides to give high initial activity and steady long-term performance. The latter can be achieved with fungicide constituents of low solubility which are gradually released to the surface during the lifetime of the paint. Matt acrylic finishes are available in a range of colours.

ENAMEL PAINTS

Enamel paints based on polyurethane or alkyd resins give highly durable impact-resistant easily cleaned hard gloss surfaces. Colours tend to be strong and bright, suitable for machinery and plant in interior and exterior locations.

MICACEOUS IRON OXIDE PAINTS

Micaceous iron oxide paints have good resistance to moisture on structural steelwork, iron railings, etc., due to the mica plates which reduce permeability to moisture vapour. A dry-film thickness of 45–50 microns is typical, thus requiring longer drying times than standard paint products. Micaceous iron oxide paints should be applied over an appropriate metal primer.

MASONRY PAINTS

Smooth- and sand-textured masonry paints are suitable for application to exterior walls of brick, block, concrete, stone or renderings. Where fine cracks are present, these can often be hidden using the sand-textured material. Usually masonry paints contain fungicides to prevent discolouration by moulds and algae. Acrylic resin-based products are predominantly water-based; however, fast-drying solvent-based systems are also produced. Mineral silicate paints form a crystalline protective layer over the masonry surface, which tends to be more durable than the organic finishes from synthetic resins.

WATER-REPELLENT AND WATERPROOFING PAINTS

Silicone water-repellent paints can be applied to porous surfaces including brick, concrete, stone and renderings to prevent damp penetration. Such treatment does not prevent rising damp, but will allow the continued evaporation of moisture within the masonry. Two-pack epoxy waterproofing systems may be applied to sound masonry surfaces to provide an impervious coating. Typical applications are to rooms where condensation causes the blistering of normal paint films; also in basements and solid external walls where penetrating water is a problem, providing that a good bond can be achieved between substrate and epoxy resin. Bituminous paints provide a waterproof finish to metals and masonry and may be used as a top dressing to asphalt or for renovating bitumen sheet roofing. Aqueous bitumen coatings, if fully protected against physical damage, can provide a vertical membrane where the external ground level is higher than the internal floor level.

EPOXY PAINTS

Epoxy ester paint coatings are highly resistant to abrasion and spillages of oils, detergents or dilute aqueous chemicals. They are therefore frequently used as finishes to concrete, stone, metal or wood in heavily trafficked workshops and factories. Many are produced as two-pack systems requiring mixing immediately before application.

Natural wood finishes

Natural wood finishes include wood stains, varnishes and oils. Wood stains are pigmented resin solutions which penetrate into the surface and may then build up a sheen finish. Varnishes are unpigmented resin solutions which are intended to create a surface film. Timber preservatives are described in Chapter 4.

WOOD STAINS

Most wood stain systems for exterior use include a water- or solvent-based preservative basecoat which controls rot and mould growth. Typical formulations include zinc or copper naphthenate, dichlorofluanid, tri-(hexylene glycol) biborate and disodium octaborate tetrahydrate. Wood stain finishes are either low-, medium- or high-build systems, according to the particular application. They usually contain iron oxide pigments to absorb the ultraviolet light which otherwise causes the surface degradation of unprotected timber. Generally, for rough sawn timber deeply penetrating wood stains are appropriate, whereas for smooth-planed timber a medium- or high-build system gives the best protection from weathering. Products are based on acrylic and/or alkyd resins.

For sawn timber, both organic solvent-based and water-based materials are available, usually in a limited range of colours. Solvent-based low-build products which are low in solids penetrate deeply, leaving a water-repellent matt finish, enhance the natural timber grain and are suitable for timber cladding. Deep penetration should eliminate the risk of flaking or blistering on the surface. Medium- and high-build products for exterior joinery offer the choice of semi-transparency to allow the grain to be partially visible, or opaque colours for uniformity. Products are available in a wide range of colours with matt or gloss finishes. The first coat both penetrates and adheres to the surface, whilst the second coat provides a continuous microporous film which is both permeable to moisture vapour and water-repellent, thus reducing the moisture movement of the timber. Additional

coats should be applied to end grain. The coating, typically 30–40 microns thick, should remain sufficiently flexible to accept natural timber movements. Low VOC products based on water-borne emulsions or high-solids solvent-borne resins are now generally available.

VARNISHES

Traditional varnishes are combinations of resins and drying oils, but most products are now based on modified alkyd resins. Polyurethane varnishes are available in matt, satin or gloss finishes, based on either water or solvent-based systems. The solvent-based systems produce the harder and more durable coatings up to 80 microns thick, suitable for exterior woodwork. Products either retain the natural wood colour, enhance it, or add colour. Screening agents to protect timber from the effects of ultraviolet light are normally included in the formulations. Urethane-modified alkyd resins are suitable for interior use and have the advantage of high resistance to scuffing and hot liquids. External weathering causes eventual failure by flaking and peeling as light passing through the varnish gradually degrades the underlying wood surface. For example, hardwood doors decorated with polyurethane varnish, protected from rain and direct sunlight by a porch, should have extended periods between maintenance. End grain should be sealed to prevent trapped moisture encouraging the development of staining fungi.

OILS

Oils such as teak oil are used mainly for internal applications. Formulations based on natural oils for exterior use are high in solids producing an ultraviolet resistant, microporous finish which may be transparent or opaque. The finish, which should not flake or crack, may be renovated by the application of a further coat.

References

FURTHER READING

ICI Paints : *ICI Dulux colour palette*. Imperial Chemical Industries plc.

Reichel, A., Hochberg, A. and Kopke, C. 2005: *Plaster, render, paint and coatings: details, products, case studies*. Basel: Birkhäuser.

INDEX

—